ELECTROMIGRATION IN ULSI INTERCONNECTIONS

International Series on Advances in Solid State Electronics and Technology (ASSET)

Founding Editor: Chih-Tang Sah

Published:

Modern Semiconductor Quantum Physics
by Li Ming-Fu

Ionizing Radiation Effects in MOS Oxides
by Timothy R. Oldham

MOSFET Modeling for VLSI Simulation: Theory and Practice
by Narain Arora

MOSFET Modeling for Circuit Analysis and Design
by Carlos Galup-Montoro & Márcio Cherem Schneider

The Physics and Modeling of MOSFETs: Surface-Potential Model HiSIM
by Mitiko Miura-Mattausch, Hans Jürgen Mattausch & Tatsuya Ezaki

Invention of the Integrated Circuits: Untold Important Facts
by Arjun N Saxena

Electromigration in ULSI Interconnections
by Cher Ming Tan

Forthcoming:

BSIM4: Theory and Engineering of MOSFET Modeling for IC Simulation
by Weidong Liu & Chenming Hu

ASSET

International Series on Advances in Solid State Electronics and Technology

Founding Editor: Chih-Tang Sah

ELECTROMIGRATION IN ULSI INTERCONNECTIONS

Cher Ming Tan

Nanyang Technological University

NEW JERSEY • LONDON • SINGAPORE • BEIJING • SHANGHAI • HONG KONG • TAIPEI • CHENNAI

Published by

World Scientific Publishing Co. Pte. Ltd.
5 Toh Tuck Link, Singapore 596224
USA office: 27 Warren Street, Suite 401-402, Hackensack, NJ 07601
UK office: 57 Shelton Street, Covent Garden, London WC2H 9HE

British Library Cataloguing-in-Publication Data
A catalogue record for this book is available from the British Library.

ELECTROMIGRATION IN ULSI INTERCONNECTIONS
International Series on Advances in Solid State Electronics and Technology

ISBN-13 978-981-4273-32-9
ISBN-10 981-4273-32-5

Typeset by Stallion Press
Enail: enquiries@stallionpress.com

Printed in Singapore by Mainland Press Pte Ltd.

To my wife (Esther) and my son (Daniel) with love and respect.

"The deeper one penetrates into nature's secrets, the greater becomes one's respect for God."

Albert Einstein

Foreword and History

I will not practice here what I preach, on limiting the Foreword to one book page, as I have managed to do for the previous four monographs of this series on device modeling. I would like to use this rare opportunity, because, I think there is an interesting story and a relevant history to tell, on how things are developed. It proves the thesis: Progress is made via generations of **natural Darwin evolution** and **artificial evolutions**-or-**intelligent designs**. Intelligent designs feed evolution, which in turn furthers design intelligence, providing the positive feedback loop that perpetually speeds up the cycle. It can be a cycle as long as many hundreds thousands of years or many human lifetimes, to the longer biological evolution times. It can become as short as one, two or three years, only about a tenth of the thirty-year human generation, such as the Gordon Moore Law. The increasing acceleration is the result of human intelligence learned and built up from many generations of evolution, the route with the positive feedback loop.

Although in the traditional or recent statistical sense, my samples are very small, they are real data, not random statistical spikes. Their impacts and consequences have affected the ~six billions as well as also all the rest of the trillion trillions evolved and evolving beings. Therefore, they cannot be just statistical random fluctuations or random spikes above the noise in the background. These examples have each one or more significant if not dominating components of artificial intelligent designs, accomplished by evolving educated and learned minds in three to five human generations, from grandparents to grandchildren, and from teachers to students.

The stories are stories of Electromigration, the subject of this book. They are described in this extended Foreword, re-titled Foreword and History. After you read this Foreword and History, I hope you will agree that the case histories you have just read show that evolution and intelligent design are in fact two of the same in a broader perspective. To wit, as we evolve, we recognize we have actually designed the evolution by our intelligence, which has sped up the evolution. The evolution then drives the intelligence to become more learned, nurtured, and grown under controlled evolution. This positive feedback closes the loop of Evolutional Intelligent Design, shortening its cycle time from billions of years to now billionth of a second.

If you are too busy to read this six-page Foreword on some of the 'ancient' histories that have occurred in less than ~ 30 years, which are the evidences of the inseparable tie between "evolution" and "intelligent design", then you should just skip forward to the book author's Preface and his Chapter 1, and then the rest of his book.

As you read this book, you will realize that it is such a complicated object, much more than the simple two-word acronym, EM. When finished, you will agree that it is an engineering masterpiece of eloquent, concise and succinct descriptions of a most important but also most engineering-complex subject. The invited young author, who has had decades of hands-on basic research and manufacturing experiences on this subject of Electromigration — EM, is Associate Professor Cher Ming Tan, presently with the Nanyang Technological University of Singapore.

This monograph is the fifth book in this series on modeling of integrated-circuit devices. Its purpose is to provide archival references, described by the model originators or authorities, on the devices and components that are to be integrated on a small, a centimeter or smaller, silicon semiconductor integrated circuit chip. It has been my intention to invite the experts to cover all the components of an integrated circuit. These include the active devices (MOS field-effect and bipolar-junction transistors), the passive devices (diodes, resistors, capacitors, and inductors), and the sensors that interface the electronics with the ambient and the user (electromagnetic and optical radiators and receivers or the antennas, biological sensors of ambient forces – pressures, flow velocities, accelerations). Not the least, we want to include the new electrical and mechanical devices when proven in applications and manufacturability. Certainly, the most important we must cover is the interconnects which are the metal conductor wires and plugs or studs, that electrically interconnect the devices and sensors, and the endurances of the interconnects during operation and shelve lives, which are the subjects of this book.

The monograph series idea came about when I was looking into the literature for references to write the keynote address. It was on the history of MOS transistor compact modeling. I was invited by the Founder of the Workshop on Compact Modeling, Associate Professor Xing Zhou of Nanyang Technology University and his international workshop program committee. It was the workshop's first keynote and it was to be presented at its fourth annual Workshop on May 10, 2005. A second purpose of this device-modeling monograph series was to provide state-of-the-art textbooks for graduate students and reference books for practicing engineers. These books would rapidly disseminate the detailed design methodologies and underlying physics, which are progressing at an ever faster rate, in the computer-aided-design of silicon semiconductor integrated circuits which contain hundreds, thousands, now billions, approaching if not already exceeding a trillion of diodes, transistors and interconnects on one small (< ~1 square centimeter) silicon die or silicon integrated circuit chip. It is also the objective of this monograph series to provide timely updates, via website and internet exchanges between the readers and authors, for immediate public dissemination, and to provide new editions when sufficient materials are accumulated.

I am especially thankful to the invited authors of the four startup volumes (Narain, Carlos+Márco, Mitiko+Hans+Tatsuya, and Arjun), and the invited authors of the later volumes (Cherming of this volume, Chenming+Weidong, Chinghsiang, Michael, Jamal, and others.). They concurred with my objectives and agreed to take up the chore of writing their books during their very busy schedules. Some have delays of one, two or even three years. Nevertheless, their monographs are still the archival records of the state of the art, and the world's authoritative contributions to the device modeling literature, because these authors are the creators and/or authorities of the models, and because the models are the industry-wide consensus models used by all circuit designers in the past and recent generations of integrated circuit designs.

The present volume is a review of the literature on Electromigration (EM) in metals. EM is the ultimate failure cause of silicon integrated circuits. It also affects, if not directly determines, the ultimate performance limit of silicon integrated circuits because the RC signal delay in digital switching circuits and resonance broadening and gain lowering in analog circuits are limited by the resistance of the metal interconnect line, R. This resistance increases due to EM, caused by the high density of electrical current through the interconnect lines. On the microscopic events leading to failure, in one failure mode, EM of the metal atoms of the metal conductor line leaves behind a void in the line, which eventually opens the line when the void grows to the cross-sectional area of the line, causing the transistors and devices to disconnect electrically. In the second mode of EM failure, the migrating metal atoms of the metal line can pile up at favored locations in the line. (Read this book to find where these favored locations are.) The pile-up of the metal atoms would grow to a sufficient size to electrically short-circuit the adjacent lines. Such shorts become increasingly frequent with increasingly closer adjacent lines as the circuit dimension, hence interline spacing, or the pitch, continually decreases into the nanometer range.

The vast engineering literature on EM is a reflection of the many chemical elements and compounds that make up the interconnecting metal lines and studs. The chemical composition is complicated further by the recent adoption of copper for its higher conductivity, in place of aluminum, to give lower R and RC delay, and to give higher signal amplification and Q or sharper pass band from RCL resonators for lower crosstalk and distortion. The complication in Cu interconnect structure arises because Cu diffuses easily into silicon to short out the devices. This must be prevented by diffusion barriers made of multi-layers of very thin films of buffer metal elements and compounds, and insulators. Additional complications come from the three-dimensional (3-D) geometries of the interconnection metal lines. All of these increase fabrication complexity and cost, and lower manufacturing yield.

For example, the 3-D interconnects (lines and studs through via's) are built into many (~10) levels over the thin silicon surface layer which contains the many silicon transistors, diodes, resistors and capacitors. **Planarization** (2-D) of each interconnects' layer is necessary to provide direct routing, in order to reduce the electrical signal delay in digital circuits through the length of interconnect metal lines. The transistors, at each of the technology generations, from micrometer to now nanometer, have always been much faster than the interconnects' RC delays. The 2-D planarization is also necessary to minimize abrupt changes in line directions, which could increase the electrical current density at the abrupt bends. The higher density would accelerate the rate of EM failure via void growth and metal pileup, directly via the microscopic interaction of the higher electric field at the bends with the migrating atom, and indirectly via space-gradients of atomic concentration (diffusion rate) or atomic flux divergence, and also space-gradients of the local lattice temperature (reaction or atomic trapping and detrapping rates). However, the chemical-mechanical polishing (CMP) or planarization technique is not only process tedious, but also a source of minute-impurity-incorporation and atomic damage or crystal defect formation from the bulk-penetrating surface roughing during the polishing step.

Furthermore, because of the many parameters that control EM, its engineering has been a factory art of recipe-driven, statistically optimized, empirical or trial-and-error task, influenced by the traditional chemical and metallurgical engineering practices of statistical control. This reminds me of a similar if not more well-known but certainly fully played-out case in a modern silicon integrated circuit factory. It is the person-specific favored fit-formula approach in modeling the dependence of the electron and hole mobility's on the concentration of the randomly located bulk and surface-interface impurities and defects, which scatter the electrons and holes, and on the local temperature and local electric field. It has been a subject of domain ownership by the manufacturing departments and individual engineers whose jobs and survivals are at stake. But, electron and hole mobility's are fundamentally and practically much simpler because they are <u>electronic</u> and **<u>ma</u>**croscopic even when the dimensions are now down to the **<u>mi</u>**croscopic nanometers, because there are still enough number of the interacting particles (electrons, holes and scattering centers) to allow a macroscopic modeling of the electron and hole mobility's. On the other hand, EM and the general phenomena of atomic migration under concentration and temperature gradients and under electric fields are much more complicated, because, they are atomic and **mi**croscopic even at the larger dimensions of the micrometer or micron era of yesteryears. This is further compounded now by the increasingly smaller dimensions into nanometers, resulting in even fewer atoms (and particles) responsible for EM. Worse, the traditional chemical-electrical-metallurgical engineers' empirical practice of thermally activated rate, given by the product of a jump or collision-frequency factor (frequently called the pre-exponential factor) and the Boltzmann factor $\exp(-E_{act}/kT)$ from the particle number variation with particle kinetic energy of a dilute concentration of scattering and scattered particles, was used by Black to fit the EM failure-time data, but its derivation has been based on a fundamentally flawed assumption of the presence of a Coulomb-like electrostatic force between the metal electrons and the metal ions, known as the electron wind force, like the rapidly moving and electrically neutral air molecules at rather dilute concentrations, that move a person, a tree, a building by the wind force. But, it ignored the fact that there are so many mobile conduction electrons in a metal, not unlike a glueish jello or an extremely high concentration of molecules in dense liquids, which could "instantaneously" "screen" the electrostatic force from the positively charged ionic core of each of the many metal atoms, preventing the charged atomic cores from "feeling" the negatively charged drifting electrons that carry the electrical current, rendering the electron-ion or electron-wind force untenable. The simple Debye screening length analysis shows that the many valence electrons in a metal (~ 10^{23} cm^{-3}) would screen the atomic ion charge in ~ 0.1A, which is much less than the inter-ion-core distance of the metal's lattice, ~ 2A, estimated from the inter-atomic core spacing or the lattice constant of the metal crystal.

As a consequence of the ppEM (pp=personal property), I was strongly discouraged by the "technology czar" of the world's leading silicon IC chip manufacturer, who happens to be also the most supportive to his former teacher among my 50 PhD students, to not enter this pp arena in 1980, because he just received a request to godfather the dispute, with a 3-foot pile of reports and stacks of 8-inch EM-test-pattern wafers. He told me that he would definitely be not able to loan me any 8-inch silicon wafers with interconnect patterns designed for EM tests since he had no ownership, unlike the generations of gate oxides (100A to 10A) and CMP silicon surfaces to investigate the interface traps on oxidized CMP silicon surfaces, for which (the CMP technology) he was the first to put into volume manufacturing for the semiconductor industry.

This EM episode has now triggered my memory on the history of these two dominating technologies (gate oxide and CMP) since EM is tightly coupled to both. It must be told since probably no one could except the actor himself who is too modest. This history also leads to the importance of CMP and hence its development with great efforts and best brains, and at substantial costs, asides from its secrecy, as a proven example of intelligently designed evolution with the characteristic time of about two years rather than the random Darwin evolution of millenniums or millionniums. There are two aspects, the manufacturing and the physics.

Let me first tell the story of EM manufacturing technology history. The CMP and gate oxide technologies were actually originated and evolved from the young-Shockley led group at Bell Telephone Laboratories (where else?) when the older Bardeen and Brattain of the group discovered the transistor "effects" in 1948 using the two point-contacts on chemically etched surface of a germanium single crystal called the 'bulk' to give the 3-terminal device that has the minimum number of leads or terminals necessary to give amplification of electrical signal. However, the point-contact transistor (PCT) of Bardeen and Brattain was plagued by instability and irreproducibility from the surface chemical treatments and even the exposure to ambient during and before operation. Such sensitivity and irreproducibility quickly closed the human's first transistor factory (Western Electric, 555 Union Boulevard, Allentown, Pennsylvania). I was lucky to see this world's first mechanically automated transistor factory, during my 1956 and first job interview before it was to close. So, Shockley invented (or "designed" via intelligence from prior learning and training) the two transistors to avoid surface effects, in 1952, the field-effect transistor (majority carrier drift current in the bulk of a semiconductor) and in 1949, the p/n junction transistor (minority carrier diffusion current also in the bulk of a semiconductor). We are still using these two transistors and their principles to this day, 60 years later. He, Shockley, designed their geometries to place the current path inside the semiconductor bulk, away from the semiconductor surfaces, to avoid the surface "effects" observed by Bardeen and Brattain in their PCT. However, the 3-D bulk device structures invented by Shockley evolved naturally to 2-D devices in a thin surface layer that is protected or passivated by a stable thermal silicon dioxide layer. Such 3-D to 2-D evolution was necessary to meet the demand of transistor operation not just for audio electronics (20-20K Hz) offered by the bulk transistors, but also at higher frequencies for the AM (550K to 1650K Hz) and FM (88M to 108M Hz) radios, and at still higher communication frequency bands, and also at higher speed for faster computer operations. The 3-D to 2-D evolution was also aimed to increase integration to put more electrical signal processing functions on one small silicon die (or chip). But fortunately (again an evolution by someone's design sometimes ago, billions of years ago) silicon is unique in having a surface oxide which is stable in the planet earth's ambient of oxygen and nitrogen, even some water vapor or hydrogen. No other chemical element's oxide has such electronic stability (tens of years or more) of one part per trillion change.

Thus, the historical path of the evolutionary invention of CMP as a practice of intelligent design, should not be surprising in its choice of routes and peoples, which I witnessed as part of the design and as the dispatch to tell its story: Shockley's BTL@1949 ➔ Shockley Transistor Corporation via Shockley@1956 ➔ Fairchild Semiconductor Corporation via Moore and Sah@1959 who previously both worked for Shockley ➔ UIUC (University of Illinois at Urbana-Champaign) via Sah@1962 ➔ Intel Corporation via Leo Yau who earned the PhD in ECE at UIUC under Sah in 1969, but delayed his arrival at Intel until 1977 via postdoctoral and junior faculty tenures ➔ USA and then world-wide semiconductor industry via Yau's Intel equipment contractor around ~1991. The Intel equipment contractor, found by Yau, was a German brother machine shop, which repeated the evolutional intelligent design cycle several times, with a period of about one year following Yau's designs. As the still going industrial practice today, it built the first ten units of the latest Yau design for Intel, then it was allowed to produce many more units of that same design for the rest of the semiconductor industry or Intel competitors for more profits at higher profit margins for the German brothers, while simultaneously producing for Intel the next generation from Yau's next design. Leo Yau escorted me through his 黄鼠狼窝 ("Shunk Works" = Yellow Rat Wolf Burrow), the Portland downtown riverfront German-brother machine shop, that was known to only a few Intel upper managers. To give me the private tour, they invoked the belief of the leak-proof teacher-student bond, in order to get a neutral witness from the academia. So, there is no doubt, at least in my mind, that the CMP technology for planarization of the interconnect lines and studs, that went into worldwide volume IC manufacturing in the 1990's, was an act of <u>evolution via intelligent design</u>. It had a total evolutional learning time of about 20 years (Yau's involvement time) or 40 years from Shockley Transistor Corporation's production in 1956 to reach volume manufacturing in ~1995 with about 10 technological increasingly more complex silicon transistor fabrication processing steps, initially bipolar then MOS integrated circuits, cumulated by Leo Yau's 黄鼠狼窝 at the Portland downtown river front. Their transition events or laboratory-jumping times are 1949-1956-1959-1962-1975-1995. No other self-proclaimed or company-designated inventors could possibly claim first or claim the CMP invention credit, regardless of their patent filing and issuing dates. Some companies, like Intel, reported by the media, do not file patent protection, relying solely on the trade-secret lead-time, since patent disclosures and claims hasten reverse engineering.

Let us next turn to the fundamental EM physics, which is a story of how I got into this topic. I had hoped to take the opportunity in 1984 when a America-born and Nanjing-China educated Materials Science graduate student sought me out to work on a PhD thesis project, even for free if I could not come up with a paid research assistantship, by continuing washing dishes at the student cafeteria at UIUC. But how could I not, for one who had ranked first in all the graduate ECE classes she had to take to make up the deficiencies due to the transfer of major from Materials Science to Electrical and Computer Engineering. Therefore, I took this up as an "intelligent design" opportunity to look into the alternatives for the fundamentally flawed Black EM formula based on the electron-wind force, which I learned from reading Rolf William Landauer's 1956-1978 articles which led him to conclude the nonexistence of the electron-wind force even using the most basic fundamental formulation of the many-body problem. The alternative theory, which I then derived, was an electron-windless formula for neutral atoms moving away from a void in the metal line, causing the growth of the void. My simple kinetics formula was based on just diffusion and trapping and detrapping of neutral atoms on the surfaces and in the surface layer of a void, with no drift current and no ions. It was a very familiar subject to me, because I was drilled by Shockley 20 years earlier via all the theoretical analyses on the electron-hole generation-recombination-trapping statistics (kinetics), which had resulted in the 1958 Sah-Shockley Physics Review article. That was the second, and last I wrote under Shockley's tutelage. However, the 1984 pp that encircled Yau, derailed my hope, from not able to get industrial wafers with EM test patterns from Leo Yau at Intel to get the experimental proof of the windless theory.

Then, another 20 years later (~1984 to 2006), one more "intelligence design" appeared. I jumped at the window of opportunity. This happened when I was invited by Professor Kok-Khoo Phua (the second time, since he also invited me to start a book series with his book publishing company, the World Scientific Publishing Company, 15 years earlier in 1991, which is now publishing this EM book) to lecture and consult during November-December 2006 at the Nanyang Technological University (NTU) in Singapore. During the one-week on-site consulting review about NTU's teaching and research programs, I met a young associate professor, Cher Ming Tan. He rang my bell when he reported his EM work to the research review committee. Upon being asked by me during the meeting and the post-meeting one-on-one discussion, he showed and gave me a copy of his 250-page review article on EM, which was about to be published by the leading international journal on materials. I immediately invited him to write a book. He agreed. The rest is history, recorded and proven by this monograph.

I would like to thank all the WSPC production staff members and WSPC copy editor Mr. Tjan Guangwei at Singapore, and the acquisition editor, VP Ms. Zhai Yubing in New Jersey, for their efforts and supports. Special thanks are due NTU Physics Professor Kok-Khoo Phua, the Founder and Chairman of WSPC, for his farsight, which is a proof of evolutionary intelligent design. I also acknowledge Dr. Jie Binbin, Professor of Physics of Xiamen University, for editorial assistance, and for his support from President Zhu Chongshi, School of Physics Dean Wu Chenxu, and Department of Physics Chair Zhao Hong of Xiamen University, Xiamen, China.

Sah Chihtang (Chih-Tang Sah)
CTSAH Associates, Florida, USA.
Department of Physics, Xiamen University, China.
May 10, 2010.

Preface

Interconnect is a vital part of integrated circuit (IC) to connect the different transistors together to form a network so that together they can perform complex functions that it is designed for. The reliability of integrated circuit therefore depends heavily on the reliability of interconnects. There have been extensive researches on IC interconnect electromigration over the past few decades. As a result, researchers in this field are often overwhelmed with many publications on the subject matter, especially for postgraduate students working in this area. This book attempts to compile the research papers as comprehensive as possible and organize them so that a clear picture on the mechanisms of electromigration and the governing factors for both Al and Cu based interconnects can be obtained. It should be useful for back end process engineers to understand the key process parameters in order to ensure good interconnect electromigration performance. It will also be useful for IC designers to know how their layout could affect the electromigration performance. The differences in the mechanisms of electromigration of Al and Cu are also discussed in depth.

While electromigration has been extensively researched, there are still many remaining challenges. Further research investigations are necessary. They are discussed in the last chapter.

Contents

Foreword vii

Preface xiii

1. Introduction 1

1.1 What is Electromigration? 1
1.2 Importance of Electromigration 2
1.3 Outlines of this Book 6

2. History of Electromigration 11

2.1 Understanding the Physics of Electromigration 11
2.1.1 Quantum mechanical theory of electromigration 11
2.1.2 Practical engineering electromigration formulation 16
2.1.3 Concept of flux divergence 20
2.2 Electromigration Lifetime Modeling 22
2.3 Electromigration Lifetime Improvement 29
2.4 Electromigration Aware IC Design 30

3. Experimental Studies of Al Interconnections 37

3.1 Introduction 37
3.2 Process-Induced Failure Physics 38
3.2.1 Microstructural inhomogeneities of metallization 38
3.2.1.1 Grain size 40
3.2.1.2 Grain size distribution 40

3.2.1.3 Texture of a metal line 41
3.2.2 Presence of impurity 42
3.2.3 Mechanical stress in the film 43
3.2.4 Presence of defect 45
3.2.4.1 Length dependence of lifetime 45
3.2.4.2 Length dependence of the standard deviation of EM lifetime distribution . 46
3.2.5 Temperature gradient 46
3.2.6 Material differences 51
3.2.7 Temperature 52
3.3 Design-Induced Failure Mechanisms 54
3.3.1 Proximity of metal lines 54
3.3.2 Inter-metal dielectric (IMD) thickness between metal lines and oxide thickness underneath the first metallization 54
3.3.3 Number of metallization levels 54
3.3.4 Use of barrier layers 55
3.3.5 Via separation length 58
3.3.6 Cornering of metal line and Step height of metal lines 59
3.3.7 Use of passivation layer 59
3.3.8 Metal width variation 61
3.3.9 Reservoir effect 61
3.4 Self-Induced Process During EM 63
3.4.1 Self-induced stress gradient 63
3.4.2 Self-induced temperature gradient 64
3.4.3 Microstructure change of interconnect 65
3.5 Electromigration Test Structure Design 67
3.5.1 NIST test structure 68
3.5.2 Test structure for multi-level metallization . . . 73
3.5.3 Test structure for bamboo structure 77
3.6 Package-Level Electromigration Test (PET) 77
3.7 Rapid Electromigration Test 79
3.7.1 TRACE method 80
3.7.2 Standard Wafer-level Electromigration Accelerated Test (SWEAT) 83

3.7.3 Wafer-level isothermal Joule-heated electromigration test (WIJET) 85
3.7.4 Wafer level constant current electromigration test (Lee, Tibel and Sullivan 2000) 87
3.7.5 Breakdown energy of metal (BEM) 87
3.7.6 Pitfalls of SWEAT 90
3.7.7 Potential pitfalls of constant current test method 99
3.7.8 Potential pitfall of breakdown energy method (BEM) 100
3.7.9 Potential pitfalls of WIJET 100
3.7.10 Summary 101
3.7.11 Highly accelerated electromigration test 101
3.8 Practical Consideration in Electromigration Testing . . 103
3.8.1 Failure criteria used in EM testing 103
3.8.2 Interpretation of the measured $\Delta R/R_0$ 105
3.8.3 Actual temperature of test strips 108
3.8.4 Test structure used 109
3.8.5 Current density used 109
3.8.6 Short length effect 109
3.8.7 Failure model used in EM accelerated testing (deviation from Black equation) 111
3.9 Failure Modes in Electromigration 113
3.9.1 Open/resistance increase 113
3.9.2 Short 115
3.10 Test Data Analysis 115
3.11 Failure Analysis on EM Failures 125
3.12 Conclusion 133

4. Experimental Studies of Cu Interconnections 143
4.1 Different in Interconnect Processing and its Impact on EM Physics 144
4.2 Process-induced Failure Physics 152
4.2.1 Interface between Cu and surrounding materials 152
4.2.1.1 Surface engineering 153
4.2.1.2 Alternative cap-layer materials 156
4.2.2 Microstructure 162

4.2.3 Presence of impurity 173
4.2.4 Mechanical stress 180
4.2.5 Barrier metal 191
4.2.6 Presence of defects 199
4.2.7 Temperature gradient 200
4.2.8 Material differences 200
4.2.9 Temperature 201
4.3 Design-Induced Failure Mechanism 202
4.3.1 Line width dependence of EM lifetime 202
4.3.2 Current crowding 208
4.3.3 Line width transition 211
4.3.4 Reservoir effect 212
4.3.5 Current direction dependence of EM lifetime . . 217
4.3.6 Blech short length effect 221
4.3.7 Via structure design 224
4.4 Electromigration Testing 225
4.5 Statistics of Cu Electromigration 228

5. Numerical Modeling of Electromigration 243

5.1 1D Continuum Electromigration Modeling 245
5.2 2D EM Modeling 246
5.2.1 Sharp interface model 248
5.2.2 Phase field model 251
5.3 Electromigration Simulation Using Atomic Flux Divergence and Finite Element Analaysis 253
5.3.1 Computation methods for Atomic Flux Divergence (AFD) 254
5.3.1.1 Formulation of AFD 254
5.3.1.2 Voiding mechanism simulation 257
5.3.1.3 Lifetime prediction 258
5.4 Monte Carlo Simulation of Electromigration 260
5.4.1 Monte Carlo simulation of the movement of atoms during EM 260
5.4.2 Monte Carlo simulation of void movement during EM 262
5.4.3 Holistic EM simulation 263
5.5 Resistance Change Modeling 264

6. Future Challenges 269

6.1 Electromigration Modeling 269
6.2 EDA Tool Development 271
6.3 Physics of Electromigration 274
6.4 Electromigration Testing 274
6.5 New Failure Mechanism for Interconnects 275
6.6 Alternative Interconnect Structure 277
6.7 Alternative Interconnect System 278

Index 285

Biography 291

CHAPTER 1
Introduction

1.1 What is Electromigration?

Electromigration is the phenomena of interconnect metal self-diffusion along an interconnection as high current density is passing through the interconnection. As a result of the metal movement, voids will be formed on some parts of the interconnection, and hillock due to the accumulation of the metal atoms will be formed on different parts of the interconnection. The presence of voids will increase the resistance of the interconnection, and the presence of hillock will cause short circuit between the adjacent interconnections if the hillock is developed side-way and short circuit between the different levels of interconnections if the hillock is developed vertically and punch through the inter-metal dielectric.

Electromigration has been a subject of scientific study for over 100 years, but the interest remained academic until it became a major failure mechanism for integrated circuits (IC) in 1959 as thin and narrow metal films are used for interconnections. Indeed, electromigration was the one of the first few failure mechanisms found in IC.

Unlike the bulk conductor which will melt from Joule heating at $\sim 10^4$ A/cm^2, the metallization in IC can sustain current densities greater than 10^7 A/cm^2 due to the good thermal contact with the silicon substrate. It is this high current density that makes the effect of electromigration becomes significant.

In addition, thin film interconnections in IC possess small grain sizes and high surface or interfacial area to volume ratios with many high mobility diffusion paths, allowing mass transport of the self-diffused metal atoms at low temperatures. When these diffusion paths intersect with one another, and each having different mobility, accumulation of atoms or voids nucleation will results at the intersection. This combination of a high driving force for electromigration and the availability of inhomogeneous high-mobility diffusion-paths network makes thin film conductor susceptible to electromigration damage.

1.2 Importance of Electromigration

With the technology scaling according to the International Technology Roadmap for Semiconductors (ITRS) 2007, the geometrical dimensions of transistors and interconnects decrease linearly. On the other hand, supply voltage scaling is saturating. These different scaling trends increase the electric fields. Furthermore, new materials, such as the porous low-k materials and Cu are introduced to meet evolving technologies requirements. As a result, effects that were considered to be second order in previous technology nodes, such as reliability, are becoming important considerations during IC design (Papanikolaou 2006). For example, electromigration continues to worsen due to smaller interconnect width and thickness as well as smaller separation between interconnections. As a result, coalesce of fewer vacancies can cause unwanted significant via resistance rise. Time-dependent dielectric breakdown happens much more often because of lower k materials with the worse breakdown behavior (McPherson 2007).

Interconnect reliability, with electromigration as the dominate failure mechanism, is the main factor that determines the circuit reliability especially when the line width becomes much narrower that renders high current density as shown by Srinivasan (Srinivasan 2004). They showed that the failure rate of a scaled 65 nm processor is more than three times higher than a similarly pipelined 180 nm processor, with electromigration (EM) and time-dependent dielectric breakdown showing the most dominant mechanisms with the advancement in the technology nodes. The increasing failure rate is believed to be due to the potentially shorter time to failure of an interconnection with narrower line width and the increasing sensitivity of the circuits to interconnect line resistances as the circuits are operating at

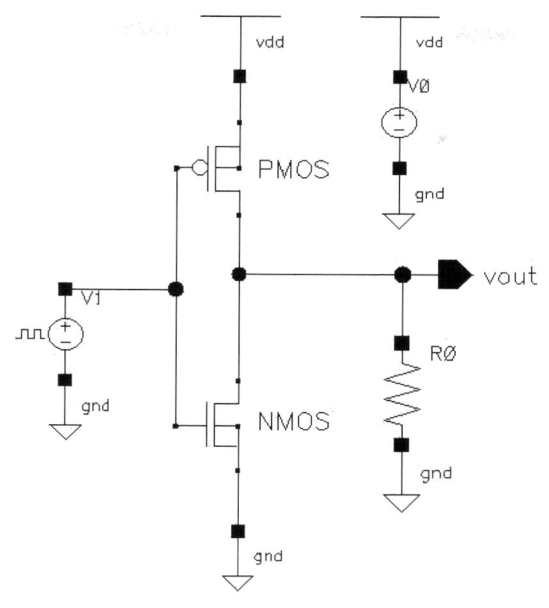

Fig. 1.1. Schematic of a simple inverter circuit to illustrate the increasing sensitivity of circuit performance to interconnects resistances.

higher frequency. For illustration purpose, we study a simple CMOS voltage inverter circuit as shown in Fig. 1.1.

We model interconnects in the circuit as resistors whose values will increase with time due to electromigration. Figures 1.2 and 1.3 show that the higher the interconnect resistance, the worse the circuit performance, i.e. longer signal delay and higher power consumption, and such impact of interconnect resistance to IC performance is greater as the operating frequency of the integrated circuits became higher.

Also, with the same simulation setup using Cadence EDA tool, when the circuit operating frequency increases (e.g. from 100 MHz to 1 GHz), the resulting temperature of the circuit increases as can be seen from the temperature extrapolation in Fig. 1.4 using ANSYS finite element modeling.

The increase in the circuit temperature as shown in Fig. 1.4 will increase the electromigration degradation rate (Cher Ming Tan; Roy October 2007), thus shorten the life time and degrade the reliability of an integrated circuit.

From the above discussion, it is obvious that the study of interconnect reliability in an integrated circuit is increasingly important as technology

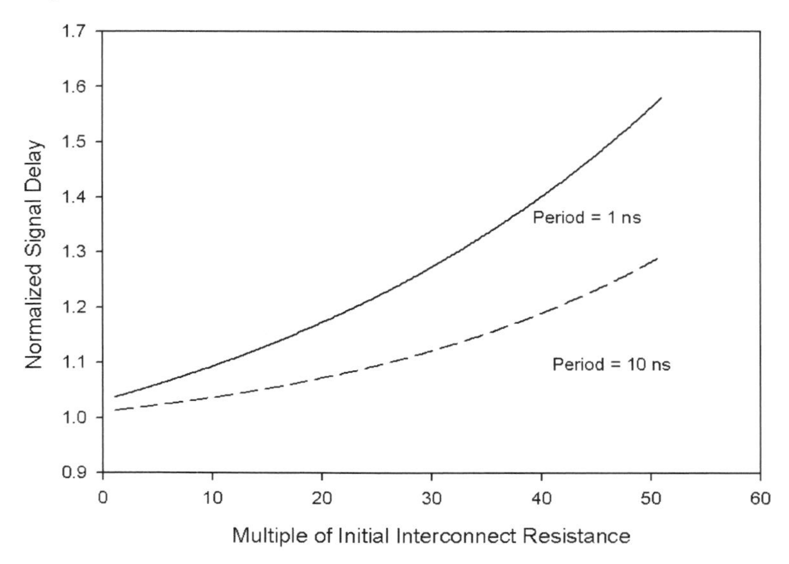

Fig. 1.2. Effect of interconnect resistance on signal delay under different operating frequency. (Initial % signal delay for 10 ns period = 0.8%; for 1 ns period = 4%).

node advances in order to ensure its reliability. This concurs with the recent work by Guo *et al.* (Jin Guo December 2008).

In fact, chip failures due to power grid issues, whether IR drop or electromigration, are already being discovered by chip designers. Since these issues are related to the number and the way components are assembled on a chip and are primarily a global phenomenon, power grid analysis is becoming a required addition to many design flows (Cooke, Goossens *et al.*). This, together with the rapid evolution of technology scaling, does not allow needed time for proper thorough evaluation of new interconnection reliability, thus additional efforts at the design stage are essential now to obtain robust and reliable chips (Guo, Papanikolaou *et al.* 2008).

In order to provide adequate knowledge on the interconnect electromigration for both the IC manufacturers and designers, a comprehensive monograph on electromigration for ULSI Interconnections covering the physics of electromigration, the various factors, both process and design related, that can influence electromigration, the testing methodologies of interconnect reliability and the modeling of electromigration is necessary. It is to these purposes that this monograph is written.

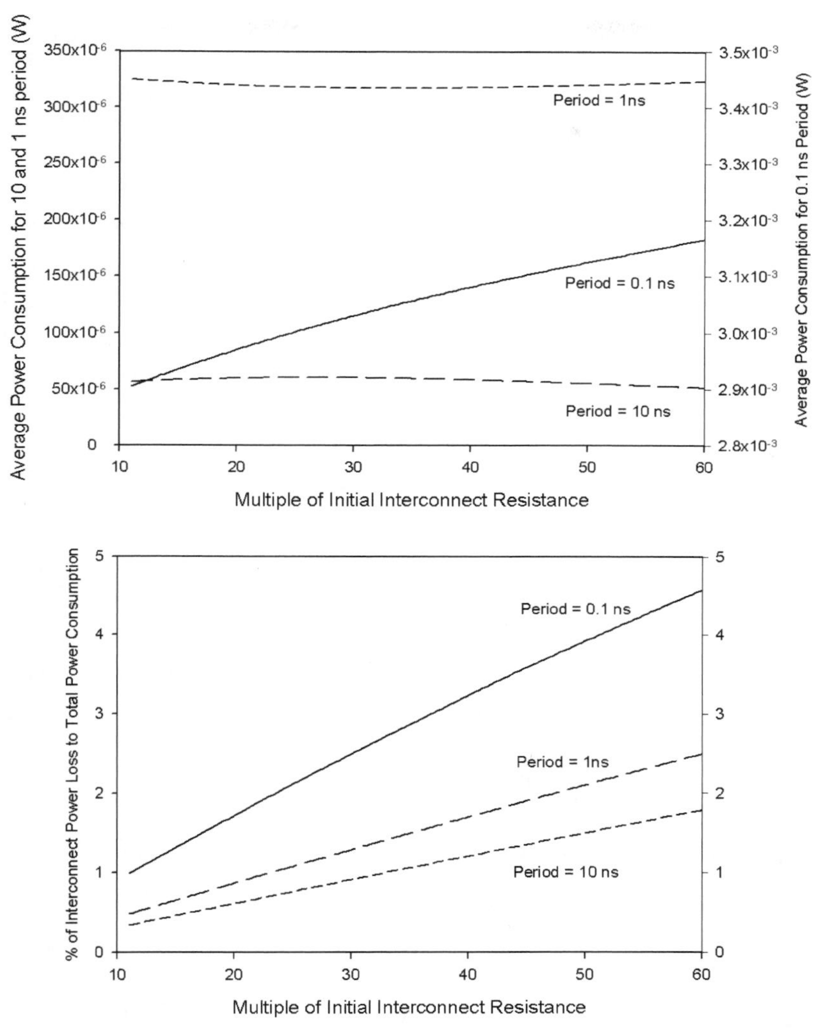

Fig. 1.3. Effect of interconnect resistance on circuit power consumption under different operating frequency. Note that the right vertical axis of the left figure is for the case of circuit operating at 1 GHz.

The modeling of electromigration is particularly important because it helps to identify the critical weak spots of an interconnection system in a short time. A typical IC layout has millions of interconnects. Analyzing the reliability problems on each of them is prohibitively time consuming.

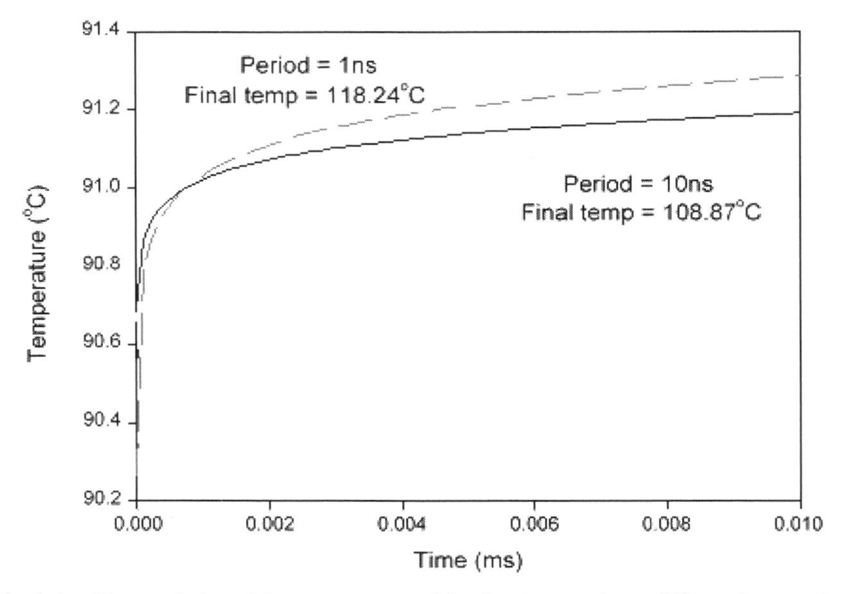

Fig. 1.4. Time variation of the temperatures of the circuit operating at different frequencies.

Therefore, the key interconnects which will be investigated will have to be identified through electromigration modeling. Also, if the modeling of electromigration can be integrated into EDA tool, a complete electromigration aware IC design can be realized.

1.3 Outlines of this Book

Since the discovery of electromigration (EM) phenomena in the 60's, the theory of EM revolved around the electron wind force as it was identified to be the sole driving force responsible for the EM failure observed in interconnects. This electron wind force formulation by Fiks (Fiks 1959) and Huntington and Grone (Huntington and Grone 1961) is a major contribution to the study of EM, and it has explained many experimental EM observations in Al interconnect. Later on, Black developed an empirical equation to relate the median time to failure with respect to the current density and temperature of the metal interconnects (Black 1967), and has been widely used till now.

With the various interconnect system developed, and a continuous shrinkage of the interconnect line width as well as the change in the interconnect material, the theory of EM is being refined and driving forces other than the electron wind force are found to be significant, and the correctness of the Black equation in predicting the median times to failure of interconnect is also become questionable. In this work, the evolution of the physical theory of EM will be explored, and both the diffusion path and driving force approaches in EM study will be discussed. The continuous interaction of the various driving forces during the EM process will be shown.

While the basic of EM is the metal atoms movement under driving forces through various diffusion paths in the metal interconnection, the changing dominance of the driving forces and the diffusion paths due to the design and process of an interconnect system complicate the understanding of the EM physics. Fortunately, with the advancement in numerical tools, our understanding of the underlying physics is enhanced, and key factors that affect the EM performances of an interconnect system can be identified. This is significant for ULSI development so that a reliable interconnect system can be built as ULSI technology is advancing.

Various numerical schemes such as finite-element-analysis, solution of coupled diffusion equations for EM etc. have been developed and will be presented in this work. The advantages, disadvantages, achievement and correlation of these models to the experimental results will be discussed. Some quantum mechanical treatments of the phenomenon and their short-comings will also be presented.

EM testing is essential for all wafer fabrication plants with interconnect back end of line (BEOL). Various test structures have been developed and used in the last two decades for EM testing, and they will be presented. As the interconnect technology is improving, the required time of EM testing is getting longer, and different type of accelerated tests are developed in both academic and industrial. Some of these accelerated tests however, do not produce results that can help in predicting the life time of an interconnect system under normal operation condition, and in some cases, they can even produce misleading information. Their problems stem from the improper test structure design, incorrect use of stress conditions and inaccurate test data analysis methods used. All these problems will be discussed in detail so that a rigorous accelerated stress testing for EM can be established.

Physical failure analysis is the postmortem of EM experiments. Methodology that can reduce the required resources for physical analysis will be presented. Non destructive techniques to detect the failure locations and their capability will be presented.

With the vast amount of knowledge on Al EM, extension of this knowledge to the present Cu interconnect will be presented, and the differences in the EM behaviors for Al and Cu will be highlighted and explained in relation to their different process technologies. The various processes and materials related improvement for interconnect EM will be reviewed. The current progress and outstanding issues on Cu EM characteristics will be addressed, and the different key factors that influence Cu EM will be discussed in details. The sources of these factors and their interaction will be explored, and their effect on the EM activation energy, void locations, void growth and void nucleation dominated mechanisms will be discussed. The understanding of these sources from design and process perspective will be helpful in the design of Cu interconnect system. Environmental related factors are also considered so that the EM performance of Cu based interconnect system can be accurately assessed.

As the ULSI technology is advancing, and the demand for ULSI in term of speed and functionality are ever increasing, the interconnect system is facing a number of different challenges ahead. Novel interconnect systems such as 3D interconnect are proposed and experimented, new interconnect materials are also being explored. The reliability of these new proposals remains unknown, and in the last section, the potential reliability problems for these new proposals will be outlined and discussed.

The structure of the book is organized as follows. We begin the description of the history of EM in Chapter 2. We then proceed to describe the various experimental studies of EM for Al and Cu respectively in Chapters 3 and 4 where the test structures used for EM, the test methodologies, the failure analysis methods for EM failures, the reported experimental results for EM as well as the data analysis of the time to failure for EM test will be discussed. With these experimental findings, a much better understanding of the physics of EM is obtained and hence a physics based numerical modeling of EM is possible, and it enables us to identify the design, process and material related factors that govern the EM of interconnects which will be useful for both the foundries and IC designers. The various numerical modeling for EM will be described in Chapter 5. Lastly, the future challenges in interconnect EM will be discussed in Chapter 6.

References

Black, J. R. (1967). *Proceedings of the 6th Annual reliability Physics Symposium*. New York, IEEE, 148–159.

Tan, C. M. and Roy, A. (October 2007). Electromigration in ULSI interconnects. *Materials Science & Engineering R*.

Cooke, L. H. and Goossens, M., *et al. Signal Integrity Effects in System-on-Chip Designs — A Designer's Perspective*.

Fiks, V. B. (1959). On the mechanism of the mobility of ions in metals. *Sov. Phys. Solid State* **1**: 14.

Guo, J. and Papanikolaou, A., *et al.* (2008). The analysis of system-level timing failures due to interconnect reliability degradation. *IEEE Transactions on Device and Materials Reliability* **8**(4): 652–667.

Huntington, H. B. and Grone, A. R. (1961). Current-induced marker motion in gold wires. *J. Phys. Chem. Solids* **20**: 76–87.

Guo, J., Stucchi, M., Croes, K., Tokei, Z. and Catthoor, F. (December 2008). The analysis of system-level timing failures due to interconnect reliability degradation. *IEEE Transactions on Device and Materials Reliability*.

McPherson, J. (2007). Reliability trends with advanced CMOS scaling and the implications for design. CICC, 405–412.

Papanikolaou, A. (2006). Reliability issues in deep deep sub-micron technologies: Time-dependent variability and its impact on embedded system design. *International Conference VLSI-SoC, IFIP*.

Srinivasan, J. A. Bose, P. and Rivers, J. A. (2004). The impact of technology scaling on lifetime reliability. *2004 International Conference on Dependable Systems and Networks*, 177–186.

CHAPTER 2
History of Electromigration

2.1 Understanding the Physics of Electromigration

2.1.1 *Quantum mechanical theory of electromigration*

The first reported work on EM was presented by Fiks in 1959 (Fiks 1959). Around the same time in 1961, Huntington and his coworkers at the Rensselaer Polytechnic Institute contributed significantly to the EM failure process (Huntington and Grone 1961) through their studies on current induced motion of surface scratches on bulk metals. They concluded that thermally activated metal ion becomes essentially free from lattice and is acted by two opposing forces in a metal. These forces are called (a) direct force and (b) 'electron wind' force. The direct force is the force experienced by the activated positive metal ion in the opposite direction to the electron flow due to the application of electric field. On the other hand, 'electron wind' force is the force experienced by metal ion in the direction of electron flow due to the momentum exchange between the moving electrons with the ion.

In the late 60s, EM threatened the existence of the integrated circuits (ICs) (Blech and Sello 1966). Electron moves with very high speed and higher momentum is transferred to metal ions when the current density is high in the interconnections in ICs. As a result of the higher momentum transfer to ion, 'electron wind' force dominates and causes an appreciable mass transport in the direction of electron flow. The resultant force of the direct and 'electron wind' forces was termed as 'electron wind force' for simplicity or simply 'wind force' in the ballistic model of EM developed by Huntington and Fiks (Fiks 1959; Huntington and Grone 1961; Huntington 1975; Verbruggen, Griessen and Groot 1986; Duryea and Huntington 1988). By the end of the 60's, EM degradation for many metals including Au, Cu, Al, Ag etc. and metal alloys was studied (Fiks 1959; Huntington and Grone 1961; Blech and Sello 1966; Hartman and Blair 1969; Kang, Burgess, Coleman and Keil 1969; Patil and Huntington 1969; Huntington 1975; Verbruggen, Griessen and Groot 1986; Duryea and Huntington 1988).

The ballistic model developed by Huntington and Fiks mentioned earlier was based on electron scattering by defect (or impurity) and the corresponding momentum transfer is from the scattering electrons to the defect. In this model, the system was defined by the electron density n_e, defect density n_d and the relaxation time τ of the electrons including all the scattering mechanisms. It was assumed that in a collision with a defect, an electron transfers all of its momentum $-e\vec{E}\tau$ to the defect. Thus in a time τ_i, which is the transport life time corresponding to the scattering by the defect under consideration (or mean time between two successive collisions), momentum transfer by the electrons is equal to $-n_e e\vec{E}\tau/n_d$ per defect. The momentum transfer per unit time, which is also the driving force, is thus given as

$$\vec{F}_{wind} = -\frac{n_e e\tau}{n_d \tau_i}\vec{E} = -\frac{n_e \rho_d}{n_d \rho} e\vec{E}, \tag{2.1}$$

where ρ and ρ_d is the total resistivity of the metal interconnect and the resistivity contribution from the defects in the metallization. At room temperature, it is estimated that

$$\left(\frac{\rho_d}{n_d}\right)\left(\frac{n_e}{\rho}\right) \approx 50.$$

Quantum mechanically, the above mentioned idea can be expressed as (Verbruggen 1988; Lodder 1989)

$$\vec{F}_{wind} = \sum_{\vec{k},\vec{k}'} \hbar(\vec{k} - \vec{k}')P_{\vec{k}\vec{k}'} f_{\vec{k}}(1 - f_{\vec{k}'}), \tag{2.2}$$

where $\vec{k}$ is the electron wavevector, $P_{\vec{k}\vec{k}'}$ is transition probability given by the generalized Fermi golden rule and $f_{\vec{k}}$ is the shifted Fermi-Dirac distribution given by

$$f_{\vec{k}} = f_{\vec{k}}^{0} + \left[-\frac{\partial f^{0}}{\partial \varepsilon} \right] (-e)\tau \vec{V}_{\vec{k}} \cdot \vec{E}. \tag{2.3}$$

Here $f_{\vec{k}}^{0}$ is the Fermi-Dirac distribution in the absence of electric field, e the electronic charge, $\vec{V}_{\vec{k}}$ the Fermi velocity, and $\vec{E}$ the macroscopic electric field.

Substitute Eq. (2.3) into Eq. (2.2), and with some mathematical manipulation, Eq. (2.1) can be reproduced. Thus a free electron quantum model basically does not give more insight details of the phenomenon and reproduces the ballistic model.

From Eq. (2.1), it can be noticed that the driving force is inversely proportional to the resistivity of the metal interconnect and directly proportional to the defect resistivity. Thus 'electron wind force' is expected to decrease with increasing temperature. Also, reducing the defect density will reduce the 'electron wind force', hence enhances the metal EM performance. This fact was confirmed experimentally for some metals, although much smaller values of the wind force were reported by Huntington (1975).

Although the ballistic model explains the basic observations, there were several questions arisen on this model probably due to its over simplified assumptions. For example, it was not clear that how ballistic model can be extended if one goes beyond free-electron approximation, especially for situation like scattering by cluster of atoms or by atom-vacancy complexes when momentum transfer must be petitioned among defects and lattice. This problem was taken into consideration in the EM formalism by Bosvieux and Friedel (1962) where the disturbance of current flow near the defect, due to the scattering of electrons by the defect was taken into account in this formalism. To maintain these disturbances, the force exerted by the lattice defect on the electrons is equal to the force on the defect by the polarized charge distribution, in a stationary state. The wind force is then given by Verbruggen (1988)

$$\vec{F}_{wind} = \int n(\vec{r}) \left[-\frac{\partial V_b}{\partial \vec{R}_i} \right] d^3 r, \tag{2.4}$$

where $n(\vec{r})$ is the actual electron density in the presence of impurity and electric field, and V_b is the bare potential of the defect located at $\vec{R}_i$. The actual electron density in an independent particle approximation can be obtained by populating the electron-scattering states $\psi_{\vec{k}}(\vec{r})$, and using Eq. (2.3), we have

$$n(\vec{r}) = \sum_{\vec{k}} f_{\vec{k}} |\psi_{\vec{k}}(\vec{r})|^2. \tag{2.5}$$

Equations (2.4) and (2.5) are the starting point of all the calculations of the wind force in metals. In order to justify Eq. (2.4), Born-Oppenheimer adiabatic approximation or Feynman-Hellman theorem is usually invoked. Equation (2.4) is also produced from the Kubo's linear response formalism (Kumar and Sorbello 1975) for driving force as shown by Sham (1975) and Schaich (1976). One of the advantages of the linear response formalism is that it considers a many-body statistical mechanics system and thus one can justify the nature of the approximation introduced in the formalism easily.

In order to compute the wind force, pseudopotential formalism was used. Sorbello (1973) calculated the wind force on vacancies and impurities for virtually all free-electron-like metals within the pseudopotential formalism. A similar technique was used by Genoni and Huntington (1977) to study the anisotropy of the effective charge number (Z^*) in Zn, Cd, Mg, etc. Their theoretical and experimental results were in reasonable agreement.

For the wind force in noble and transition metals (Gupta 1982; Gupta, Serruys, Brebec and Adda 1983) evaluated Eq. (2.4) within muffin-tin formalism and satisfactory results were obtained especially for noble metals. However, the values of Z^* in Nb for vacancies and impurities are an order of magnitude smaller. Using a finite cluster of muffin-tin potentials with the preservation of multiple scattering, Brand and Lodder presented satisfactory results, though their results were dependent on the cluster size (Lodder and Brand 1984; Brand and Lodder 1986).

Although there were reasonable agreements between theory and experiment, the situation was still not satisfactory due to the validity of the approximations made to evaluate Eq. (2.4) and thus further refinement of the model was required.

Landauer and Woo (1974) noticed that two contributions will be missed if Eq. (2.4) is evaluated in Born approximation. These two missing terms are the residual resistivity dipole (RRD) field and the carrier density modulation (CDM) effect.

The importance of the RRD field for driving force in EM was shown by Bosvieux and Friedel (1962). It is to be noted that RRD field is of second or higher order in potential which gives a significant corrections to the wind force in the case of strong scattering. The details computation of RRD field can be found in (Landauer 1975; Sorbello and Dasgupta 1977; Sorbello 1981). Later on, exact solution of the scattering problem was presented, and it was shown by Schaich (1976) that there is no need for an order-by-order calculation of the corrections.

The CDM field reduces the direct force in EM. The CDM field arises from the local disturbance in the potential due to the presence of a lattice defect. Continuity of current will require a change in the velocity of the carriers and consequently a change in conductivity near a defect which implies a local change in electric field. For a defect which creates an attractive potential, the CDM field is in the opposite direction from the background field. Thus CDM field reduces or screens the direct force in EM. Das-Peierls EM theorem (Das and Peierls 1975) and the work presented by Landauer (Landauer 1975; Landauer 1977) are the most appealing work on CDM field. These workers considered the momentum balance of electrons in the presence of defect, impurity and background scattering.

Apart from the above mentioned refinement in ballistic model which is based on the electron wind force, Landauer (1989) concluded that the electron wind force is untenable even considering the most fundamental and complete many body quantum transport theory. [C. T. Sah, (1996) Fundamentals of Solid-State Electronics — Solution Manual, World Scientific Publishing Company: 174–176. C. T. Sah, (2006) Personal communication with Prof Sah (2008), and C.T. Sah and B.B. Jie (2008)] Sah and Jie obtained a resistance formula due to EM considering atomic diffusion and trapping on void surface without the electron wind force, and they managed to produce the time to failure equation similar to that of the Black's equation. The three most important unique features of the Sah's model are the time dependence of the resistance of a wire, its stress current dependence, and the thermal and electric field dependent capture and emission rates of the migrating atoms and their vacancies, and also their thermal activation energies that are a combination of the bond breaking and migration energies. The former two can be readily verified by experiments, both resistance versus time measurements and photograph records, while the latter, by varying material composition, both the host and the impurities, including the differences among the different metals and the impurities such as H, N, O, F and others that are used to control and enhance the EM resistance of the conductor wires and interconnections of integrated circuits. These fundamental physics experiments have not been performed, which would settle the fundamental forces underlying EM, and also would create engineering designs with highly increased endurance against EM.

Recently, Braun, Soe, Flipse and Rieder (2007) performed experiments and showed that metal atoms on metal surface can be moved when a small electric field and current are applied. Upon resonantly trapping an electron in atom, the antibonding state becomes occupied and the bond of the atom to the surface becomes less strong and can then hop to the next adsorption site. And when an electric field is present, it determines the direction of the atom hopping. This phenomena could shed light on the nature of electron wind force. In short, after 50 years of investigation, the fundamental microscopic nature of the electromigration remains unsettled.

2.1.2 *Practical engineering electromigration formulation*

Despite the abovementioned theories of the EM phenomenon, in practice, the evaluation of the interconnection EM reliability uses simpler equations. The resultant of 'electron wind' force and 'direct force' is given by (for simplicity, some times the resultant force is also called 'wind force') (Verbruggen 1988)

$$\vec{F}_{em} = \vec{F}_{wind} + \vec{F}_{d}, \tag{2.6}$$

where the direct force is given by

$$\vec{F}_{d} = Z_d^*|e|\vec{E}, \tag{2.7}$$

and Z_d^* is a parameter related to the valance of the metallic atom. In the absence of scattering process, Z_d^* is the valence (Z) of the metallic atom.

Using Eqs. (2.1), (2.6), (2.7) and putting the atomic density N equal to n_e/Z, the total force is given by

$$\vec{F}_{em} = \left(Z_d^* - Z\left[\frac{\rho_d}{n_d}\right]\left[\frac{N}{\rho}\right]\right) \cdot |e| \cdot \vec{E}, \tag{2.8}$$

which is again simplified as

$$\vec{F}_{em} = Z^* \cdot |e| \cdot \vec{E} = Z^* \cdot |e| \cdot \rho \cdot \vec{j}, \tag{2.9}$$

where

$$Z^* = Z_d^* - Z\left[\frac{\rho_d}{n_d}\right]\left[\frac{N}{\rho}\right], \tag{2.10}$$

and it is called the 'effective charge number' which is the measure of ion-electron interaction. Lower value of this number implies lesser momentum exchange from electron to ion. The relation $\rho = E/j$ has been used in Eq. (2.9) in which j represents the current density in the conductor.

The nature of the contribution of the 'direct' and 'electron wind' forces in EM can be understood by the determination of Z^* as a function of temperature. In order to do this, one can express Eq. (2.10) as

$$Z^* = Z_d^* + K/\rho(T). \tag{2.11}$$

Verbruggen (1988) obtained the experimental results of Z^* in V, Nb and Ta and described several typical cases of effective charge dependency on electrical resistivity. Fractional values of Z_d^* are obtained instead of $+1$, and this could be due to the CDM effect as discussed earlier. He also found that for Hydrogen in V and Nb, the electron-wind force is in the same direction as direct force, where the opposite is true for the case in Ta, indicating the significance of the host atoms of the conductors to their EM behavior.

The drift velocity of metal atoms during EM is given by the Nernst-Einstein equation as

$$\vec{v}_d = D\frac{\vec{F}_{em}}{k_B T}, \tag{2.12}$$

where the atomic diffusion coefficient D is given by

$$D = D_0 \mathrm{Exp}\left(-\frac{E_a}{k_B T}\right). \tag{2.13}$$

E_a is the activation energy associated with the diffusion process, k_B is the Boltzmann's constant and T represents the temperature. Using Eqs. (2.9), (2.12) and (2.13), the drift velocity can be expressed as (Blech and Kinsbron

1975)

$$\vec{v}_d = \frac{D_0}{k_B T}|e| \cdot \rho \cdot Z^* \cdot \vec{j} \cdot \text{Exp}\left(-\frac{E_a}{k_B T}\right). \tag{2.13}$$

In the early days, EM drift velocities were obtained using TEM (Blech and Meieran 1969) and SEM (Weise 1972) techniques. More sophisticated and direct measurement of EM drift can be found in (Blech and Kinsbron 1975; Blech 1976).

The diffusion of metal atoms during EM is governed by the Fick's diffusion laws with additional terms. Clement and Lloyd contributed significantly in the solution of EM related diffusion equations (Shatzkes and Lloyd 1986; Clement and Lloyd 1992). The vacancy flux (which is opposite to the atomic flux) for EM mass transport is given by (Shatzkes and Lloyd 1986; Clement and Lloyd 1992)

$$J = -D \cdot \nabla c + \frac{cD}{k_B T} F_{em}, \tag{2.14}$$

where c is the vacancy concentration. The first term on the right-hand side of the equation represents the flux due to concentration gradient, and the second term represents the flux due to the electron wind force.

With the Fick's second law in diffusion given as

$$\frac{\partial c}{\partial t} = -div J. \tag{2.15}$$

Equation (2.14) becomes

$$\frac{\partial c}{\partial t} = div(D \cdot \nabla c) - div\left(\frac{cD}{k_B T} F_{em}\right). \tag{2.16}$$

Assuming EM diffusion takes place only along the length (x-direction) of a conductor of length l, Eq. (2.16) reduces to

$$\frac{\partial c(x,t)}{\partial t} = D\left\{\frac{\partial^2 c(x,t)}{\partial x^2} - a\frac{\partial c(x,t)}{\partial x}\right\}, \tag{2.17}$$

with

$$a = \frac{F_{em}}{k_B T}.$$

The parameter 'a' has the unit of cm^{-1}, and it is considered as the characteristic length for EM (Kircheim and Kaeber 1991).

The boundary and initial conditions for Eq. (2.17) can be summarized as follows:

Boundary conditions

(a) $J(0, t) = 0,$
(b) $c(-\infty, t) = c_0,$
(c) $J(-l, t) = 0,$
(d) $c(-l, t) = c_0,$

Initial condition

(e) $c(x, t = 0) = c_0,$

where c_0 is the equilibrium vacancy concentration. The boundary condition (a) and initial condition (e) is common to all cases. Thus there are three cases arise when considering other boundary condition (b), (c) or (d) respectively.

Analytical solutions to Eq. (2.17) with boundary condition (b) (semi-infinite solution) at $x = 0$ and with boundary condition (c) (finite solution) for all x respectively were reported (Shatzkes and Lloyd 1986). In such cases, the time-to-failure (t_f) was expressed as (Shatzkes and Lloyd 1986)

$$t_f = BT^2 j^{-2} e^{\frac{E_a}{k_B T}}. \tag{2.18}$$

Notice the T^2 proportionality term in Eq. (2.18) which does not appeared in the Black's equation as described in the next section. The variation of $\ln(t_f / T^2)$ as a function of $1/k_B T$ was also shown to be a straight line, and its slope gives the activation energy. Note also that the Sah formula provides the whole range of two exponents of T and J separately depending on materials and also stress fluence, not provided (2.18) but observed by all experiments.

Clement and Lloyd (1992) presented the numerical solutions of Eq. (2.17) with the three boundary conditions as shown in Fig. 2.1. They found that all the solutions coincide with each other outside the saturation regime, and the deviation of the solution from the limiting semi-infinite solution is only significant near the saturation regime.

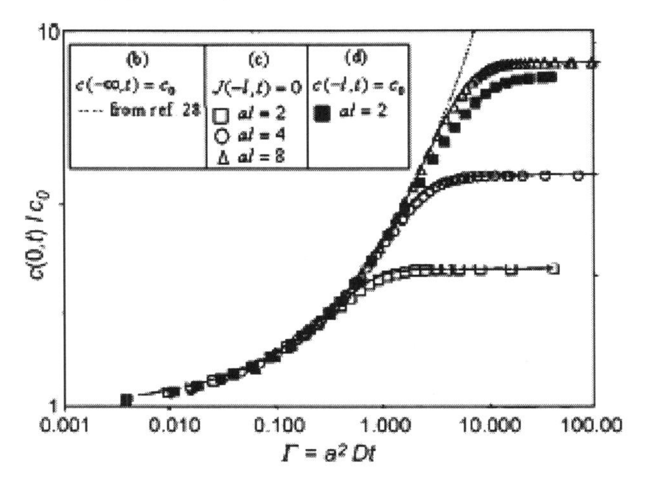

Fig. 2.1. Solutions of Eq. (2.17) for different boundary conditions (Reprinted with permission from Clement and Lloyd 1992. Copyright 1992, American Institute of Physics).

The solution with boundary condition (c) corresponds to the conservation of the concentration of vacancies and the volume of the conductor is not permitted to change. This type of situation may arise for a conductor with strong thick passivation and confined by hard surrounding materials so that the conductor volume is not permitted to change.

2.1.3 *Concept of flux divergence*

Another useful concept in EM analysis is the atomic flux divergence. It is obvious that atoms mass transport cannot induce failure in metallization unless there is non-zero atomic flux divergence (AFD). Thus for a given section of interconnect, if $\nabla J = 0$, the material entering into and flowing out from the section is exactly the same. On the other hand, when $\nabla J \neq 0$, mass accumulation (i.e. when $\nabla J < 0$) or depletion (i.e. when $\nabla J > 0$) occurs and this results in EM failure.

A simple relation between failure time (t_f) and AFD ($\nabla \bullet J$) can be found as follows. Lets ΔN be the number of atoms accumulated in a small interconnect segment of length Δl for a time period of Δt due to the presence of the flux divergence, then for an infinitesimal time period, we have

$$\Delta N = \lim_{\Delta t \to 0} \delta \cdot h \cdot \Delta t \cdot \Delta l \cdot (\nabla \bullet J), \tag{2.19}$$

where δ and h are the effective diffusion path width and film thickness (or diffusion path thickness) respectively. Multiplying Eq. (2.19) by atomic volume (Ω), the growth rate of the volume V of mass depletion or accumulation can be written as (Christou 1994; Roy and Tan 2006)

$$\frac{\partial V}{\partial t} = \delta h \Omega \Delta l (\nabla \bullet J). \tag{2.20}$$

Integrating Eq. (2.20) yields

$$\int_0^{V_c} \partial V = \Omega \Delta l \int_0^{t_f} \delta h (\nabla \bullet J) \partial t, \tag{2.21}$$

where V_c is the critical volume of mass depletion or accumulation required for failure to occur. As there are many interacting factors that produce the resultant AFD, the variation of AFD with time is complex and they do not remains constant over a given period of time once the mass depletion or accumulation begin. Also, because of the increasing inhomogeneity of the interconnect during EM, the term ($\nabla \bullet J$) is a function of time and cannot be taken outside the integration. Likewise, δ and h are also function of time.

In order to have a qualitative relationship between the time to failure and AFD, the concept of average is used, and Eq. (2.21) is reduced to (Roy and Tan 2006)

$$t_f = \frac{V_c}{\Omega \Delta l \overline{\delta h \nabla \bullet J}}. \tag{2.22}$$

Equation (2.22) gives an important statement i.e. the failure time is inversely proportional to the AFD. So homogeneous metallization is highly desirable to reduce AFD, which in turn improves the interconnect EM life-time.

There are many causes by which AFD can occur in the metallization. One of the major causes of AFD in polycrystalline interconnection was reported to be the non-uniform grain structure and their orientation in the metallization (Attardo, Rutledge *et al.* 1971; Rosenberg and Ohring 1971; Schoen 1980; Nikawa 1981; Harrison 1988; Marcoux, Merchant *et al.* 1989). A well-known structural inhomogeneity was found at the junction of three grains, the so-called triple point, and it is shown schematically in Fig. 2.2.

With the concept of AFD, practical EM formulation of actual ULSI interconnect can be performed. This will be elaborated in the later chapters.

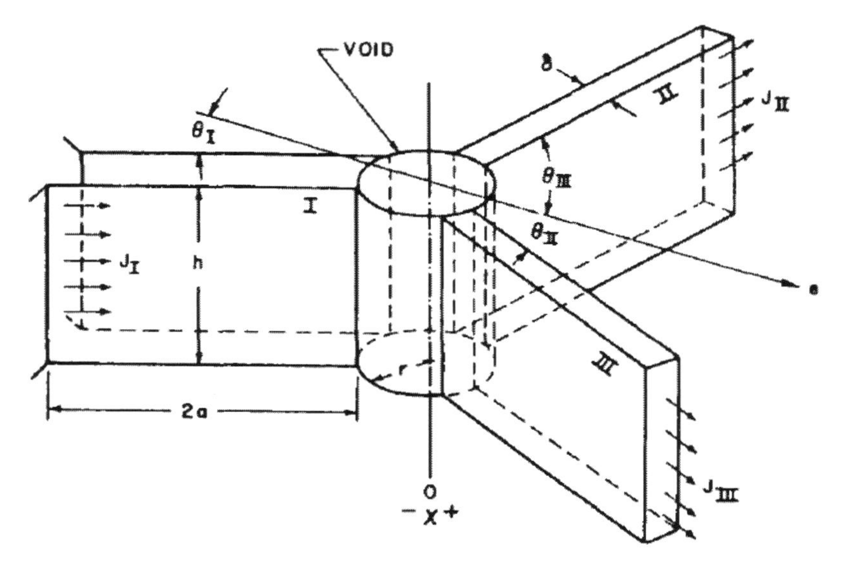

Fig. 2.2. Schematic of atomic flux at triple point (Reprinted with permission from Rosenberg and Ohring 1971. Copyright 1971. American Institute of Physics).

2.2 Electromigration Lifetime Modeling

In order to extrapolate the time to failure of interconnect due to EM to normal operating condition, Black (1969) carried out series of EM experiment on Al based metallization systematically in the late 60's. Despite the argument on the presence of electron wind force as mentioned earlier, he formulated a semi-empirical equation for median time to failure of a group of identical conductor undergone same EM stressing as follows basing on the concept of electron wind force:

$$t_{50} = Aj^{-2} \exp\left(\frac{E_a}{k_B T}\right), \tag{2.23}$$

where

t_{50} = median time to failure,
A = a constant which contains a factor involving the cross-sectional area of the film,
j = current density in the conductor,
E_a = activation energy,

k_B = Boltzman's constant, and
T = conductor temperature.

This equation is known as Black equation and is used widely in the EM study of VLSI interconnects.

Black conducted many EM experiments to understand the impact of conductor width and thickness and conclude that the constant 'A' in Eq. (2.23) is given by $A = Bwt$, where 'B' is a characteristics constant, 'w' and 't' are the conductor width and thickness respectively. It was thus found that EM degradation follows Arrhenius type relationship with temperature. The failure distribution for an identical group of conductors undergone same EM stressing was found to follow 'log-normal' distribution (Black 1969).

With this Black's equation, one can performed accelerated EM test at higher temperature and higher current density, and the median time to failure obtained from the test can then be extrapolated to the normal operating condition easily which will be discussed in Chapter 3. Due to its ease of extrapolation and its accuracy, the Black's equation becomes the most widely used equation for the extrapolation of EM lifetime.

The current density exponent in the Black's equation is 2, and this is the consequence of the time it takes for stress to build up to the point of failure, and it is referred to as nucleation-dominated failure. In many cases, the current density exponent is not observed to be 2. As such, a generalized Black equation is produced as follows

$$t_{50} = Aj^{-n} \exp\left(\frac{E_a}{k_B T}\right). \tag{2.24}$$

If n is not equal to 2, then the failure is not nucleation dominated. If $n > 2$, the failure process can be either dominated by temperature gradients or by the Blech length effect. If $n < 2$, failure is growth dominated, where the time elapsed following nucleation for damage in the form of voids or hillocks is a significant portion of the time to failure. If $n = 1$, this implies either the nucleation time vanishes or there was pre-existing damage in the structure (Lloyd and Rodbell 2006). Recent experiments have provided strong support for this view (Yokogawa and H. 2005) and have confirmed that the activation energies for both nucleation and growth are similar.

However, if the value of n is changing, the unit for A will also be changed, and this does not make physical sense. Note that the Sah formula

(Sah 1996; Sah and Jie 2008) covers the entire range of T and J and their power exponents, without such empirical separations assumed by Lloyd. Lloyd recently proposed that the time to failure should be partitioned into nucleation time and growth time as shown below (Lloyd 2007).

$$t_{50} = t_{nuc} + t_g = \left(\frac{AkT}{j} + \frac{B(T)}{j^2} \right) \exp\left(\frac{E_a}{kT} \right), \quad (2.25)$$

where A and B are constants that contain geometric information, such as the size of the void required for failure. $B(T)$ also has a temperature dependence that depends on which failure model is chosen (Shatzkes and Lloyd 1986; Korhonen, Borgesen *et al.* 1993; Clement and Thompson 1995).

With this modified equation, extrapolations to use condition will be different from a power law as in the case of conventional Black equation. Knowing how to deal with this accurately is challenging because both the power law and the nucleation and growth models in Eq. (2.25) will provide a nearly equally excellent fit to the data at higher current densities, but the extrapolation to lower current densities will differ substantially. Fortunately, the lifetime extrapolated with the Black equation is shorter and hence it is conservative. But this may also be too restrictive in some cases. Lloyd provided a procedure for the extrapolation using Eq. (2.25) and the detail of the procedure can be found in (Lloyd 2007).

Recently, Li, Tan and Raghavan (2009) revisited the Black's equation and outlines the inherent assumptions made in the equation as follows.

(a) There are only two stress factors during EM, temperature and current density.
(b) The pre-exponential term is a temperature independent constant.
(c) There is no interaction between temperature and current density/ other stress factor.
(d) The overall effective activation energy is independent on current density/other stress factor.

By carefully examine each assumptions, they showed that all the assumptions are no longer valid for interconnection goes below 350 nm (Li, Tan and Raghavan 2009). In order to verify their finding experimentally, they perform an accelerated package level EM tests on an aluminum Via-line structure where the via-filling material is Tungsten (W), the barrier metal is TiN/Ti composition and the TARC layer is Ti. The width of the Al line and the via diameter are 0.288 um. The final anneal temperature of the wafer is at 410°C. The test temperatures are 250°C, 200°C and 150°C at the same stress current density of 3.5 MA/cm^2. The test continues until a 1% change in the resistance of the metal interconnection is observed. The time to failure data are analyzed and extrapolated to low stress condition (100°C & 3.5 MA/cm^2) based on the Black's equation and the unbiased statistical method without any assumption on the physical model. Such a statistical method can be applied only if the acceleration is true linear acceleration, and the detail of the statistical method can be found in (Elsayed 1996; Meeker and Escobar 1998). True linear acceleration is a term in reliability statistics that represents invariant of the failure mechanism under the different stress conditions. Therefore, one can perform accurate extrapolation to any operating condition using the linear relationship between stress level and the lifetime of the product without any bias, i.e. no assumption on the underlying physical model is needed. This lifetime can be Mean Time to Failure, or Median Time to Failure etc.

From the probability plot of the time to failure data at the three test temperatures, we can see that they are parallel as shown in Fig. 2.3, indicating that they are indeed the true linear acceleration.

Plotting the ln(MTF) vs. stress level (i.e. temperature), our results appear as straight line in the plot as shown in Fig. 2.4 as expected from a true linear acceleration. The prediction results due to the Black's equation and the unbiased statistical method are summarized in Table 2.1, and we can see that the Black's equation can over-estimate the lifetime by more than 100% when extrapolated to the normal operating condition.

The overestimation of the Black's equation is believed to be due to the neglection of the effect of thermo-mechanical stress in the interconnect system. At high testing temperature, the stress in the metal line is moderate as the testing temperature is closer to the stress free temperature, which was set at the final annealing temperature of the wafer. However, at low testing temperature, the stress in the metal line is significant and becomes another stress factor during EM testing. The extrapolation from high testing temperature condition thus underestimates the acceleration effect due to the thermo-mechanical stress and overestimates the EM lifetime at the lower testing temperature condition.

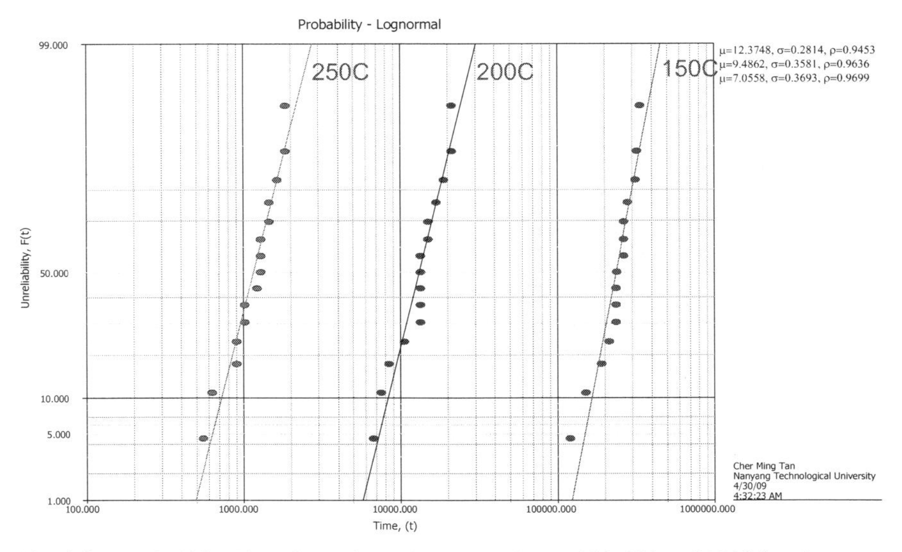

Fig. 2.3. Probability Plot of Accelerated EM Test data at 250, 200 and 150°C and current of 4.5 mA.

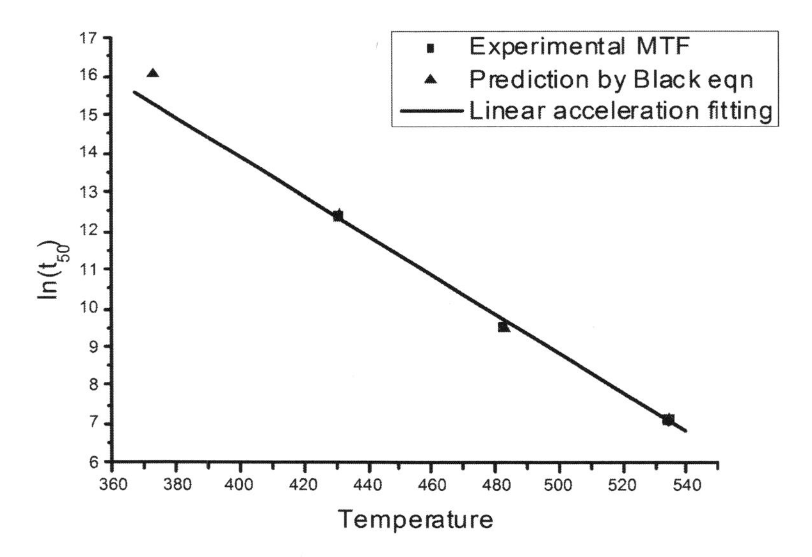

Fig. 2.4. The result of experiments and predictions.

Table 2.1. Results of predictions by the Black's equation. Value of t_{50} at 100°C is obtained from Fig. 2.4 based on the principle of true linear acceleration.

Test current density ($j = 3.5$ MA/cm^2)	Stress condition	Extrapolated condition		
	200°C	250°C	150°C	100°C
	Reference	1.16×10^3	2.42×10^5	4.41×10^6
Black's equation	13226.79	6.16%	3.72%	111.24%

Besides the use of the Black equation for extrapolation, numerical modeling has also been used with limited success. The prediction of the lifetime of a test structure depends on the method to determine the realistic time in the simulation and the specified condition/criterion beyond which the elements do not exist anymore. In the work by Sasagawa, Naito, Saka and Abe (1999) and Sasagawa, Nakamura, Saka and Abe (2002), the calculation process of the numerical simulation for the lifetime prediction is carried out repeatedly until the metal line fails where the entire line width is either occupied by the elements whose temperature exceeds the melting point of the metal due to Joule heating and/or the thickness of the elements become smaller than a pre-defined infinitesimal threshold value. Based on their experimental measurement and the hypothesis of the effective width of the slit void, they determined the threshold value should be 2×10^{-3} times the initial film thickness. By assigning one calculation step to a realistic time, the lifetime of the test structure can thus be predicted. The method of prediction is found to have a good agreement with their experimental observations. Although the method of prediction is found to have a good agreement with their experimental observations, the assignment of the "realistic time" to the calculation step is not rigorous theoretically, and its extension to general EM test is questionable.

Different prediction approaches is employed by Dalleau *et al.*'s study (Dalleau and Weide-Zaage 2001; Dalleau, Weide-Zaage and Danto (2003); Weide-Zaage, Dalleau, Danto and Fremont (2007)). The time dependent evolution of the local atomic concentration is expressed as

$$div(J_{EWM} + J_{th} + J_{str}) + \frac{\partial N}{\partial t} = 0. \tag{2.26}$$

Based on the above atomic flux equations, they derived the approximated Atomic Flux divergence (AFD) for the respective driving forces as follows:

$$div(\vec{J}_{EWM}) = \left(\frac{E_A}{k_B T} - \frac{1}{T} + \alpha\frac{\rho_0}{\rho}\right) \cdot \vec{J}_{EWM} \cdot \nabla T, \tag{2.27}$$

$$div(\vec{J}_{th}) = \left(\frac{E_A}{k_B T} - \frac{3}{T} + \alpha\frac{\rho_0}{\rho}\right) \cdot \vec{J}_{th} \cdot \nabla T + \frac{NQ^* D_0}{3k_B^3 T^3} j^2 \rho^2 e^2 \exp\left(-\frac{E_A}{k_B T}\right), \tag{2.28}$$

$$\begin{aligned} div(\vec{J}_{str}) = {} & \left(\frac{E_A}{k_B T} - \frac{1}{T}\right) \vec{J}_{str} \nabla T + \frac{2EN\Omega D_0 \alpha_l}{3(1-\upsilon)k_B T} \\ & \times \exp\left(-\frac{E_A}{k_B T}\right)\left(\frac{1}{T} - \alpha\frac{\rho_0}{\rho}\right)\nabla T^2 \\ & + \frac{2EN\Omega D_0 \alpha_l}{3(1-\upsilon)k_B T} \exp\left(-\frac{E_A}{k_B T}\right)\frac{j^2 \rho^2 e^2}{3k_B^2 T}. \end{aligned} \tag{2.29}$$

With Eqs. (2.27)–(2.29), Eq. (2.26) can be re-written as

$$Nf + \frac{\partial N}{\partial t} = 0, \tag{2.30}$$

where f is a function including the different physical parameters. The theoretical evolution of the atomic density is then given by Eq. (2.31)

$$N = N_0 e^{-ft}, \tag{2.31}$$

and N_0 is the initial value of the atomic concentration. In their simulation, the element is considered to be empty of material when the atomic concentration reduced to 10% of the initial concentration ($N/N_0 = 10\%$). By solving Eq. (2.31), the time required to delete one element can be calculated. As the void growth is simulated by the deletion of groups of element, time to failure of the structure can be extracted by accumulating the time needed for the deletion of all the elements. However, difference is found between lifetime measurements and calculations.

Recently, Li, Tan and Hou (2007) demonstrate a methodology to incorporate the Monte Carlo method into Finite element analysis to simulate the EM process from void nucleation to void growth, considering the thermodynamics process of EM (various diffusion paths) and the kinetic of

EM (various driving forces). They used the method to derive an expression for the Median time to failure of EM failure governed by the void nucleation as in narrow interconnects (Li, Tan and Raghavan 2009), and their results are found to agree well with the experimental data, even to the extrapolated data. Chapter 5 will describes the method in a greater detail.

2.3 Electromigration Lifetime Improvement

With extrapolation, the lifetime of interconnection at normal operating conditions were found to be inadequate and extensive works on EM were performed to understand the EM so as to extend the lifetime.

Only the major improvement works for Al is described below. Other improvement works for both Al and Cu will be described in the subsequent chapters.

Soon after the introduction of the Black's equation, it was reported that a dielectric passivation layer on the top of Al metallization layer can improve the EM reliability which showed higher activation energy as compared to the bare metallization (Spitzer and Schwartz 1969). Later on, passivation thickness dependent EM life-time was also observed (Lloyd and Smith 1983).

In 1976, Blech and his coworkers at Bell Lab found that the EM mass transport ceases when the conductor length is below a certain length for a given EM stress condition (Blech 1976; Blech and Herring 1976; Blech and Tai 1976). The length of the conductor below which EM mass transport was not observed is called critical length for a given EM stress condition. In order to explain the critical length in EM, Blech proposed the 'Back flow' of atoms opposite to the electron flow direction due to the presence of back stress in the interconnect when voids and hillocks are formed. This phenomenon is known as 'Blech effect' or 'Short length effect'.

With continuous shrinking in interconnection line-width as IC technology follows the Moore's Law, further research activities move toward the improvement of the EM life-time. Alloying by Cu and/or Silicon doping into Al interconnections was found to have a better EM-life time as compared to pure Al (Blech 1977; Vaidya, Fraser and Lindenberger 1980), and Al with 0.5% Cu was found to have the best EM resistant, and is commonly adopted in wafer fabrication industry that use Al. Similarly, incorporation of hydrogen into the metallization during processing was reported to be beneficial for EM life-time improvement as well (Meyer 1980).

As the interconnect line width is continuously decreasing, Vaidya, Kinsbron and their coworkers conducted line-width dependence of EM tests at Bell Labs in 1989 (Kinsbron 1980; Vaidya, Sheng and Sinha 1980), and they observed that when the grain size of the metal film is comparable to the film width, the EM life-time becomes better, and this is due to the significant reduction in the grain boundaries mass flow of atoms and the number of triple points in interconnections as the grain structures of the interconnections are becoming bamboo structures.

In order to eliminate contact pitting due to Si-Al reactions, a barrier material for metallization was proposed in IC technology to protect inter diffusion between Al and Si in the late 70s and implemented in the 80's (Fu, K-Y and Fuji, T. 12–13 June 1989; Fu, K-Y and Fuji, T. 12–13 June 1989; Hoang, R. A. Coy *et al.* 12–13 June 1990; Singlevich and Bordelon 12–13 June 1990; Ghate, Blair and Fuller 1977; Ghate and Blair 1978; Ghate 1982; Sinke, Frijlink *et al.* 1985). Refractory materials such as Ti, W, etc. was proposed as barrier material, since these materials can withstand very high temperature without EM. The other reasons of adapting these refractory materials as barrier layer include low diffusivity in SiO_2, ease of purity, high yield strength, compatibility with the standard metallization deposition process, closeness of lattice constant with that of Al etc. A dramatic improvement in the EM performance was observed from the introduction of barrier material with special care in nitridation, oxidation, chamber vacuum break etc. of the various layers in the multilayered structure (Singlevich and Bordelon 12–13 June 1990; Huang, Shofner *et al.* 2006; Sinke, Frijlink *et al.* 1985; Ting, Hong *et al.* 1996; Tan, Roy *et al.* 2005). This improvement is because the barrier material serves as an alternative current path even after void formation due to EM. It can also melt the Al and eliminate the void in the Al and resume connection. The effect of the barrier metal will be further elaborate in the next chapter.

2.4 Electromigration Aware IC Design

While thorough electrical testing can help to ensure IC performances, time-dependent reliability issues are difficult to filter in testing. With the importance of EM for IC reliability, inclusion of electromigration consideration in IC design has been practiced by many workers. An automatic wire widening method had been developed by Heng (1998) for EM reliability enhancement in 2D layout; Lienig (2005) described a 2D physical design methodologies for current density verification and current-driven layout de-compaction to address electromigration. In general, EM checking is conducted during design verification by referencing the current density limitation in an EM design rule and compare it with the calculated average current of the target interconnect. However, the influence of a maximum instantaneous current is unchecked in most design. Also, as will be shown later, current density is no longer the sole driving force for electromigration in interconnects when the line width decreases below 200 nm (Li Aug 2007), and interconnection electromigration reliability depends critically on their surrounding materials (Li 2007), the 2D modeling with the current density as the only consideration mentioned above is no longer adequate.

Although Jerke *et al.* (Jerke January 2004) built a quasi 3D model of the interconnect via region to verify the irregularities for current densities determination in arbitrarily shaped custom-circuit layouts, their model consider only the via instead of the entire interconnect system in an IC, and thus it is of limited application.

Recently, Tan and He (2009) developed a 3D circuit model for the interconnection reliability investigation and they have shown that the weak spots in interconnections are not those as determined based on the current density alone. A lot more work is needed in order to accurately assess the interconnection reliability for a given IC layout.

References

Attardo, M. J., Rutledge R. and Jack R. C. (1971). Statistical metallurgical model for electromigration failure in aluminum thin-film conductors. *Journal of Applied Physics* **42**: 4343–4349.

Black, J. R. (1969). Electromigration — A brief survey and some recent results. *IEEE Trans. Electron Devices* **16**(4): 338–347.

Blech, I. A. (1976). Electromigration in thin aluminum films on titanium nitride. *Journal of Applied Physics* **47**: 1203–1208.

Blech, I. A. (1977). Copper electromigration in aluminum. *Journal of Applied Physics* **48**: 473–477.

Blech, I. A. and Herring, C. (1976). Stress generation by electromigration. *Appl. Phys. Lett.* **29**: 131–133.

Blech, I. A. and Kinsbron, E. (1975). Electromigration in thin gold films on molybdenum surfaces. *Thin Solid Films* **25**: 327–334.

Blech, I. A. and Meieran, E. S. (1969). Electromigration in thin Al films. *Journal of Applied Physics* **40**(2): 485–491.

Blech, I. A. and Sello, H. (1966). A Study of failure mechanisms in silicon planar epitaxial transistor. *Physics of Failure in Electronics* **5**: 496.

Blech, I. A. and Tai, K. L. (1976). Measurement of stress gradients generated by electromigration. *Applied Physics Lett.* **30**: 387–389.

Bosvieux, C. and Friedel, J. (1962). Sur l'electrolyse des alliages metalliques. *J. Phys. Chem. Solids* **23**: 123.

Brand, M. G. E. and A. Lodder (1986). Electromigration in niobium. A finite-cluster-model study. *Phys. Stat. Solidi B* **133**: 119–125.

Braun, K. F., Soe, W. H., Flipse, C. F. J. and Rieder, K. H. (2007). Electromigration of single metal atoms observed by scanning tunneling microscopy. *Applied Physics Lett.* **90**: 023118-1.

Christou, A. (1994). *Electromigration and Electronic Degradation*, New York, John Wiley & Sons.

Clement, J. J. and Lloyd, J. R. (1992). Numerical investigations of the electromigration boundary value problem. *Journal of Applied Physics* **71**: 1729–1731.

Clement, J. J. and Thompson, C. V. (1995). Modeling electromigration-induced stress evolution in confined metal lines. *Journal of Applied Physics* **78**: 900.

Dalleau, D. and Weide-Zaage, K. (2001). Three-dimensional voids simulation in chip metallization structures: a contribution to reliability evaluation *Microelectronics Reliability* **41**(9–10): 1625–1630.

Dalleau, D., Weide-Zaage, K. and Danto, Y. (2003). Simulation of time depending void formation in copper, aluminium and tungsten plugged via structures. *Microelectronics Reliability* **43**: 1821.

Das, A. K. and Peierls, R. (1975). The force of electromigration. *J. Phys. C: Solid State Phys.* **8**, 3348–3352.

Duryea, T. W. and Huntington, H. B. (1988). The driving force for electromigration of an atom adsorbed on simple metal surface. *Surface Science* **199**: 261–281.

Elsayed, E. A. (1996). *Reliability Engineering*, Addison Wesley Longman, Inc.

Fiks, V. B. (1959). On the mechanism of the mobility of ions in metals. *Sov. Phys. Solid State* **1**: 14.

Fu, K.-Y., Travis, E., Sun, S. W., Grove, C. L., Pyle, R. E., P. F. and S. P. (12–13 June 1989). Ti-thickness dependent electromigration resistance of the Al-Cu-Si/TiNx/TiSiy barrier contact system. *IEEE Proc. of VMIC*, 439–446.

Fujii, T., Okuyama, K., Moribe, S., Torii, Y., Katto, H. and Agatsuma, T. (June 1989). Comparison of electromigration phenomenon between aluminum interconnection of various multilayered materials. *IEEE Proceedings of VMIC*, 477–483.

Genoni, T. C. and Huntington, H. B. (1977). Transport in nearly-free-electron metals. IV. Electromigration in zinc. *Physical Review B* **16**: 1344–1352.

Ghate, P. B. (1982). Metallization for very-large-scale integrated circuits. *Thin Solid Films* **93**: 359–383.

Ghate, P. B. and Blair, J. C. (1978). Electromigration testing of Ti:W/Al and Ti:W/Al-Cu film conductors. *Thin Solid Films* **55**: 113–123.

Ghate, P. B., Blair, J. C. and Fuller, C. R. (1977). Metallization in microelectronics. *Thin Solid Films* **45**: 69–84.

Gupta, R. P. (1982). Theory of electromigration in noble and transition metals. *Physical Review B* **25**: 5188–5196.

Gupta, R. P., Serruys, Y., Brebec, G. and Adda, Y. (1983). Calculation of the effective valance for electromigration in niobium. *Physical Review B* **27**: 672–677.

Harrison, J. W. (1988). A simulation model for electromigration in fine-line metallization of integrated circuits due to repetitive pulse currents. *IEEE Trans. Electron Devices* **35**: 2170–2179.

Hartman, T. E. and Blair, J. C. (1969). Electromigration in thin gold films. *IEEE Trans. Electron Devices* **16**(4): 407–410.

Heng, Z. C. (1998). *A fast minimum layout perturbation algorithm for electromigration reliability enhancement.* 1998 IEEE International Symposium on Defect and Fault Tolerance in VLSI Systems Proceedings.

Hoang, H. H., Coy, R. A. and McPherson, J. W. (12–13 June 1990). Barrier metal effects on electromigration of layered aluminum metallization. *IEEE Proceedings of VMIC*, 133–141.

Huang, J. S., Shofner, T. L., *et al.* (2006). Direction observation of void morphology in step-like electromigration resistance behavior and its correlation with critical current density. *Journal of Applied Physics* **89**: 2130–2133.

Huntington, H. B. (1975). *Diffusion in Solids; Recent Developments.* A. S. N. a. J. J. Burton. New York, Academic Press, Inc.

Huntington, H. B. (1975). Effect of driving forces on atom motion. *Thin Solid Films* **25**(2): 265–280.

Huntington, H. B. and Grone, A. R. (1961). Current-induced marker motion in gold wires. *J. Phys. Chem. Solids.* **20**: 76–87.

Jerke, G. L. (2004). Hierarchical current-density verification in arbitrarily shaped metallization patterns of analog circuits. *IEEE Transactions on Computer-Aided Design of Integrated Circuits and Systems.*

Kang, K. D., Burgess, R. R., *et al.* (1969). A Cr-Ag-Au metallization system. *IEEE Trans. Electron Devices* **16**(4): 356–360.

Kinsbron, E. (1980). A model for the width dependence of electromigration lifetimes in aluminum thin-film strips. *Applied Physics Lett.*: 968–970.

Kircheim, R. and Kaeber, U. (1991). Atomistic and computer modelling of metallizatioin failure of integrated circuits by electromigration. *Journal of Applied Physics* **70**: 172–181.

Korhonen, M. A., Borgesen, P., *et al.* (1993). Microstructure based statistical model of electromigration damage in confined line metallizations in the presence of thermally induced stresses. *J. Appl. Phys.* **74**: 4995.

Kumar, P. and Sorbello, R. S. (1975). Linear response theory of the driving forces for electromigration. *Thin Solid Films* **25**: 25.

Landauer, R. (1975). Sources of conduction band polarization in the driving force for electromigration. *Thin Solid Films* **26**: L1–L2.

Landauer, R. (1989). Comment on Lodder's 'extact' electromigration theory. *Solid State Commun.* **72**(9): 867.

Landauer, R. and Woo, J. W. F. (1974). Driving force in electromigration. *Physical Review B* **10**: 1266–1271.

Landauer, R. (1975). The Das-Peierls electromigration theorem. *J. Phys. C: Solid State Phys.* **8**: L389–L392.

Landauer, R. (1977). Geometry and boundary conditions in the Das-Peierls electromigration theorem. *Physical Review B* **16**: 4698–4702.

Li, W., Tan, C. M. and Hou, Y. (2007). Revisit to the finite element modeling of electromigration for narrow interconnects. *Journal of Applied Physics* 033705-1-7.

Li, W., Tan, C. M. and Raghavan, N. (2009). Predictive dynamic simulation for void nucleation during Electromigration in narrow interconnects in integrated circuits. *Journal of Applied Physics* **105**: 014305.

Lienig, J. J. (2005). Electromigration-aware physical design of integrated circuits. *18th International Conference on VLSI Design Proceedings.*

Lloyd, J. R. (2007). Black's law revisited — Nucleation and growth in electromigration failure. *Microelectronics Reliability* **47**: 1468–1472.

Lloyd, J. R. and Rodbell, K. P. (2006). Chapter 7, Reliability. *Handbook of Semiconductor Interconnection Technology*. (eds) G. C. Schwartz and K. V. Srikrishnan, Taylor & Francis.

Lloyd, J. R. and Smith, P. M. (1983). The effect of passivation thickness on the electromigration lifetime of Al/Cu thin film conductors. *J. Vac. Sci. Technol.* **A1**: 455.

Lodder, A. (1989). The driving force in electromigration. *Physica A* **158**: 723–739.

Lodder, A. and Brand, M. G. E. (1984). Electromigration in transition-metal hydrides: a finite-cluster-model study. *J. Phys. F: Metal Phys.* **14**: 2955–2962.

Marcoux, P. J., Merchant, P. P., *et al.* (1989). A new 2D simulation model of. Electromigration. *Hewlett-Packard J.*: 79–84.

Meeker, W. Q. and Escobar, L. A. (1998). *Statistical Methods for Reliability Data*, John Wiley & Sons Inc. USA.

Meyer, D. E. (1980). Effects of hydrogen incorporation in some deposited metallic thin films. *Journal of Vacuum Science and Technology* **17**: 322–326.

Nikawa, J. (1981). Monte Carlo Calculations based on the generalized electromigration failure model. *Proc of IEEE International Reliability Physics Symposium*: 175–181.

Patil, H. R. and Huntington, H. B. (1969). Electromigration and associated void formation in silver. *J. Phys. Chem. Solids* **31**: 463–474.

Rosenberg, R. and Ohring, M. (1971). Void formation and growth during electromigration in thin films. *Journal of Applied Physics* **42**: 5671.

Roy, A. and Tan, C. M. (2006). Experimental investigation on the impact of stress free temperature on the electromigration performance of copper dual damascene submicron interconnect. *MIcroelectronics Reliability* **46**(9–11): 1652–1656.

Sah, C. T. (1996). *Fundamentals of Solid-State Electronics — Solution Manual*, World Scientific Publishing Company, 174–176.

Sah, C. T. (2006) Personal communication. Electron-windless-driftless model was constructed and electric-current-stress-time-depenendent resistance formula was derived by Prof Sah and experimentally verified by an PhD student at the University of Illinois in Urbana-Champaign in 1985.

Sah, C. T. and Jie, B. (2008). The driftless electromigration theory (diffusion-generation-recombination-trapping theory). *Journal of Semiconductors* **29**(5): 815–821.

Sasagawa, K., Naito, K., Saka, M. and Abe, H. (1999). A method to predict electromigration failure of metal lines. *Journal of Applied Physics* **86**: 6043.

Sasagawa, K., Nakamura, N. Saka, M. and Abe, H. (2002). Governing parameter for electromigration damage in the polycrystalline line covered with a passivation layer. *Journal of Applied Physics* **91**: 1882.

Schaich, W. L. (1976). Theory of driving force for electromigration. *Physical Review B* **13**: 3350–3359.

Schoen, J. M. (1980). Monte Carlo calculations of structure-induced electromigration failure. *J. Appl. Phys.* **51**: 513–521.

Sham, L. J. (1975). Microscopic theory of the driving force in electromigration. *Physical Review B* **12**: 3142–3149.

Shatzkes, M. and Lloyd, J. R. (1986). A model for conductor failure considering diffusion concurrently with electromigration resulting in a current exponent of 2. *Journal of Applied Physics* **59**: 3890.

Singlevich, S. G. and Bordelon, M. D. (1990). A study of aluminum sputter deposition parameters and TiW barrier surface effects on electromigration. *IEEE Proceedings of VMIC*, 371–373.

Sinke, W., Frijlink, G. P. A., *et al.* (1985). Oxygen in titanium nitrade diffusion barriers. *Applied Physics Lett.* **47**: 471–473.

Sorbello, R. S. (1973). A pseudopotential based theory of the driving forces for electromigration in metals. *J. Phys. Chem. Solids* **34**: 937–950.

Sorbello, R. S. (1981). Residual-resistivity dipole in electron transport and electromigration. *Physical Review B* **23**: 5119–5127.

Sorbello, R. S. and Dasgupta, B. (1977). Local fields in electron transport: Application to electromigration. *Physical Review B* **16**: 5193–5205.

Spitzer, S. M. and Schwartz, S. (1969). The effects of dielectric overcoating on electromigration in aluminum interconnects. *IEEE Trans. Electron Devices* **16**: 348–350.

Tan, C. M. and He, F. (2009). *ULSI Interconnect Reliability Study Using 3D Circuit Model.* submitted.

Cher Ming Tan, Yuejin Hou and Wei Li (August 2007). Revisit to the finite element modelling of electromigration for narrow interconnects. *Journal of Applied Physics* **102**: 033705.

Tan, C. M., Roy, A., *et al.* (2005). Effect of vacuum break after the barrier layer deposition on the electromigration performance of aluminum based line interconnects. *Microelectronics Reliability* **45**: 1449–1454.

Ting, L., Hong, Q.-Z. and H. W.-Y (1996). Reduction in flux divergence at vias for improved electromigration in multilayered AlCu interconnects *Applied Physics Lett.* **69**: 2134–2136.

Vaidya, S., Fraser, D. B. and Lindenberger, W. S. (1980). Electromigration in fine-line sputter-gun Al. *Journal of Applied Physics* **51**: 4475–4482.

Vaidya, S., Sheng, T. T. and Sinha, A. K. (1980). Linewidth dependence of electromigration in evaporated Al-0.5%Cu. *Applied Physics Lett.* **30**: 464–466.

Verbruggen, A. H. (1988). Fundamental questions in the theory of electromigration. *IBM J. Res. Dev* **32**: 93–98.

Verbruggen, A. H., Griessen, R. and d. Groot, D. G. (1986). Electromigration of hydrogen in vanadium, niobum and tantalum. *J. Phys. F. Met. Phys.* **16**: 557–575.

Weide-Zaage, K., Dalleau, D., Danto, Y. and Fremont, H. (2007). Dynamic void formation in a DD-copper-structure with different metallization geometry. *Microelectronics Reliability* **47**: 319.
Weise, J. (1972). Quantitative measurements of the mass distribution in thin films during electrotransport experiments. *Thin Solid Films* **13**: 169–174.
Yokogawa, S. (2005). Scaling impacts on electromigration in narrow single-damascene Cu interconnects. *Japanese Journal of Applied Physics* **44**: 1717–1721.

CHAPTER 3
Experimental Studies of Al Interconnections

3.1 Introduction

In this Chapter, we will present the data from the Aluminum interconnects electromigration studies. As mentioned in Chapter Two, the vacancy flux in an interconnect due to electromigration is given by

$$J = -D\vec{\nabla}c + \frac{Dc}{k_B T} Z^* e\rho \cdot \vec{j}. \tag{3.1}$$

Consider a control volume as shown in Fig. 3.1, three situations are possible:

(a) $J_{\text{in}} > J_{\text{out}}$.
In this case, vacancy accumulation will be resulted which renders void nucleation and growth that lead to open circuit eventually.

(b) $J_{\text{in}} < J_{\text{out}}$.
In this case, metal atom accumulation will happen that renders the formation of hillock or extrusion which in turn lead to short circuit.

(c) $J_{\text{in}} = J_{\text{out}}$.
Depending on the condition, void shape might be changed, hence causing open circuit. However, if no significant void shape change occur, nothing will happen.

Therefore, the first condition for electromigration failure to occur is the presence of a divergence of vacancy flux somewhere in the line that

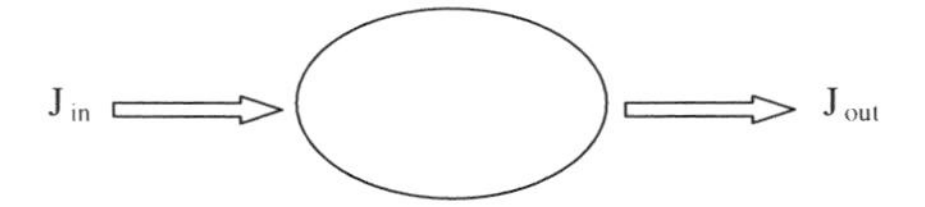

Fig. 3.1. Control volume for vacancy flux.

allows voids or extrusions to form. Hence, understand the factors that affect the vacancy flux and its divergence will enable us to understand the failure mechanisms of electromigration.

From Eq. (3.1), one can easily see that the factors that affect the vacancy flux are the following:

(a) diffusivity
(b) vacancy density
(c) temperature
(d) factors that affect the variation of electron wind-force, including current density
(e) resistivity

It is the spatial variation of these factors that will lead to vacancy flux divergence. The mechanisms that cause the spatial variations are the failure mechanisms of electromigration. These mechanisms can be process-related or design-related. Table 3.1 summarizes the study of these mechanisms in Al interconnections.

Besides these process or design-induced failure physics for electromigration, stress developed during the process of electromigration will also decelerate or accelerate the EM process. They are termed as self-healing or self-induced electromigration, and will be discussed later.

Extensive research works have been done to understand the mechanisms of the spatial variation shown in Table 3.1, and they will be described in detail as follows.

3.2 Process-Induced Failure Physics

3.2.1 *Microstructural inhomogeneities of metallization*

Baerg and Wu (Baerg and Wu Mar 1991) showed that the EM median time to failure (MTF) was directly proportional to the grain size while the standard deviation of the lifetime distribution was directly proportional to the grain size distribution. For diffusion along the grain boundary which is

Table 3.1. List of known electromigration failure mechanisms.

S/No.	Factor that affect the vacancy flux	Mechanisms of the spatial variation of the factor in the left	Process-related	Design-related
1	Diffusivity	a. Microstructural inhomogeneities of metal lines	Y	Y
		b. Temperature gradient	Y	Y
		c. Impurity distribution	Y	
		d. Mechanical stress gradient	Y	Y
		e. Material difference	Y	
		f. Defect density distribution	Y	
		g. Surface contamination	Y	
		h. Multi-level metallization	Y	Y
		i. Metal width uniformity		Y
2	Temperature	a. Underneath oxide thickness	Y	Y
		b. Metal width	Y	Y
		c. Resistivity	Y	
		d. Current density		Y
3	Defect density	—	Y	
4	Resistivity		Same as (1)	
5	Electron wind-force	a. Metal width	Y	Y
		b. Metal thickness	Y	Y
		c. Metal step height	Y	Y

the dominant diffusion path for Al during EM, the diffusion flux is given by (J.E. Sanchez and J.W. Morris 1991)

$$J_i = (ND_i\delta/dk_BT)F_i, \tag{3.2}$$

where N is the atomic density, D_i is the boundary diffusivity, δ is the effective boundary width, d is the grain size, k_B is the Boltzmann's constant, T is absolute temperature, F_i is the resolved EM driving force along the boundary.

Since the flux depends on the grain size and effective boundary width, grain size distribution play an importance role in flux divergence. Also, the diffusivity is different with different grain orientation, being smaller in films with [111] orientation (Attardo and Rosenberg 1970), and the effective boundary width is also dependent on the grain orientations and the degrees of mixture of various orientation, hence the texture of the film is also an importance element in flux divergence as elaborated below.

3.2.1.1 *Grain size*

EM lifetime has been shown to significantly increase as the linewidth becomes comparable or smaller than the average grain size (J.E. Sanchez and J.W. Morris 1991). This is because with larger grain size, the number of triple points for flux divergence is reduced. When grain size becomes larger than the linewidth, "bamboo" structure is resulted, and no triple point is available, causing a drastic improvement in EM lifetime. Diffusion then proceeds along the interface. The tendency for the grain boundaries to align themselves normal to the lengths of the lines during annealing is found to be strong when the linewidth of Al reach approximately 1 μm (Walton, Frost and Thompson 1992).

The transformation of a thin film strip to a bamboo structure proceeds at an exponentially decreasing rate, and that the rate is inversely proportional to the square of the strip width. There exists a maximum strip width to thickness ratio above which the transformation to a bamboo structure is expected to fail to proceed to completion. For strips with smooth sidewalls, this critical ratio is estimated to be between 2.1 and 3.0 (Walton, Frost and Thompson 1992).

For interfacial diffusion, since interface could contain various type of defects and contamination, the standard deviation in the EM lifetime distribution is found to increase for "bamboo" structure (J.E. Sanchez and J.W. Morris 1991). In fact, it is found that surface contamination can greatly reduce the stress required for void nucleation, thus reduces the EM lifetime (Gleixner and Nix 1996).

3.2.1.2 *Grain size distribution*

A physically abrupt change in grain size will cause corresponding changes in the diffusivity, effective boundary width, and grain size, which leads to flux divergence, and induces voiding (Attardo and Rosenberg 1970; Agarwala, Patnaik and Schmitzel 1972). Experiments have shown that a large standard deviation of a continuous grain size distribution implies more local microstructural discontinuities, and hence a lower EM lifetime (Knorr, Rodbell and Tracy 1991).

Change in grain size can also occur at steps in interconnections. It was confirmed that the presence of steps definitely degrades the EM lifetime (Kisselgof, Elliott, Maziarz and Lloyd 1991). The degradation was also found to be due to the effect of grain size divergence associated with thinner metal over step (Strausser, Euzent, Smith, Tracy and Wu 1987). Therefore, the degradation is dependent on the step height. In fact, the correlation of EM lifetime with step height is much better than with the measured step coverage (Kisselgof, Elliott, Maziarz, and Lloyd 1991).

3.2.1.3 *Texture of a metal line*

For highly oriented films, EM lifetime improvement is observed as compared with less textured film with mixed grain orientation (Attardo and Rosenberg 1970). This can be understood as the less textured film has larger diffusivity variation and larger number of diffusion paths available for vacancy diffusion. Therefore, the standard deviation of the EM lifetime distribution is also drastically reduced for good texture film. In other word, good texture film has longer mean lifetime and fewer early failures.

A suggested formula that neatly summarizes the dependence of mean lifetime on grain size (S), the standard deviation statistical spread in grain size (sigma sub gs), and the preferred crystallographic orientation is given by that the mean time to failure $MTTF = (BS/\sigma_{gs^2})\log(I_{111}/I_{200})^2$. Here the I's are the corresponding X-ray intensities for plans of indicated indicesand B is a proportionality constant. Here we can see clearly that more nearly equiaxed large grains that have a strong [111] orientation (texture) in the firm plane are desirable (Knorr D.B., Rodbell K.P., Tracy D.P., Texture and microstruture effects on electromigration behavior of aluminum metallization, Proceeding of Materials Reliability Issues in Microelectronics Symposium, p. 21–26, 1991)

3.2.2 *Presence of impurity*

Control of the Al microstructure and precipitate distribution (also known as alloying) have been the leading methodologies for improving Al EM lifetime. The effect of precipitate is as follows:

(a) Addition of impurities, e.g. copper, along the grain boundary will reduce the aluminum atom grain boundary diffusion (Agarwala, Digiacomo and Joseph 1976).
(b) However, the presence of impurity also resulted in interphase boundaries, and thus a higher boundary energies than typical Al-Al boundaries. These higher boundary energies would be more potent catalytic sites for heterogeneous void nucleation and growth (Agarwala, Digiacomo and Joseph 1976). Thus, the EM lifetime will be reduced.
(c) For copper precipitate, (b) will not happen as the dissolution of the copper precipitate causes the surrounding Al grain to grow to replace the precipitate volume by migration of the phase boundaries (Shaw, Hu, Lee and Rosenberg 1995). Many of the smaller grains are eliminated, and grain structure adjacent to the precipitate is coarsened, as was observed under TEM. This resulted in a significant change in the local diffusion paths, in particular, the reduction of the number of diffusion paths and triple points.

Therefore, with copper addition, the EM lifetime increases. The increment in lifetime is larger with increasing copper concentration. The longest lifetimes observed was with 16% copper inclusion (solid solubility limit of copper in aluminum is only 0.05 wt%). However, care must be taken that for Al line with Cu impurity, pre-annealing of the line will help Cu atoms to accumulate at the grain boundary forming the Cu precipitates, thus reduces the incubation time and renders shorter EM lifetime as compared with the samples without pre-annealing treatment (Mazumder, Yamamoto, Maeda, Komori and Mashiko 2001).

For multilevel metallization, flux divergence occur at the line/stud contact (Endicott, Bouldin and Miller 1992; Oates, Nkansah and Chittipeddi 1992; Hu and Small 1993; Hu, Small and Ho 1993). Cu is found to be important in determining the rate of mass depletion at the line/stud contact. The depletion of Al is preceded by an incubation period where Cu in Al has to first be swept out a critical distance from the cathode end of the line. The damage formation process is controlled by the Cu electromigration along grain boundaries. The length of incubation time in Al(Cu) alloys is roughly proportional to Cu concentration (Hu, Small and Ho 1993), and it can be as long as 20 hours under accelerating condition.

The MTF values can be increased by a factor of 10 to 100 by the addition of Cu. This is because if more than 0.5% Cu is added to the Al line, grain boundary and interface (IF) diffusivities are decreased by a factor of 10 to 100, decreasing the flux divergence (Brown, Korhonen, Borgesen and Li 1994).

3.2.3 *Mechanical stress in the film*

The time required to form a void is given by (Lloyd 1991):

$$t_I = \left(\frac{w}{\delta}\right)\left(\frac{A}{\Delta D}\right)\left(\frac{T}{j}\right)^2, \tag{3.3}$$

where w is the conductor stripe width, δ is the effective grain boundary width, T is temperature, j is the current density, ΔD is the change in vacancy diffusivity in the grain boundary. A is given by

$$A = 2C_f\left(\frac{k_B}{Z^*e\rho}\right)^2, \tag{3.4}$$

and

$$C_f = [\exp(\sigma_f - \sigma_a)\Omega/k_BT - 1]\exp(-\sigma_a\Omega/k_BT). \tag{3.5}$$

Here σ_f is the stress required to form a void, Ω is the vacancy volume, and σ_a is an applied stress.

From Eqs. (3.3) to (3.5), we can see that an applied stress plays a dual role in the EM lifetime (Lloyd 1991), namely it changes the kinetics of void formation and it determines the amount of stress that needs to be "supplied" by electromigration for void formation.

The normal component of stress on the grain boundary will alter the activation energy for diffusion. In order for a vacancy to move from its present site to the next, enregy must be supplied to move the adjacent atoms out of the way in order to make room for the motion. If the applied stress is a tensile stress, it will help this process, and if it is compressive, it will hinder the process. Therefore, tensile stress can have a profound effect on the electromigration failure kinetics by considerably reducing the time required for an EM-induced void to appear.

Also, immediately after the void formation, the stress remaining will be "felt" by the surrounding grain boundaries since the surface of a void is stress free. This means that the void shape will become narrower and longer and consume relatively more of the stripe width for a given volume of vacancies than would be anticipated from equilibrium considerations. Under such circumstances, fewer vacancies will be required to condense to form an open circuit, thus shorten the EM lifetime (Hong and Crook 1985; Lloyd 1991; Walton, Frost and Thompson 1992).

In Eq. (3.4), the applied stress includes the stress in the interconnect due to the mismatch of the thermal expansivities of the metal and its surrounding. Due to the high temperature processing of the metallization which is in the range of 350–400°C, the interconnects are generally in the state of tensile when they are cooled down to room temperature and even at the EM test temperature. Hence, the presence of this thermal mismatch induced stress will affect the EM life. Lloyd (Lloyd 1999) studied interconnect EM performance for different lots with different processing temperatures, and he found approximately five times improvement in EM life-time by reducing the deposition temperature from 390 to 325°C on the interconnect lines.

Along the same argument, the differential thermal expansions of the metal and its passivation material will also have significant effect on the EM life. Doan *et al.* (Doan, Lee, Lee, Flinn and Bravman 2001) performed experiments with different passivation materials, and they found substantial increases of the lifetime if compliant dielectric polymer was used as passivation as compared with the conventional oxide. This is because a critical amount of tensile stress is required to form a void as shown in Eqs. (3.3)–(3.5). For a compliant passivation material, a much larger amount of material must be removed in order to produce the same amount of tensile stress. As a result, the time to void nucleation is much longer as compared with the conventional case of SiO_2 or Silicon nitride. Similar results were also shown by Usui *et al.* (T. Usui, Watanabe, Ito, Hasunuma, Kawai and Kaneko 1999) for Al-Cu line where they compared the top dielectric materials of silicon dioxide, silicon nitride and a low-k organic spin-on glass.

While compliant top dielectric suppress void nucleation, it is susceptible to unconstrained void growth, i.e. it is poor in slowing down the void growth due to its lower effective bulk modulus and hence weaker Blech effect (Doan, Lee, Lee, Flinn and Bravman 2001). The concept of Blech effect will be discussed in the subsequent section of self healing.

Another example of the effect of passivation on the stress in Al film which in turn affect its EM lifetime is shown by Learn and Shephered (Learn and Shephered 1971) who observed 7 to 14 times of EM lifetime increase when the Al film was anodized. The drift velocity was also found to decrease with increasing anodizing thickness. This is believed to be due to the change in the self diffusivity of Al as a result of the compressive stresses imposed by the anodized layer (Ross, Drewery, Somekh and Evetts 1989). Similarly, an oxygen plasma treatment applied after patterning was also found to reduce EM mass transport, and EM lifetime increased as a function of the plasma treatment time, due to the same reasoning of the compressive stress (Wada, Higuchi and Ajiki 1985).

Therefore, the thickness, composition, mechanical properties, and processing of dielectric materials can qualitatively and quantitatively change the interconnect EM lifetime. The net impact depends on the details of the structure, and this explains the widely conflicting reports on the role of the dielectric layer for interconnect EM performances (Lloyd 1982; Teal, Vaidya and Fraser 1986; Foley, Ryan, Martin and Mathewson 1998; T. Usui, Watanabe, Watanabe, Ito, Hasunuma, Kawai and Kaneko 1999).

3.2.4 *Presence of defect*

Defect produces flux divergence. The lifetime of a conductor will be determined by the element containing the most severe defect because it results in the largest flux divergence. It is not the average value of the severities of defects, but the most severe among all the defects that determined the EM lifetime. The presence of defect produces the following.

3.2.4.1 *Length dependence of lifetime*

If the density of defects, i.e. the number of defects per unit length, is constant, the total defect population increases with an increase of the conductor length, and therefore, a longer stripe has a higher probability of containing more severe defects. Thus, a decrease of lifetime at longer lengths is expected as shown by Agarwala *et al.* (Agarwala, Attardo and Ingraham 1970).

3.2.4.2 *Length dependence of the standard deviation of EM lifetime distribution*

When interconnect length increase, the total number of defects increases, and thus the probability of finding severe defects in each sample will be higher, and hence the time to failure distribution has a smaller standard deviation as observed by Agarwala *et al.* (Agarwala, Attardo and Ingraham 1970).

Since the length dependence of t_{50} and σ is directly related to the defect distribution in the film, it is expected that the dependence of t_{50} and σ on length will shift with changing process variables. Low temperature (250°C) aging for as long as 10 hr has been found to increase lifetimes as much as 10 times (Nemoto and Nggami 1994).

3.2.5 *Temperature gradient*

Because of the temperature dependence of diffusion, there will be a maximum in the divergence of the diffusivity that leads to a severe flux divergence near the point of inflection of the temperature profile (Lloyd 1994). This temperature profile could be due to joule heating which will not heat the line uniformly (Lloyd 1994). The cause of non-uniform joule heating may be due to the following:

(a) The variation of resistivity of the line, or
(b) The variation in heat conduction due to the non-uniform thickness of the dielectric layer underneath the metallization line or
(c) The non-uniform distribution of the current density, for example, at the corner of the line or the variation in the conductor width or at the line/via intersection.

The effect of oxide thickness on the test strip temperature is as shown in Fig. 3.2 (Schafft 1987). The effect is more significant at higher current density.

From Fig. 3.2, one can see that a step in oxide thickness can produce significant temperature gradient, thus higher step height results in lower lifetime of EM. The step in oxide thickness can be of design intent, and it can also be due to process deficiency.

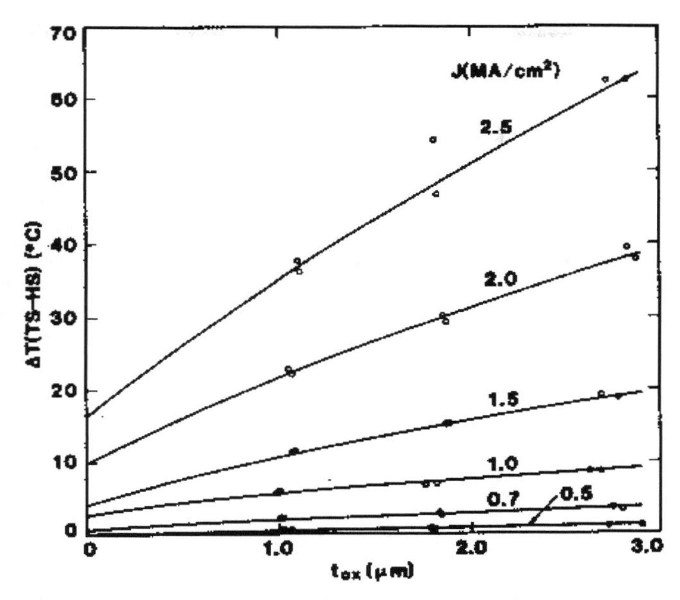

Fig. 3.2. Al EM test structure to heat sink temperature difference vs. silicon dioxide thickness. Each data point represents the average of the temperature differences of 6 test structures in a package and an average oxide thickness on the chip. (Schafft, 'Thermal Analysis of electromigration test structures', IEEE Trans. Electron Devices, 34(3), p. 664), Copyright © 1987 IEEE.

Current crowding will occur when a line has a corner or there is a line width transition along a straight line. The right-angle corner exhibits the highest level of current density and the highest gradient along the line A-A' as shown in Fig. 3.3. It is this current density gradient, not the current crowding itself that enhanced electromigration (Gonzalez and Rubio 1997). This is illustrated in Table 3.2.

Finite element analysis is performed on Al line structure with wide transition. The width ratio is 2 and the finite element model is as shown in Fig. 3.4. With a current density of 1 MA/cm^2 in the wide line segment, the temperature gradient and current density distributions can be computed as shown in Fig. 3.5. One can see the formation of the temperature gradient due to the non-uniform current density distribution.

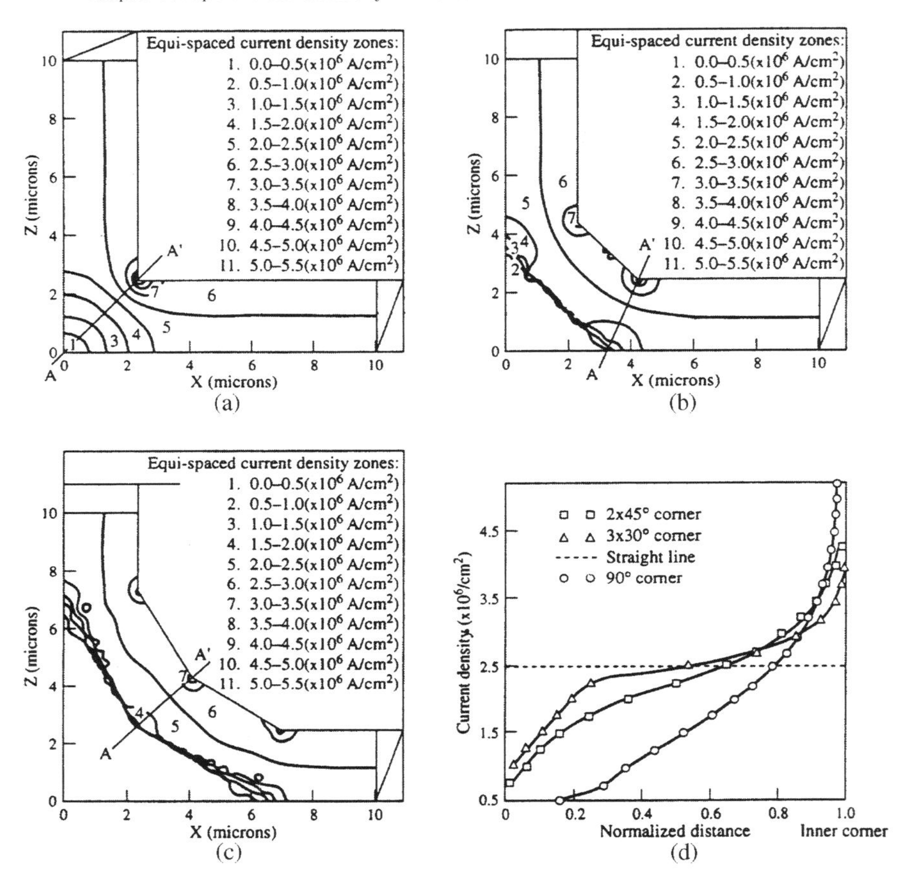

Fig. 3.3. Current density distribution in (a) right-angle corner, (b) $2 \times 45^\circ$ corner, and (c) $3 \times 30^\circ$ corner. (d) Current density values along the line A-A' for the previous three shapes as compared with the straight line current density (dash line). Reprinted from Microelectronics Reliability, 37(7), Gonzalez and Rubio, 'Shape effect on electromigration in VLSI interconnects', p. 1073, Copyright © 1997, with permission from Elsevier.

For a larger line width, the non-uniformity will be even more severe as shown in Table 3.3.

Another cause for the temperature gradient to exist is the line-via structure. Even for the Al via, the current density non-uniformity and the

Table 3.2. Normalized to $T = 200°C$. Median Time to fail (t_{50} or MTF) and standard deviation (t_{50} or DTF) for the different shapes used in the experiment, and for metal 1 (0.66 μm thick) and metal 2 (0.99 μm thick).

Metal 1					Metal 2				
Name	Shape	MTF (h)	DTF	Sample no.	Name	Shape	MTF (h)	DTF	Sample no.
R1	┌	3.74	1.820	7	R5	┌	33.05	0.914	6
R2	╭	5.34	1.503	7	R6	╭	77.18	1.67	4
R3	└┐	2.78	1.277	7	R7	└┐	40.27	0.948	4
R4	│	2.33	1.558	8	R8	│	71.29	1.135	5

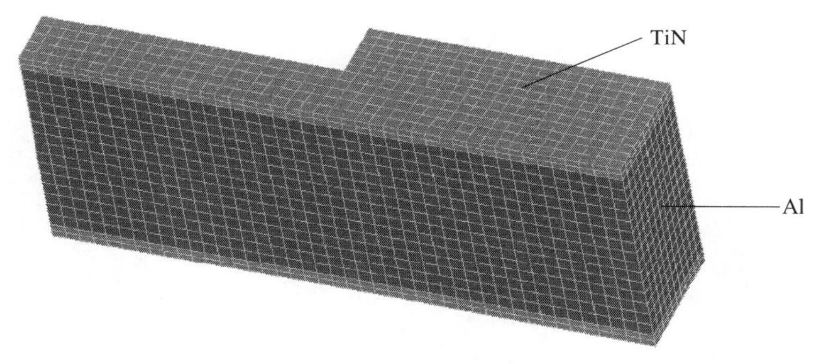

Fig. 3.4. Finite element model for width transitional structures. Wide line is two times wider than that of narrow line.

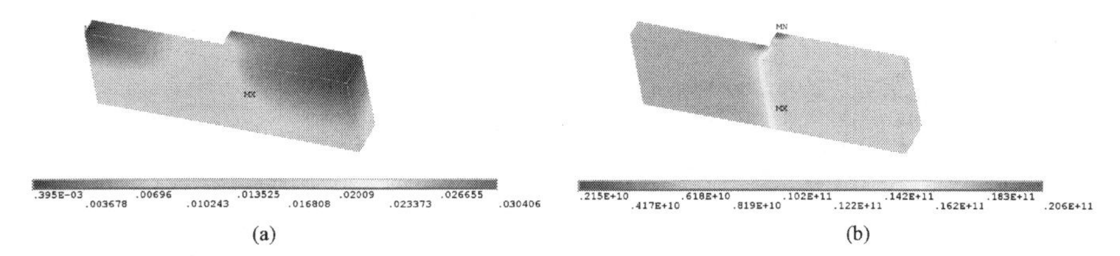

Fig. 3.5. Representative distributions for width transitional structures with line width of 0.28 μm for narrow line. (a) Temperature gradient distribution in K/μm (b) current density distribution in A/cm^2.

resulting temperature gradient distribution will be more severe than the line with width transition. Figure 3.6 shows the finite element model and Fig. 3.7 shows the distributions. Table 3.4 shows the maximum temperature gradient and the current density.

Table 3.3. Max temperature gradient and current density due to line width transition, keeping the width ratio of 2.

Width of the narrow segement (μm)	Max temperature gradient (K/μm)	Max current density (MA/cm^2)
0.28	0.030406	2.06
0.4	0.040448	2.18
0.6	0.050531	2.38

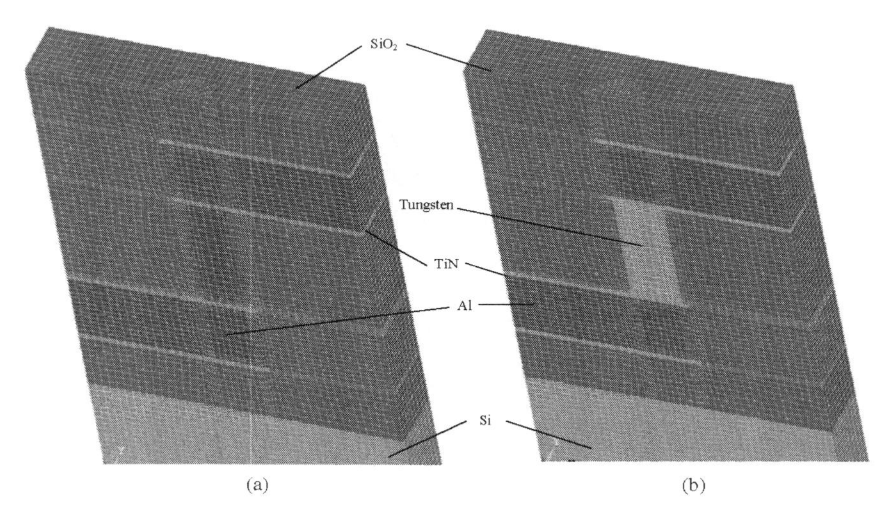

Fig. 3.6. Finite element model for Al line-via structures with (a) Al via and (b) TiN via.

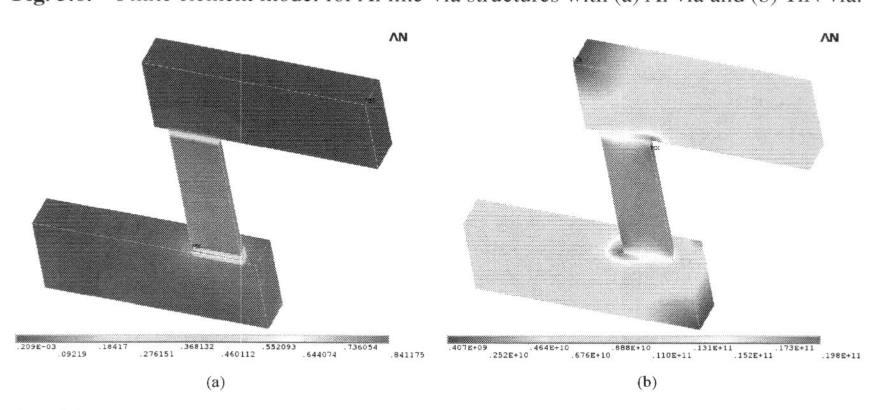

Fig. 3.7. Representative distributions for Al via structures with 0.28 μm line width. (a) Temperature gradient distribution (b) current density distribution.

Table 3.4. Max temperature gradient and current density due to line-via structure with Al via.

Line width (μm)	Max temperature gradient (K/μm)	Max current density (MA/cm^2)
0.28	0.841175	1.98
0.4	1.521	2.79
0.6	2.95	4.15

3.2.6 *Material differences*

The situation of material difference in an interconnect system is usually occur in a multi-level Al interconnect structure where W via is commonly employed. With the presence of the W stud, three fundamental characteristics, namely the discontinuity of mass transport at Al/W interfaces, the discontinuity of Cu supply at Al-Cu/W interfaces, and current crowding at studs are responsible for the decrease in the EM lifetime which is found to be less than half of that of Al-Cu straight line (Kwok, Tan, Moy, Estabil, Rathore and Basavaiah 1990).

Finite element analysis for Al interconnect structure with W via is performed with the structure as shown in Fig. 3.6(b). Similar current density distribution as shown in Fig. 3.7 is obtained, but the maximum current density and maximum temperature gradient are more severe than the case of Al via. The comparison results are shown in Table 3.5.

Figure 3.8 shows a typical failure of an Al–Cu interconnect at the W plug via area where Cu accumulation and Al depletion as a result of EM are clearly seen at the anode and the cathode end, respectively (Kawasaki, Gall,

Table 3.5. Comparison of max temperature gradient (T_g) and current density (J_s) due to line-via structures with Al and W via.

	Al/Tungsten		Al/Al	
Line width (μm)	Tg(K/μm)	Js(MA/cm^2)	Tg(K/μm)	Js(MA/cm^2)
0.28	1.457	3.43	0.841175	1.98
0.4	2.699	4.85	1.521	2.79
0.6	5.472	7.24	2.95	4.15

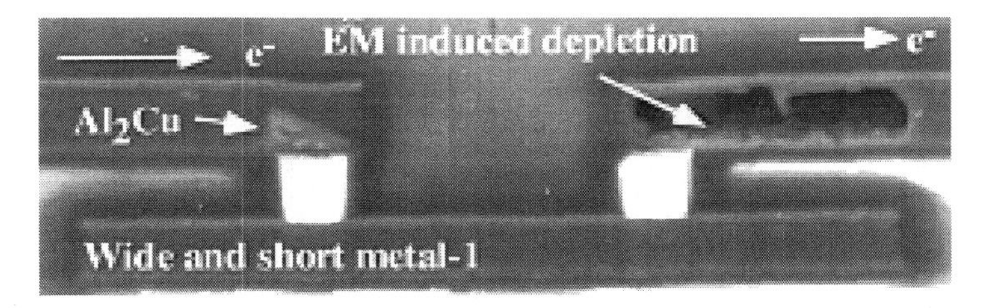

Fig. 3.8. Cross-sectional SEM micrograph of a typical EM failure site in a W plug structure. Cu accumulation (Al_2Cu) and Al depletion due to EM are seen at the anode and cathode end of the Al lines, respectively. Reprinted from Thin Solid Films, 320, Kawasaki, Gall, Jawarani, Hernandez and Capasso, 'Electromigration failure model: Its application to W plug and Al-filled vias', pp. 45–51, Copyright © 1998, with permission from Elsevier.

Jawarani, Hernandez and Capasso 1998). Since Cu is known to suppress Al migration (Rosenberg 1971), EM induced Al voids do not form or grow as long as a sufficient amount of Cu is present within the critical length, and this is attributed to the stress-induced Al backflow (Blech 1976), the discontinuity of Cu supply at Al-Cu/W interface is therefore a major factor in shortening the EM life of a Al interconnect structure.

It is found that the discontinuity of Cu supply at Al-Cu/W interfaces contributes most significantly to the reduction in the electromigration resistance of W stud chains (Kwok, Tan, Moy, Estabil, Rathore and Basavaiah 1990), and the flux of Cu away from the Al-Cu/W interfaces is enhanced by the current density at the interfaces.

Amazawa *et al.* (Amazawa and Arita 1991) and Dixit *et al.* (Dixit, Paranjpe, Hong, Ting, Luttmer, Havemann, Pual, Morrison, Littau, Eizenberg and Sinha 1995) used Al plugged via instead of W-plugged via, and they found that EM lifetime was improved by several order of magnitudes and electrical resistance was also found to be reduced. Thus, as expected, the change in the material can significant reduce the EM lifetime.

3.2.7 *Temperature*

As temperature increases, the fluxes also increase. Hence, the effect of the flux divergence will be "amplified".

Also, at higher temperature, thermal expansion mismatch can be more severe, and the resulting thermo-mechanical stresses could induce the formation of hillock and whisker. The hillock and whisker formation processes may locally enhance diffusivities by dramatically increasing vacancy concentrations, which in turn result in severe flux divergence (J.E. Sanchez 1986). It was shown that the largest voids in EM tested interconnects were found typically adjacent to significant hillocking (J.E. Sanchez, Lloyd and Morris 1990).

Electromigration mass flux divergence can create vacancy supersaturation in the grain boundary which will flow to and annihilate at the surface via the boundary or the adjacent lattice. How they flow, which is temperature dependent, determines the damage morphology (Lloyd and Koch 1992) and this has important implication in the validity of accelerated testing.

The ratio of number of diffusing vacancies into the lattice to those diffusing along the grain boundary is as shown in Fig. 3.9. At low temperature where grain boundary diffusion is predominating, failure sites are characterized by crack-like features, whereas at higher temperatures where lattice diffusion is predominating, more area around the grain boundary will be missing. As these two mechanisms have different Ea, care must be taken when extrapolate the test data at accelerated stress condition to normal operating condition (Lloyd and Koch 1992).

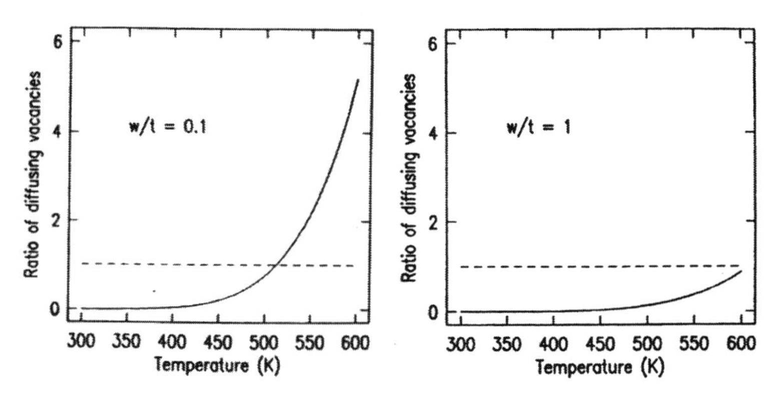

Fig. 3.9. Ratio of diffusing vacancies along grain boundary over that through the lattice. w is the average grain size, t is the thickness of the Al/Cu line, which is 1 μm in this case. Reprinted with permission from Lloyd and Koch, 'Electromigration-induced vacancy behavior in unpassivated thin films', Journal of Applied Physics, vol. 76(6), p. 3231. Copyright © 1992, American Institute of Physics.

3.3 Design-Induced Failure Mechanisms

3.3.1 *Proximity of metal lines*

The proximity of metal lines will affect the joule heating dissipation, and hence the temperature of an interconnect. Also, close proximity may result in early EM failure due to extrusion.

3.3.2 *Inter-metal dielectric (IMD) thickness between metal lines and oxide thickness underneath the first metallization*

We have shown earlier in Fig. 3.2 that the temperature of the metal lines is significantly depends on the oxide thickness under them. If low-k is used to replace the SiO_2 as IMD, the line temperature will be even higher due to the poorer thermal conductivity of the low-k material as shown in Fig. 3.10.

3.3.3 *Number of metallization levels*

Due to the presence of IMD and the corresponding heat conductivity through this IMD, when the number of metallization level increases, the line temperature will also increase as shown in Fig. 3.11 (Shen 1999).

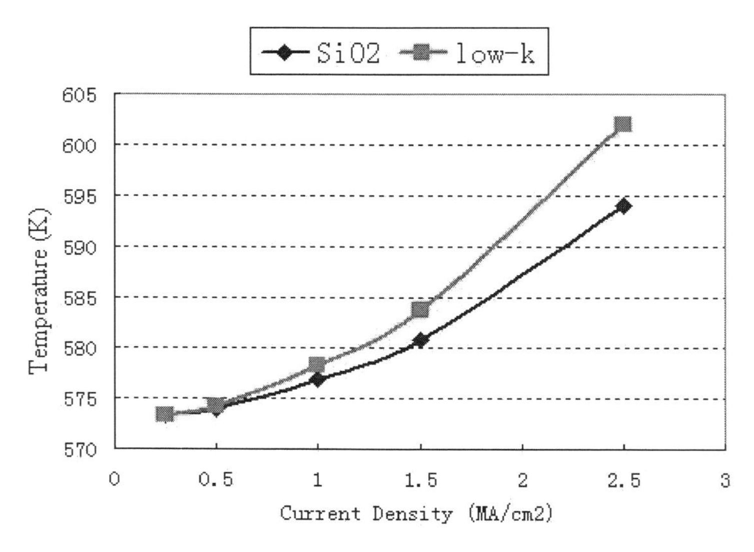

Fig. 3.10. Increase in line temperature due to Joule heating.

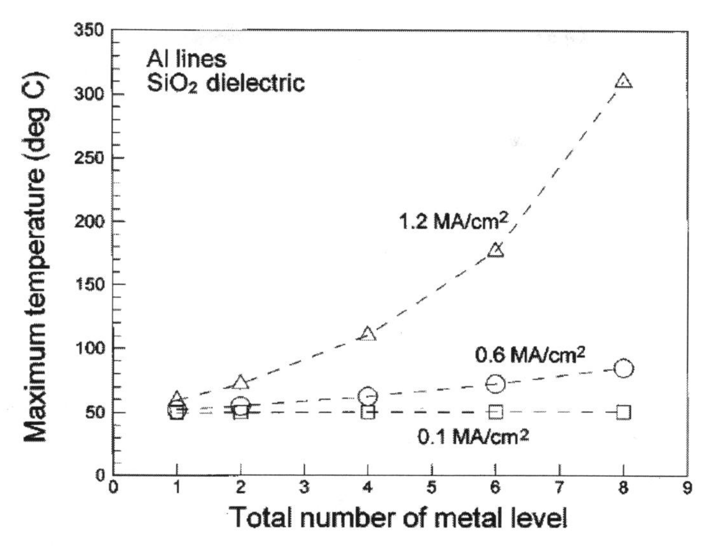

Fig. 3.11. Variation of the maximum temperature with the total number of metal levels. Reprinted with permission from Shen, 'Analysis of Joule heating in multilevel interconnects', Journal of Vacuum Science and Technology B, vol. 17(5), pp. 2115–2121. Copyright © 1999, American Institute of Physics.

3.3.4 *Use of barrier layers*

Mutilayer metallization structures with TiN as barrier and/or capping layer has been shown to be very effective in improving electromigration resistance (Hinode and Homma 1990; Fujii, Okuyama, Moribe, Torii, Katto and Agatsuma June 1989). This is because of the following two reasons:

(a) The EM resistance for TiN is found to be very high, being immune to EM at 340°C at current density of 1×10^7 A/cm^2. In fact, even at current density of 1.8×10^7 A/cm^2 and 630°C, the failure of TiN is still not due to EM (Tao, Cheung and Hu 1995).

(b) The local heating in TiN at the region where Al interconnect is broken could melt the Al and form the connection again, thus producing a self-healing effect. This is known as current by-pass effect (Ondrusek, Nishimura, Hoang, Sugiura, Blumenthal, Kitagawa and McPherson 1988).

The texture of the barrier metal can also affect the EM lifetime of Al film with barrier. It is found that when Ar plasma treatment is performed on the Ti underlayer, EM life-time is improved by one order of magnitude. The reason for this improvement is found to be the better texture quality in the case of Ar plasma treated Ti underlayer (Amazawa and Arita 1991; Dorgelo, Vroemen and Wolters 2001). EM life-time of sputtered Al alloy [Al(1%)Si(0.5%)Cu] interconnect is found to be dependent on whether pure Ar or contaminated Ar is used during deposition. It is noticed that when deposition was performed in pure Ar, EM life-time is about six times longer.

It is also found that O_2 incorporation into the TiN layer is needed to effectively protect the interdiffusion between Ti and Al (Atakov 1990). He reported that TiN underlayer reduces Al-alloy EM resistance but it improves EM resistance when TiN layer is exposed to air prior to Al deposition. As a result, chamber vacuum break prior to Al deposition was proposed to create an effective diffusion barrier layer.

But Avinun *et al.* (Avinum, Barel, Kaplan, Eizenberg, Naik, Guo, Chen, Mosely, Littau, Zhou and Chen 1998) studied the detail of the Al texture quality on various types of TiN underlayer, and they observed that upon air exposure, the proportion of the Al grain in the preferred $\langle 111 \rangle$ orientation reduces and thus it would expect to have a shorter EM life-time as compared with the non air exposure scheme. Thus a conflicting message is being generated about the suitability of vacuum break process.

Tan *et al.* (Tan, Roy, Tan, Ye and Low 2005) investigated the effect of air exposure prior to Al deposition. They carried out moderate and highly accelerated EM tests and noticed that for sample with air exposure prior to Al deposition, the EM life-time and its variations decreases when tested at highly accelerated stress while negligible variations of these quantities are observed when tested at moderate accelerated stress. Through physical analysis, they have also shown that upon air exposure, the top surface of the bottom TiN layer becomes rough and there exists TiO_2. With the aid of the physics based finite element modeling of EM, they explained the differences in the experimental observations at highly and moderate accelerated EM tests and predicted that the interconnect EM performance at normal operating should not be affected by the vacuum break. Thus the intentional chamber vacuum break prior to Al deposition should be preferred in the TiN/Al(Cu)/TiN interconnection fabrication process.

The presence of barrier metal can also block the vacancies flow. As a result, vacancies accumulation occur at the cathode side, and resistance will increases. Koizumi and Hiraoka (Koizumi and Hiraoka 1995) shows

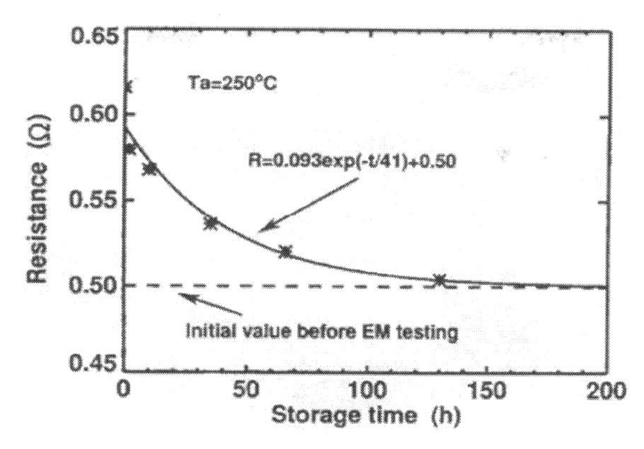

Fig. 3.12. Decay in the line resistance during storage at an ambient temperature of 250°C after EM testing. (Koizumi H., and Hiraoka K., 'The blocking barrier effect on aluminum electromigration due to titanium layers in multilayered interconnects of LSI's', IEEE Electron Device Lett., 16(7), p. 298), Copyright © 1995 IEEE.

this phenomena for Ti barrier. After EM testing, where the resistance is increased to 20%, the resistance was observed to decay when the sample was stored at high temperature of 250°C as shown in Fig. 3.12. This decay is considered to be due to the diffusion of accumulated excess vacancies.

A multilayered metal line has a number of interfaces between different metals which will affect the line EM, and there are many factors that are originated in these hetero-interfaces of the multilayered metal lines and affect the metal line differently. These factors are the interdiffusion at heterointerfaces, A1 grain size, A1 crystallographic orientation, refractory metal mechanical stress, and the contact resistance of heterointerfaces. Onoda *et al.* (Onoda, Kageyama, Tatara and Fukuda 1993) found that the contact resistance between Al and TiN is an important factor for layered metal line lifetime and high contact resistance between over-layered TiN and A1 suppresses the current bypass flow mentioned earlier when it is one order of magnitude higher compared with that of Al/Al direct contact.

To improve the contact resistance, Onoda *et al.* (Onoda, Kageyama, Tatara and Fukuda 1993) found that if a thin Ti layer is inserted in between TiN and Al through successive deposition without breaking the vacuum, the contact resistance decreases, and the current bypass effect through the over-layered TiN can be expected. Furthermore, the Ti can diffuse into Al as detected using the RBS measurement, and since Ti is one of the impurities in Al that can enhance its EM lifetime (Fischer and Neppl 1984; Dirks,

Tien and Towner 1986; Towner, Dirks and Tien 1986; Hosoda, Yagi and Tsuchikawa 1989), the lifetime of TiN/Ti/Al is found to be significantly larger than TiN/Al.

3.3.5 *Via separation length*

When the via separation along a metal line is short, Blech-length effect can occur and it will retard the atomic diffusion during EM, and hence enhances the EM lifetime of interconnect. This Blech-length effect will be discussed in the next section on Self-induced process during EM.

This concept has been used extensively in the design and layout of IC with Al-based interconnection with W plugs which presented a blocking boundary.

With this Blech-length effect, the increase in the interconnect resistance will reach a saturation as shown in Fig. 3.13 (Filippi, Wachnik, Eng, Chidambarrao, Wang and White 2002). Hence, if a circuit can tolerate such an increase in interconnect resistance, the circuit will not fail due to electromigration.

Such a resistance saturation during electromigration test was studied extensively by Filippi *et al.* (Filippi, Wachnik, Eng, Chidambarrao, Wang

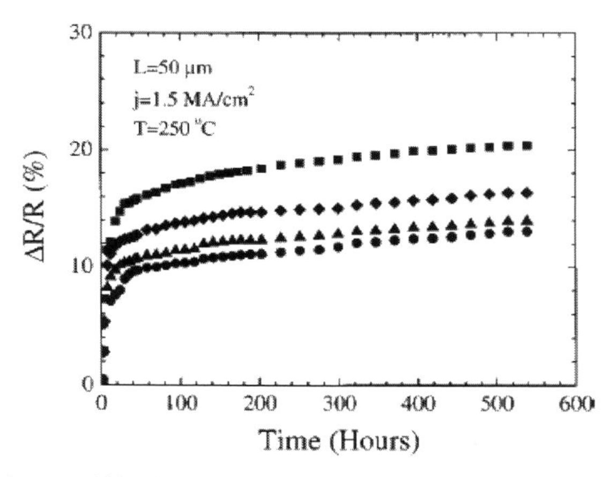

Fig. 3.13. Resistance shift (%) vs time for the 50 μm-long samples. The test conditions are 1.5 MA/cm^2 and 250°C. Only four out of 12 samples are shown. Reprinted with permission from Filippi, Wachnik, Eng, Chidambarrao, Wang and White, 'The effect of current density, strip length, stripe width, and temperature on resistance saturation during electromigration testing', Journal of Applied Physics, vol. 91(9), pp. 5878–5795. Copyright © 2002, American Institute of Physics.

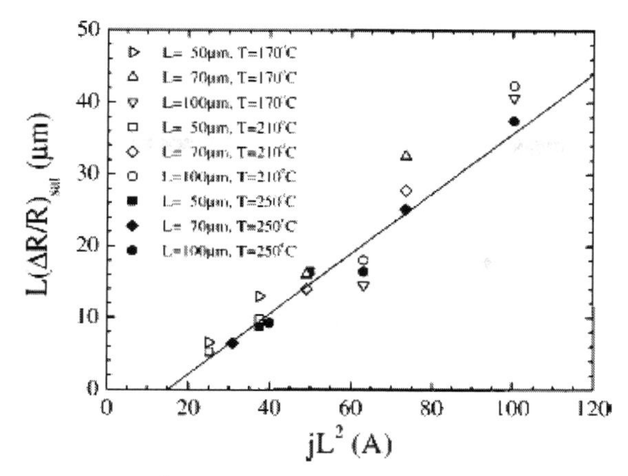

Fig. 3.14. $L(\Delta R/R)_{\text{sat}}$ vs jL^2 for the 0.33 μm-wide samples tested at 170, 210, and 250°C. The symbols correspond to the average values. Note that some of the symbols overlap because the data for these stress cells are nearly equal; namely, (■□) at $jL^2 = 25.1$ A, and (♦◇) at $jL^2 = 49.1$ A. The solid line is the least-squares fit based on the 250°C data. Reprinted with permission from Filippi, Wachnik, Eng, Chidambarrao, Wang and White, 'The effect of current density, strip length, stripe width, and temperature on resistance saturation during electromigration testing', Journal of Applied Physics, vol. 91(9), pp. 5878–5795. Copyright © 2002, American Institute of Physics.

and White 2002) where they examined the effect of current density, interconnect length and width as well as temperature on the resistance saturation. They found that the maximum resistance change follows a linear relationship with jL regardless of the line length and temperature as shown in Fig. 3.14. However, such a relationship depends on the line width as the effective bulk modulus of the interconnect depends on the line width as shown in Fig. 3.15.

3.3.6 *Cornering of metal line and Step height of metal lines*

The shape of metal line and step height will affect the temperature gradient as has been discussed in Sec. 3.2.5.

3.3.7 *Use of passivation layer*

The presence of passivation layer has been found to increase Al EM lifetime (Black 1969; Ainslie *et al.* 1972; Wada *et al.* 1973; Lloyd and Smith 1983). By comparing SiO_2 passivation and SiN passivation, Wada *et al.*

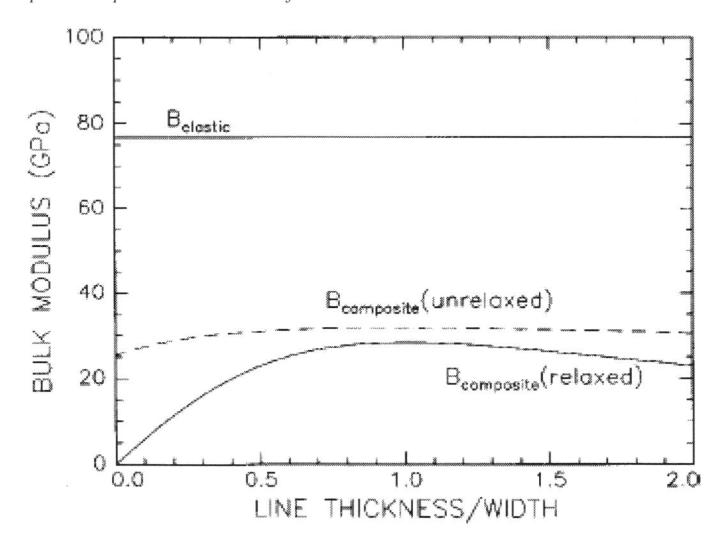

Fig. 3.15. The elastic bulk modulus, B_{elastic}, and effective composite bulk moduli before, $B_{\text{composite}}$ (unrelaxed), and after, $B_{\text{composite}}$ (relaxed), shear stress relaxation for different aspect ratios of interconnect lines. Reprinted with permission from Filippi, Wachnik, Eng, Chidambarrao, Wang and White, 'The effect of current density, strip length, stripe width, and temperature on resistance saturation during electromigration testing', Journal of Applied Physics, vol. 91(9), pp. 5878–5795. Copyright © 2002, American Institute of Physics.

(Wada, Sugimoto and Ajiki 1987) found that SiN passivation layer prevents aluminum hillock formation due to electromigration because the mechanical strength of SiN passivation layer is greater than that of SiO_2 passivation layer. However, when a micro-crack is presence in the passivation layer, such an inhibition of hillock growth will be diminished, and it indeed accelerates EM.

On the other hand, Nishimura *et al.* (Nishimura, Okuda, Ueda, Hirata and Yano 1990) studied the effect of stress in the passivation layer on the electromigration lifetime of the vias. They found that the median time to failure for the via chain with the slope angle of 55° is about 1/18 times shorter than that with the slope angle of 85°, despite a better step coverage of Al film in the via with slope angle of 55°. This was attributed to the thicker passivation thickness on the via which implied a higher compressive stress in the passivation. Their results are shown in Fig. 3.16.

Besides the stress effect in passivation layer on the EM lifetime, it is also found that hydrogen that evolved from the silicon nitride layer

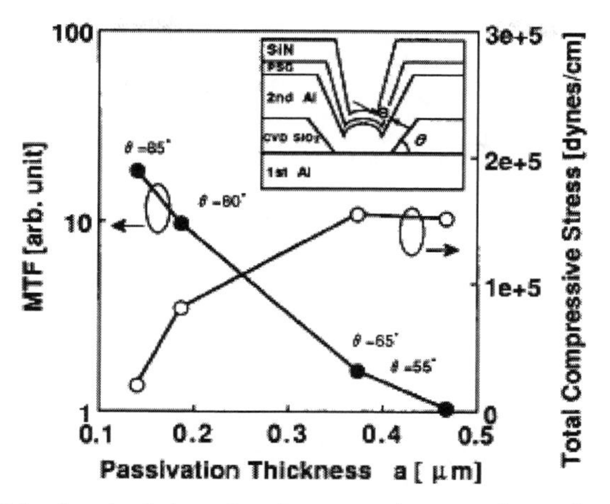

Fig. 3.16. MTF for the via chain and total compressive stress in passivation layer versus passivatin thickness (SiN/PSG) in 1.2 μm diameter via. (Nishimura, Okuda, Ueda, Hirata and Yano, 'Effect of stress in passivation layer on electromigration lifetime for Vias'. Proc. IEEE Symposium on VLSI Technology, p. 31–32), Copyright © 1990 IEEE.

during processing will also served as nucleation sites for stress- and electromigration-induced voids in Al films, rendering them to have voids formed away from the interconnect sidewall in contrast to the results from other studies of void nucleation in passivated aluminum lines. The presence of the hydrogen was confirmed by the nuclear reaction analysis performed by Lee *et al.* (Lee, Bravman, Doan, Lee and Flinn 2002).

3.3.8 *Metal width variation*

The variation in the interconnect line width will cause current crowding and has been discussed in the earlier section on temperature gradient.

3.3.9 *Reservoir effect*

The reservoir is a protrusion of multi-level Al–0.5%Cu interconnects around W via as shown in Fig. 3.17. Even though the reservoir does not carry an electric current during electromigration test, the lifetime of the multi-level interconnect systems can be prolonged by the reservoir, which is called 'reservoir effect'. Many experimental works (Fujii, Koyama and Aoyama 1996; Skala and Bothra 1998; Dioin 2000; Dion 2001; Le, Ting, Tso and Kim 2002) have verified the reservoir effect by demonstrated that

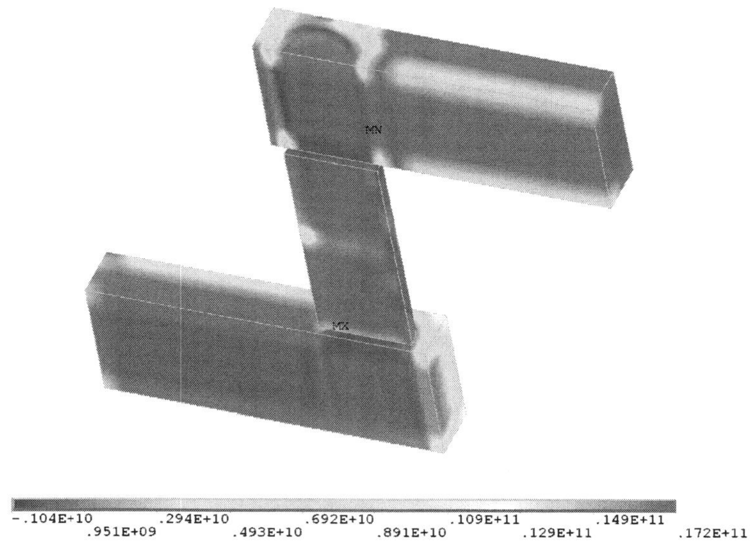

Fig. 3.17. Total AFD distributions for Al via structures with 0.28 μm line width. Unit: atoms/$\mu m^3 \cdot s$.

electromigration lifetime is dominated by the reservoir area and the number of contacts/vias.

The reasons for the lifetime enhancement effect due to reservoir are related to the hydrostatic stress and its gradient in the reservoir region. First of all, the maximum atomic flux divergence occur at two sites in the reservoir regions as shown in Fig. 3.17 through our finite element analysis. Without the reservoir, the two maximum AFD sites coincide and situated on top of the via, thus shorten the time to failure. In the presence of the reservoir length, the two sites are separated, and thus lengthen the time to failure. When the reservoir length reaches a certain value, the contribution of AFD at the corner of the reservoir to open circuit at the line-via interface no longer significant, and the time to failure is dominated by the maximum AFD directly on top of the via, hence lifetime enhancement due to reservoir length cease. A detail analysis can be found in (Tan, Hou and Li 2007).

Furthermore, in the presence of the reservoir, the higher mechanical stress is now shifted to the reservoir instead of on top of the via, and the stress at the top of the via is reduced. Since the equilibrium vacancy concentration depends on the mechanical stress as in Eq. (3.6) (Clement and Lloyd 1992; Korhonen, Borgesen, Tu and Li 1993; Clement and Thompson

1995; Clement 1997), void growth rate will be larger in the reservoir while the growth rate is reduced at the top of the via, thus enhances the EM lifetime.

$$c_c = c_0 \exp(f'\Omega\sigma/kT). \tag{3.6}$$

Here c_c is the equilibrium vacancy concentration in the presence of mechanical stress, c_0 is the equilibrium concentration in the absence of stress, $f' = 1 - f$ and f is the average vacancy relaxation factor; σ is the hydrostatic stress.

3.4 Self-Induced Process During EM

During electromigration, some self-induced processes are operative. These processes could be enhancing the failure, they could also heal the degradation which enhance the lifetime.

3.4.1 *Self-induced stress gradient*

Stress gradients can be produced by the regions of mass accumulation (compressive stress) and mass depletion (tensile stress) during EM. These stress gradients may give rise to mechano-diffusion fluxes in the direction opposite that of the EM fluxes, thus reducing the net EM flux. (Blech and Herring 1976; Ross, Drewery, Somekh and Evetts 1989). The net flux will be the lowest when the hillock and the void are closest (J.E. Sanchez and J.W. Morris 1991). Therefore, the net drift velocity for aluminum is given by (Schreiber 1985):

$$v_d = \left(j\rho e Z^* - \frac{\Delta\sigma\Omega}{L}\right)\frac{D_o}{kT}\exp\left(-\frac{E_a}{kT}\right), \tag{3.7}$$

where $\Delta\sigma$ is the back-flow stress, Ω is the atomic volume and L is the conductor length. Blech showed that the net drift velocity will become zero for a critical product of current density and conductor length $(jL)_c$, referred to as the threshold product.

Blech (Blech 1976) assumed that the back stress gradient is constant over the entire length, and hence jL should be a constant for a given temperature. However, Filippi *et al.* (Filippi, Biery and Wachnik 1995) found that when the length of conductor is long, even with the condition of jL equal to the critical value, EM still prevail. This is because only after some stress time, the back stress gradient will be developed and reach a constant over the entire length. If the length is long, the time required will be long, and substantial electromigration damage will have already occurred before the time is reached.

3.4.2 *Self-induced temperature gradient*

When the line width is small, and the structure becomes bamboo, the presence of void can produce gradients in current density and temperature around a void as shown in Fig. 3.18 (O. Kraft and E. Artz, "Numerical simulation of electromigration-induced shape changes of voids in bamboo lines", Applied Physics Lett. 66(16), p. 2063–2065, 1995). These gradients together with the electron flow could change the shape of the void without changing its volume as metal atoms are brought by electrons wind (which have enhanced diffusion rate on the void surface) from one side of the void to the other side as illustrated in Fig. 3.19 (O. Kraft and E. Artz, "Numerical simulation of electromigration-induced shape changes of voids in bamboo lines", Applied Physics Lett. 66(16), p. 2063–2065, 1995). This results in a slit-like crack in the metal line, and greatly reduces the EM MTF of the metal line.

The void shape change shown in Fig. 3.19 is due to the varying diffusion rates of atoms along the void surface. Atoms move faster at higher temperature, and hence as they are diffusing from the right side of the void due to the driving force, they are diffusing faster and faster, leaving vacancies accumulation on the right side and top of the void. After they pass

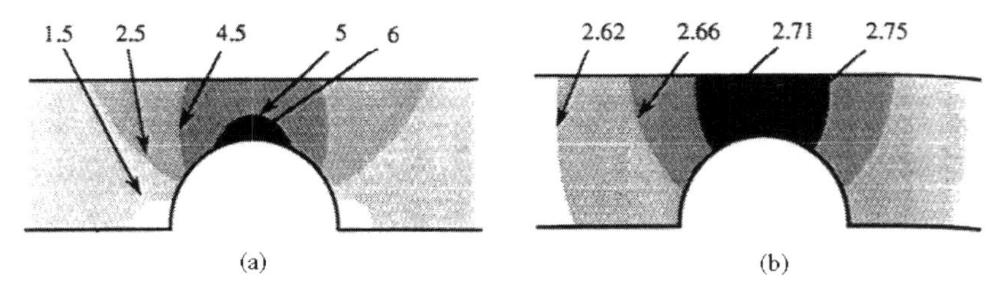

Fig. 3.18. Results of the finite element calculation: (a) current density distribution in the vicinity of a semi-circular void with a radius of 0.6 μm in a 1 μm wide Al line and an applied current density of 2 MA/cm^2, and (b) resulting temperature distribution. Reprinted with permission from Kraft and Arzt, "Numerical simulation of electromigration-induced shape changes of voids in bamboo lines", Applied Physics Lett. 66(16), p. 2063–2065, Copyright © 1995, American Institute of Physics.

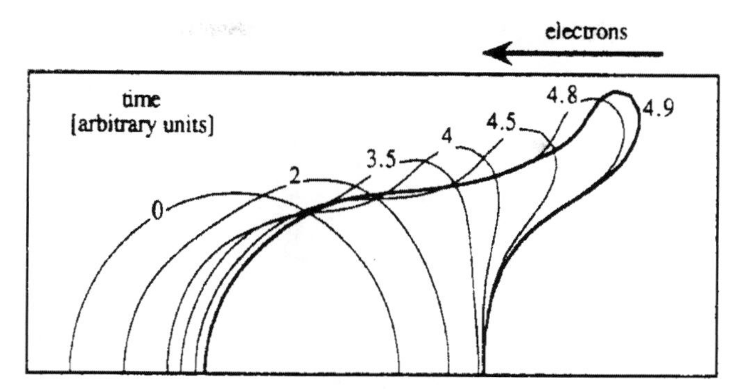

Fig. 3.19. (a) Time sequence showing a void shape change as calculated by the numerical simulation. The time is given in arbitrary units. The height of the frame is equal to the line width used in the simulation. Reprinted with permission from Kraft and Arzt, "Numerical simulation of electromigration-induced shape changes of voids in bamboo lines", Applied Physics Lett. 66(16), p. 2063–2065, Copyright © 1995, American Institute of Physics.

through the tip of the void, they are diffusing slower and slower due to the decreasing temperature they are experiencing, causing atoms accumulation on the left side of the void.

3.4.3 *Microstructure change of interconnect*

There have been numerous reports showing that interconnect film thinning precedes void formation (Rosenbaum and Berenbaum 1968; Blech and Meieran 1969; Berenbaum 1971; Ohring 1971; Rosenberg and Ohring 1971; Rosenberg, Mayadas and Gupta 1972). When electron wind force approaches a grain boundary, a region on the cathode side tends to thin preferentially (R.W. Vook, Cheng and Park 1993). Figure 3.20 shows an example of film thinning after EM stress (Vook 1994).

While most of the thinning are observed at the triple point, some observations show that thinning can also occur along a grain boundary (Vook 1994). Also, when the electron wind force is approximately parallel to a grain boundary, grain boundary grooving often occurred (R.W. Vook, Cheng and Park 1993).

Many experiments involving observations of film thinning show the presence of surface diffusion processes. Others have observed entire voids migrating against the electron wind down the interconnection (Blech 1976; Poate, Tu and Mayer 1978; Madden, Maiz, Flinn and Madden 1991). Thus it

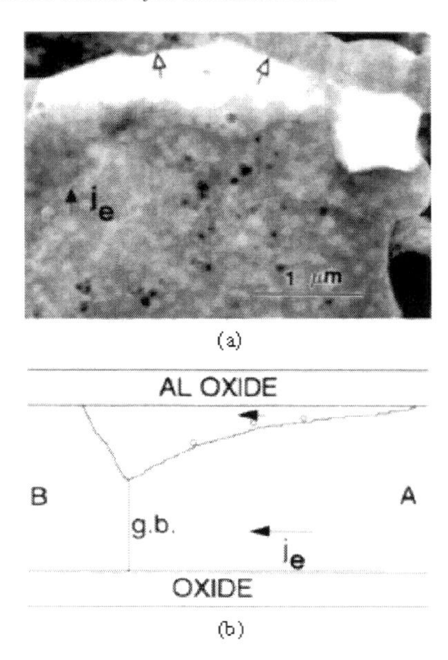

Fig. 3.20. Preferential thinning on the cathode side of a grain boundry in Al: (a) TEM micrograph; (b) sketch illustrating the phenomenon. The solid arrow gives the direction of electron flow. Reprinted from Materials Chemistry and Physics, 36, Vook, 'Transmission and Reflection electron microscopy of electromigration phenomena', pp. 199–216 Copyright © 1994, with permission from Elsevier.

is believed that the thinning and void migrations are due to surface diffusion mechanism (Ho 1970). With such a thinning and void migrations, the non-uniformity of current density will be enhanced, thus increase the atomic flux divergence and shorten the time to failure. This explains the rapid increase in the resistance of an interconnection after void nucleation.

Interestingly, when the current direction is reversed, regrowth takes place between two bounding oxide layers in Al (Vook 1994). Thus the EM lifetime of interconnects under AC current is expected to be lengthen.

While it is agreed that surface diffusion is the main mechanism for the film thinning, one found that in some cases, some kinds of atoms on a given substrate drift to the anode while others drift to the cathode. They are at variance with the bulk studies where atoms always drift toward the anode as a result of momentum transfer by the electron wind to the migrating atoms. Hence, it is clear that surface electromigration is yet to be well understood (Vook 1994). In fact, with the Cu interconnection, such surface electromigration is expected to play an even dominating role for the Cu EM as will be seen in the next Chapter.

3.5 Electromigration Test Structure Design

The complexity of electromigration process and its dependence on the interconnect structure can be seen from the above discussion. Therefore, to quantitatively assess the EM reliability of an interconnect process and structure, actual EM tests must be performed, and in particular, due to the requirement of high reliability of the interconnects, accelerated life test is needed. Numerical models can only give qualitative merit of the backend process and the interconnect structure.

In order to ensure repeatability and accuracy of an EM test, and in view of the sensitivity of the EM process with respect to the interconnect structure, standard test structures are developed so that the failure mechanisms observed under accelerated life test are the same as that under normal operating conditions. Only with this assurance can the test results obtained at accelerated life test be extrapolated to the normal operating conditions.

Two common structures for electromigration testing are (i) a long line with large pads at either end as shown in Fig. 3.21 for a single interconnect level, and such structure is also known as NIST (National Institute of Standards and Technology) structure; and (ii) an interconnect line with W-studs at the line end ("Via-fed" structure) for multi-level interconnect. The effects of (a) flux divergences at the line end, (b) interfacial diffusion, and

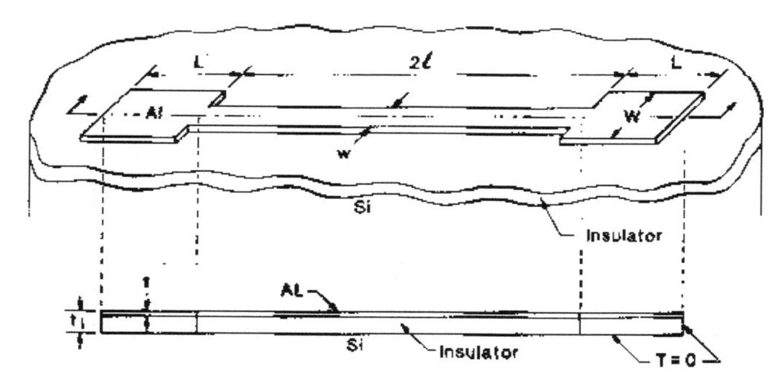

Fig. 3.21. NIST test structure for EM (Schafft 1987). (Schafft, 'Thermal Analysis of electromigration test structures', IEEE Trans. Electron Devices, 34(3), p. 664), Copyright © 1987 IEEE.

(c) Cu-depletion generally result in different lifetimes for the two types of test structures under the same testing conditions (Venkatrkrishnan and Ficalora 1992). Besides, structure design can affect the temperature gradients and joule heating during the EM test, which in turn affect the accuracy and reproducibility of the reliability assessment of VLSI interconnect. Let us now look into more detail of the two types of test structures.

3.5.1 *NIST test structure*

For NIST structure, Schafft (Schafft 1987) identified the various factors that will affect the accuracy and repeatability of the test results. His evaluation was made with the following assumptions:

(a) The far end of the end-contact segment is maintained at the silicon substrate temperature
(b) No temperature gradient exist along the top surface of the silicon substrate lying beneath the test structure
(c) Effect of temperature dependence for the thermal conductivity of the insulator and of the metallization can be ignored in comparison with that due to the metallization resistivity
(d) No passivation layer over the metallization
(e) No adjacent metallization near the interconnect under test

Under the above-mentioned assumptions, the following factors must be considered in test structure design as identified by Schafft (Schafft 1987):

(a) Thickness of the underlying insulator

We have seen the effect of oxide thickness on the test strip temperature in Fig. 3.2. The effect is more significant at higher current density.

Besides the importance of oxide thickness, the importance of thickness uniformity can also be seen. The step in oxide thickness can also produce significant temperature.

(b) Thermal conductivity of the underlying insulator

Calculation showed that the shape of the curves in Fig. 3.2 is sensitive to the value of thermal conductivity used for oxide and that only a narrow range of thermal conductivity values would give a reasonable fit to the experimental data. For example, the value of 0.0096 W/cm°C gave the best fit to the data while a value of 0.010 W/cm°C will result in a noticeably deviation of the fit. This implies that the temperature of the test structure is highly sensitive to the thermal conductivity of the underlying oxide.

In general, the poorer the thermal conductance between the metal line and the silicon substrate, the larger the temperature of the test line will be for a given current density.

Since the thermal conductivity of oxide is a function of the deposition process and of subsequent processing, test structure fabrication is critical. This make EM lifetime sensitive to process condition.

(c) Presence of passivation layer and adjacent metal line

The presence of passivation layer and adjacent metal line will affect the temperature and temperature profile of the test strips as have been seen in the previous section. These factors need to be considered in the data interpretation of the test results. It also indicates the importance of direct measurement of the test line temperature, temperature profile and film stress.

(d) Strip Width/oxide thickness Ratio

The temperature of the test line of infinitely long is given by (Schafft 1987)

$$T = T_a + \frac{J^2 \rho_o}{\frac{K_i}{tt_i}[1 + 0.88t_i/w] - J^2 \rho_o \beta}, \tag{3.8}$$

where T_a is the ambient temp. J is the current density, ρ_o is the resistivity of the metallization, β is the TCR, t is the metallization thickness, t_i is the insulator thickness, w is the test line width, and K_i is the thermal conductivity of the insulator.

If t_i/w is very small, i.e. either the oxide thickness is very small or width of the test strip is large, 1D heat flow will be resulted. The temperature variation with oxide thickness for the 1D heat flow will be more significant as shown in Fig. 3.22. Figure 3.22 also shows the importance of the lateral heat dissipation. Therefore, it again indicates the importance of direct measurement of the test line temperature and temperature profile.

(e) Length of the test line

The shape of the temperature profiles of test lines is shown in Fig. 3.23 for three 400 μm long test lines. A thermal interaction length l_θ is used to characterize the shape of the temperature profiles of test lines and is defined as the distance from the end of the test line at which the line reaches 90% of

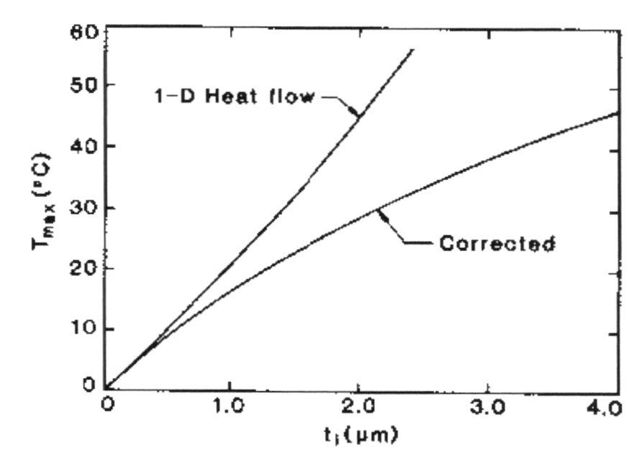

Fig. 3.22. Maximum test line temperature vs. insulator (silicon dioxide) thickness for conditions where 1D heat flow is assumed and where deviations from 1D heat flow are included. Here $J = 2.5\,\text{MA/cm}^2, t = 1\,\mu\text{m}, \rho_0 = 3.14\times10^{-6}\Omega\cdot\text{cm}, \beta = 3.9\times10^{-3}{}^\circ\text{C}^{-1}$, $K_i = 0.0096\,\text{W/cm.}^\circ\text{C}$, and $w = 4\,\mu\text{m}$. (Schafft, 'Thermal Analysis of electromigration test structures', IEEE Trans. Electron Devices, 34(3), p. 664), Copyright © 1987 IEEE.

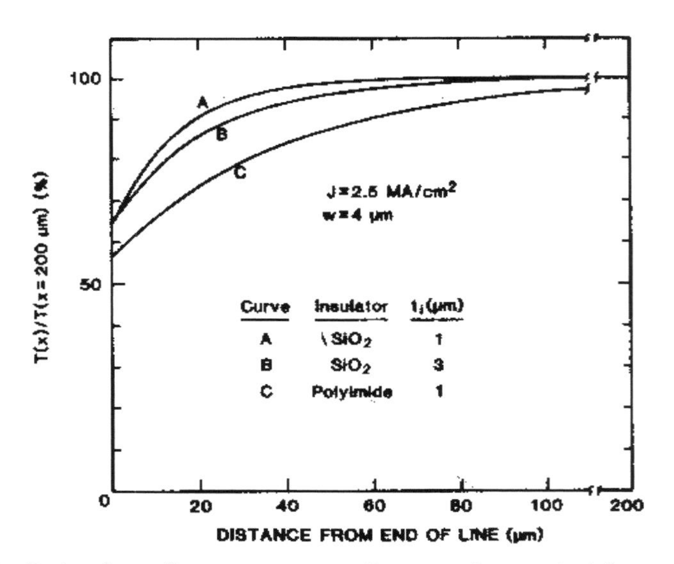

Fig. 3.23. Ratio of test line temperature at distance x from end of line to max. line temperature vs. distance x for different insulator materials and thicknesses. The length L of the end contact segment is $120\,\mu$m. (Schafft, 'Thermal Analysis of electromigration test structures', IEEE Trans. Electron Devices, 34(3), p. 664), Copyright © 1987 IEEE.

the maximum temperature of an infinitely long line. From Fig. 3.23, one sees that the poorer thermal conductance of the substrate results in a longer l_θ.

Therefore, to avoid conditions where the test line will have a large variation in temperature along its length and thereby make difficult an analysis of a stressed test, the length of the test line should be longer. In general, test lines much shorter than 400 μm should not be used.

(f) End segment length

The profile of the test line is essentially unaffected as long as the end-contact segment L is greater than approximately 100 μm. Changing the insulator from the oxide to the much less thermal conducting polyimide shifts the threshold value of L to above 100 μm. (Schafft 1987)

Therefore, to ensure that the end segment has a negligible effect on the temperature profile of the test line, L should be longer than 100 μm for the case of oxide.

(g) Width ratio (width of end contact/width of test strip)

This ratio can affect the temperature gradient in the test line. The closer the ratio is to unity, the less its effect on the temperature profile near the ends of the test line. But if this ratio is reduced sufficiently, electromigration failures will begin to occur in the end segment rather than in the test line due to high temperature gradient at the interaction of the end segment and the test line. A safe lower limit for the width ratio is two. Figure 3.24 shows the temperature gradient at the end of the test line for a current density of 2.5 MA/cm^2.

Figure 3.24 also shows that with poorer thermal conductor, the magnitude of the gradient increases. Increasing the power dissipation by increasing the linewidth to 4 μm will also increase the gradient.

On the other hand, if the end-segment width is too large, it can became a source or reservoir which replenshied the depleted material in the test strip. In fact, some depletion of the reservoir as it narrowed towards the test strip has been noted (Shingubara, Nakasaki and Kaneko 1991). This gives an upper limit for the width ratio.

(h) Current density

The degree of joule heating has an important effect on the temperature gradient at the end of the test line. The magnitude of the effect is illustrated in Fig. 3.25.

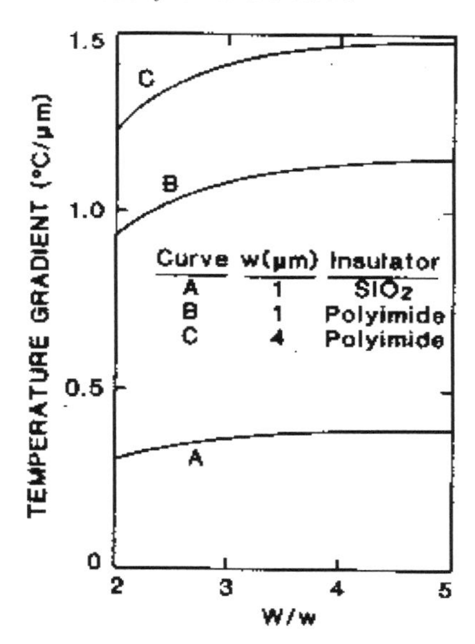

Fig. 3.24. Temperature gradient at the end of test line vs. the ratio of the width of the end-contact segment to that of the test line. The current density is 2.5 MA/cm^2. (Schafft, 'Thermal Analysis of electromigration test structures', IEEE Trans. Electron Devices, 34(3), p. 664), Copyright © 1987 IEEE.

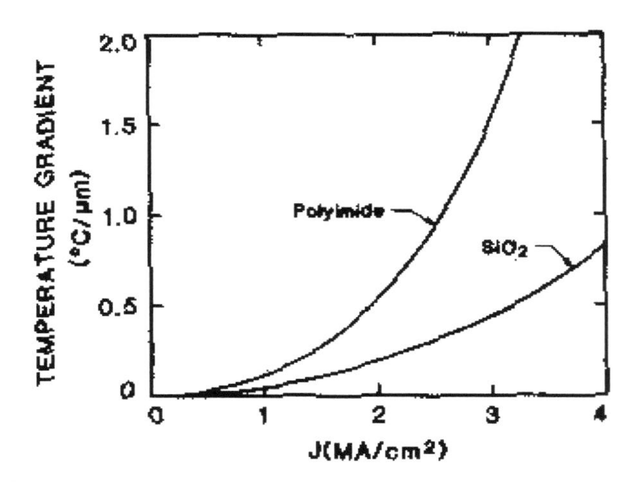

Fig. 3.25. Temperature gradient at the end of the test line versus current density for a silicon dioxide and for a polyimide insulator. The linewidth and insulator thickness are 1 μm. (Schafft, 'Thermal Analysis of electromigration test structures', IEEE Trans. Electron Devices, 34(3), p. 664), Copyright © 1987 IEEE.

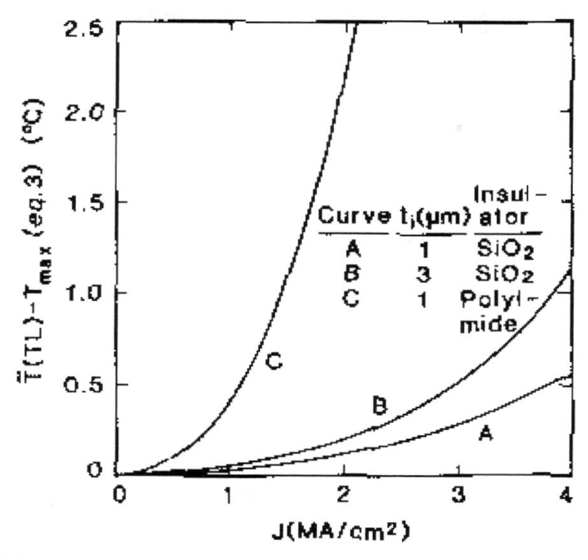

Fig. 3.26. Difference between the mean test line temperature and the maximum test line temperature versus current density. (Schafft, 'Thermal Analysis of electromigration test structures', IEEE Trans. Electron Devices, 34(3), p. 664), Copyright © 1987 IEEE.

Joule heating also results in a large difference between the mean temperature and the peak temperature of the test line as shown in Fig. 3.26.

The consideration in the test structure design discussed above is for the single metallization. It is useful for the characterization of the metal deposition and annealing processes. For multi-level metallization, current flows between different levels can results in void formation at the interface between W-plug vias and Al test as shown earlier in Sec. 3.2.6. Hence, to assess the reliability of such interconnect structure, "via-fed" test structure needs to be used.

3.5.2 *Test structure for multi-level metallization*

The types of test structures used for the drift-velocity type failure are as shown in Fig. 3.27 (Ting and Graas 1995). For the "via-fed" structures, the observed log-normal sigma is smaller as compared with the NIST structure because the "drift-velocity-type" failure having a weak dependence on microstructure variations on the test lead (Ting and Graas 1995). The results are shown in Table 3.6. For "via-fed" test structures, the weakest link is

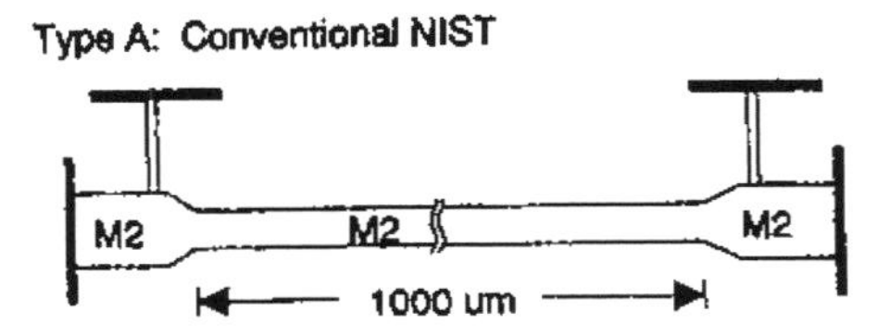

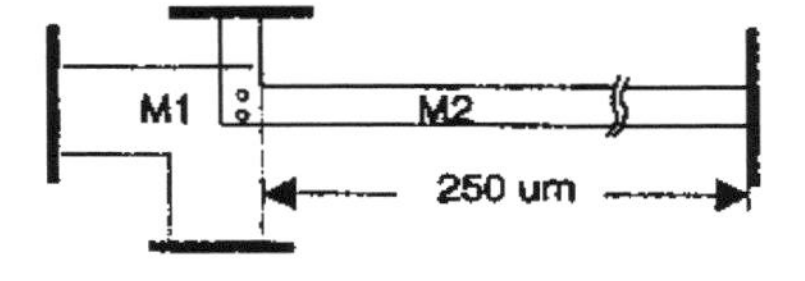

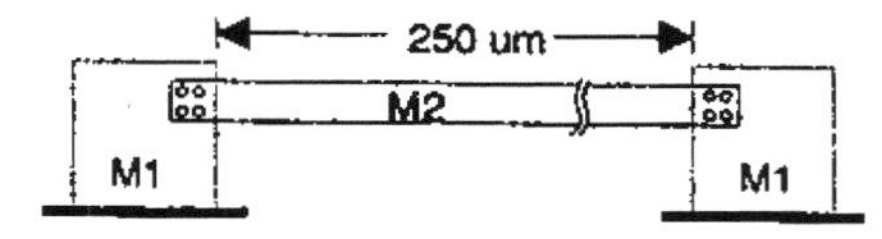

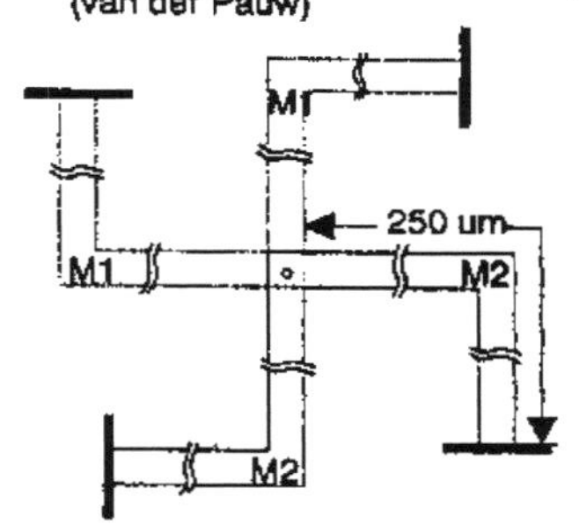

Fig. 3.27. Layout schematics of EM test structures studied. In the EM tests, the direction of the electron flow was from left to right. (Ting and Graas, 'Impact of test structure design on electromigration lifetime measurement'. Proc. of International Reliability Physics Symposium, p. 326–332) Copyright © 1995 IEEE.

located at the heterogeneous interface between W-plug vias and the Al test lead, which enables a large atomic flux divergence due to drifting of Al atoms away from the interface. In the same token, the lifetime dependence on test lead linewidth is much smaller for "via-fed" test structure as was observed experimentally.

Table 3.6. Comparison of lifetimes measured on different type test structures for W = 3.05 μm. Except for type A, all "via-fed" test structures has a single via for current feeding (Ting and Graas 1995).

Type	Test structure scheme	Median lifetime (hrs)	Lognormal sigma
A	Conventional NIST	1091	0.42
B	"via-fed" with a short voltage tap lead	107	0.23
C	"via-fed" without voltage tapping	39	0.30
D	"via-fed" with a long voltage tap lead	40	0.21

In the same way as NIST test structure, the "via-fed" structure has some design aspects that will affect the EM lifetime. These aspects are as follows.

(a) Design of kelvin-type voltage taping (Ting and Graas 1995)

From Table 3.6, one can see that among the various type of "via-fed" test structures, the design of voltage tapping for Kelvin-type measurements can significantly affect lifetime measurement. Use of a short voltage tap lead enabled the bond pad at the end of the short lead to play a sourcing role and replenished the depleting atoms that migrated away from the vias/lead interface. This led to delay in void formation and an increase in lifetime. The experimental results in Table 3.6 showed that replacement with a long voltage tap lead of 250 μm (type D) yielded a comparable lifetime as obtained by using he test structure without inclusion of voltage tapping (type C).

(b) Number and layout of vias

The effect on lifetime measurement for varying the number of vias used in current feeding was found to be generally small by Ting *et al.* (Ting and Graas 1995). Small changes in the measured lifetime were obtained for varying the number of vias from 1 to 9 (Ting and Graas 1995).

On the other hand, the layout of multiple vias can affect the lifetime significantly due to the reservoir effect. Skala *et al.* (Skala and Bothra 1998) studied 6 different type of vias layout from M1 to M2 as shown in Fig. 3.28, and they found that the t_{50} of the interconnect can be affected by the layout of the via as shown in Table 3.7.

(c) Current density uniformity (Ting and Graas 1995)

As have been seen earlier, current crowding can have an effect on the EM lifetime of an interconnect. Ting *et al.* (Ting and Graas 1995) found that

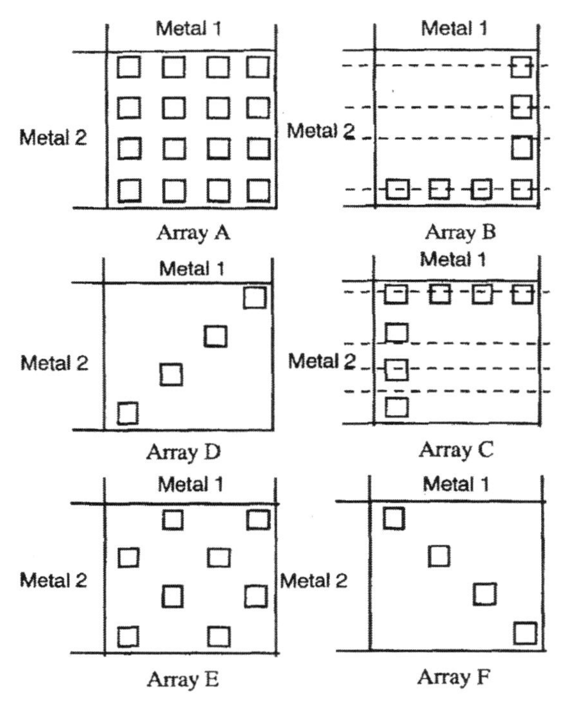

Fig. 3.28. Layout of different metal 1/metal 2 via array interconnects. (Skala and Bothra, 'Effects of W-plug via arrangement on electromigration lifetime of wide line interconnects'. Proc. IEEE International Technol. Conf., p. 116–118) Copyright © 1998 IEEE.

Table 3.7. Summary of t_{50} and sigma for via arrays shown in Fig. 3.28. Failure criteria is 10% increases in resistance (Skala and Bothra 1998).

Via array	t_{50} (hours)	Sigma
A	160	0.169
B	87.6	0.091
C	123.5	0.103
D	133	0.103
E	137.2	0.162
F	105	0.095

the failure times at large stress currents in the "via-fed" structure are much smaller compared with those obtained for the uniform profile. At small stress currents, both uniform and non-uniform current density profiles gives similar results.

3.5.3 *Test structure for bamboo structure*

To apply the NIST test structure to Bamboo metal lines, problem arises as the end-contact segments have average grain size smaller than the line width, giving rise to alternate 'bamboo' sections and so-called 'cluster' sections, with connecting grain boundaries. This caused the end-segments to show a worse resistance to electromigration and a shorter lifetime, even if the current density flowing through them was half of the value in the test line (Munari, Vanzi, Scorzoni and Fantini 1995).

Test on passivated 0.9 μm wide bamboo structure revealed that the EM test was terminated due to the end-segment breaks. Most of the failures detected were located along the end-segments, and their locations were quite randomly along their whole length of the end-segments. There were not likely to be related to thermal gradients since they were not detected at the sites where the temperature variations were expected to be the highest (Munari, Vanzi, Scorzoni and Fantini 1995). In this regard, in real devices, 'bamboo' lines are not the most sensitive part of the structure. The weak points are the connections between wide and narrow lines. Different structures should be designed in order to highlight these weaknesses (Munari, Vanzi, Scorzoni and Fantini 1995).

3.6 Package-Level Electromigration Test (PET)

As mentioned earlier, in order to assess the EM of an interconnect, accelerated EM test is necessary with a test structure as described above. The most reliable method of EM testing is the PET.

In 1968, Rosenberg *et al.* (Rosenberg and Berenbaum 1968) first introduced the idea of monitoring resistance to follow the structural damage during accelerated EM testing from the rate of resistance change (RRC) as given as follows.

$$\frac{1}{R(t=0)}\frac{dR(t)}{dt} = A\exp\left(-\frac{E_a}{k_B T}\right). \tag{3.9}$$

In the above expression, all symbols carry the usual meaning. Equation (3.9) for RRC during EM induced mass transport is also studied by many other researchers (Hummel, DeHoff and Geier 1976; Pasco and Schwarz 1983; Lloyd and Koch 1988; Maiz and Segura 1988; Scorzoni, Cardinali, Baldini and Soncini 1990). The activation energy is extracted from the ratio of the RRC at two different temperatures (say at T_1 and T_2) from the following expression

$$\frac{\dot{R}_1}{\dot{R}_2} = \exp\left[-\frac{E_a}{k_B}\left(\frac{1}{T_1} - \frac{1}{T_2}\right)\right]. \tag{3.10}$$

Silicon oil baths were used to change instantaneously the sample temperature in steps of 25°C between 60 to 150°C and activation energy was found to be 0.5 to 0.6 eV using Equation (3.10) for Al strips interconnects with line width of 7.6 μm (Rosenberg and Berenbaum 1968).

In today PET method, processed wafer is diced and each chip is die-attached to package with die-attach epoxy that can withstand high temperature of 450°C. The typical size of the chip is a few mm^2. Wire bonding is then performed on the test dice and gold wire is normally used. The packaged samples are then loaded into oven for EM test. A constant current density of several MA/cm^2 and a constant oven temperature (>150°C) are used as EM stressing parameters. The electrical resistance of every sample is monitored continuously. Once the change in resistance of a sample increased to a pre-selected level, the tester stops sending current through that sample, but the thermal stress (fixed by oven temperature) still continues on the failed sample until the test terminates.

In the PET method, current and temperature are normally taken as independent of each other. For high current density test, the additional temperature rise due to Joule heating is also estimated by using TCR data of the sample under test. The TCR data is obtained before the EM stressing. In order to obtain sample TCR, the sample resistance is measured at pre-selected temperatures (maximum temperature is equal to test temperature) by sending a short current pulse of magnitudes ~0.1 mA.

PET method allows us to determine all the Black's parameters (i.e. E_a, n and A). To determine all the parameters, the EM tests are performed at 2 different temperatures and 2 different current densities. One set of test will be at T_1, J_1; another set at T_2, J_1; and the last set at either T_1, J_2 or T_2,

J_2. Thus a total of 3 different stress conditions are required. With these 3 stress conditions and the respective MTFs at the stress conditions, one can determine the activation energy and the current exponent as (JEDEC 1995):

If $t_{50,1}$, $t_{50,2}$ and $t_{50,3}$ correspond to the MTF at the three stress conditions, J_1 is the stressed current density for the first two stress conditions, and J_2 is the stressed current density for the third stress conditions, and T_1, T_2 and T_3 are the actual line temperatures at the three stress conditions (although the test temperature for the first and third stress conditions are set to be the same, due to the different current densities and hence different degree of Joule heating, the actual line temperatures for the two stress conditions are different), using the MTF from the first two stress conditions, and based on the Black's equation, we have

$$\frac{t_{50,1}}{t_{50,2}} = e^{\frac{E_a}{k}\left(\frac{1}{T_1}-\frac{1}{T_2}\right)}$$

where k is the Boltzmann's constant. Thus activation energy E_a can be determined. Using the result from the first and third stress conditions, we have

$$\frac{t_{50,1}}{t_{50,3}} = \left(\frac{J_1}{J_2}\right)^n e^{\frac{E_a}{k}\left(\frac{1}{T_1}-\frac{1}{T_3}\right)}$$

and the current exponent n can be determined. One can also use the second and third stress conditions to obtain n in a similar manner.

Though PET takes the longest test time in comparison to other EM test methods, it is free from crude approximation. However, like all other methods, it assumes the invariability of failure mechanism in accelerated test and in actual field of operation.

As interconnect technology becomes more reliable, the Time to Failure of interconnect get longer. Hence, various versions of EM tests are developed in industry to shorten the test time as will be discussed in the next section.

3.7 Rapid Electromigration Test

Generally, electromigration test is performed in package level in order to obtain accurate results. However, industries like to speed up the testing so that decision can be made in a short time. Two approaches are commonly employed to meet the goal, namely the use of fast wafer level electromigration test and the use of highly accelerated conditions.

The fast wafer level electromigration test can save the time and cost of packaging, and hence a larger sample size is possible, making it attractive for electromigration evaluation of thin-film metallization for ULSI today. Fast wafer level techniques can provide useful electromigration information within a few seconds to a few hours. The different fast wafer level electromigration tests developed include the TRACE method (Pasco and Schwarz 1983), Standard Wafer-level Electromigration Acceleration Test (SWEAT) (Root and Turner 1985), the Wafer-level Isothermal Joule-heated Electromigration Test (WIJET) (Jones and Smith 1987), the Constant Current method (Snyder 1994), and the Breakdown Energy of Metal (BEM) method (Hong and Crook 1985).

The major disadvantage of wafer level testing, however, is that the temperature and current cannot be controlled independently, and for this reason, little trust is placed in lifetime projections produced by these methods. In addition to the coupled parameters, the results are different for each method (Lee, Tibel and Sullivan 2000). Also, these techniques were developed with speed in mind. Hence, 'overstressing' of the metallization where highly accelerated stress conditions are commonly used, and this could result in failure mechanisms different from those present in real life applications, making their lifetime predictions meaningless (Foley, Molyneaux and Mathewson 1999). In this section, the various types of Fast Wafer Level (FWLR) electromigration test methods will be described in detail from the perspective of electromigration physics and reliability statistics, so that the best feasible FWLR can be identified.

3.7.1 *TRACE method*

In 1983 Pasco and Schwarz (Pasco and Schwarz 1983) proposed a resistometric technique known as TRACE (temperature-ramp resistance analysis to characterize electromigration) for fast EM testing. In this method, a temperature ramp is applied to a packaged sample while the test sample is subjected to a high current density. By measuring the resistance change during the experiment within an elapsed time $\sim$1 h, the parameters A and Ea of Eq. (3.9) are extracted. This method is subsequently applied to wafer level EM characterization in air with some improvements of the thermal model (Finetti, Ronkainen, Blomberg and Suni 1986; Finetti, Suni,

Armigliato, Garulli and Scorzoni 1986; Finetti, Suni, Santi, Bacci and Caprile 1987; Finetti, Scorzoni, Armigliato, Garulli and Suni 1988).

The interconnect resistance change during EM testing is taken as an increase in the lattice defect concentration and Matthiessen's rule is introduced in the proposed model for EM damage. Matthiessen's rule splits the sample resistance into two parts as follows:

$$R(t) = R_0[1 + \alpha(T - T_0)] + R_i(t), \tag{3.11}$$

where R_0 is the component of resistance related to the scattering of electron waves by phonons in the lattice (ideal resistance), R_i is the part of the resistance independent of temperature originating from the presence of impurities (residual resistance) and alpha (write in greek symbol) is the TCR. The weak variation of R_0 and TCR with temperature is neglected by considering a small range of temperature change. It is also assumed that a relative resistance change up to $<10\%$ is only due to the influence of R_i, related to the scattering from impurities, though there are experimental evidences against this hypothesis (Schwarz and Felton 1985).

Nevertheless, with the introduction of ramping temperature condition $T = T_0 + bt$, where T_0 is the starting temperature and b is the heating rate (usually in the range of 0.5 to 5 K/min), Eq. (3.11) can be written as

$$R(t) = R_0[1 + \alpha bt] + R_i(t). \tag{3.12}$$

The normalized resistance change ΔR after time t, considering change in R_i is ΔR_i during the same elapsed time, is given by

$$\frac{\Delta R_i}{R(0)} = \frac{\Delta R}{R(0)} - \alpha bt, \tag{3.13}$$

where $\Delta R/R(0)$ is measured during EM testing. Thus the EM component of the normalized resistance change in the TRACE technique is the total resistance change minus the linear baseline containing temperature component.

Integrating Eq. (3.13) using the temperature ramping condition $dt = dT/b$ (ΔR_i should be independent of T), the following relation can be obtained:

$$T^{-2}\frac{\Delta R}{R(0)} = \left(\frac{Ak_B}{bE_a}\right)\exp\left(-\frac{E_a}{k_BT}\right). \tag{3.14}$$

The value of activation energy and the constant A are extracted by plotting left hand side of Eq. (3.14) (in logarithmic scale) as a function of $(1/k_BT)$.

The slope gives the value of activation energy and the value of the constant is derived from the intercept of the above mentioned plot.

Though the value of activation energy is found to be in good agreement with the expected value, a slight linear dependence of activation energy on the heating rate (b) is noticed. This problem is resolved by introducing an error factor (κ) into Eq. (3.13) and rewritten it as

$$\frac{\Delta R_i}{R(0)} = \frac{\Delta R}{R(0)} - \kappa \alpha b t. \tag{3.15}$$

The reason for introducing the above mentioned error factor and its derivation can be found in (Schwarz and Felton 1985). During the TRACE test, κ varies in the range of 1 to 1.1, and it varies as a function of stress condition especially.

The sample temperature is assumed to follow the oven temperature, plus a constant over-temperature ΔT arisen from Joule heating. It is not very hard to understand that ΔT is not constant when temperature of the sample holder increases. Thus another error is introduced in the last term of Eq. (3.15) of the same order of magnitude as the error factor κ. This error is ignored for the TRACE technique used in (Finetti, Suni, Armigliato, Garulli and Scorzoni 1986; Finetti, Suni, Santi, Bacci and Caprile 1987; Finetti, Scorzoni, Armigliato, Garulli and Suni 1988; Scorzoni, Neri, Caprile and Fantini 1991).

It is assumed and justified by Rosenberg *et al.* (Rosenberg and Berenbaum 1968) that EM damage is responsible for the first percent of relative resistance change and as a result Eq. (3.9) is used up to a relative resistance change in the range of 5 to 10%. During EM damage, the failure process should proceed under well-defined conditions of temperature, temperature distribution and current density before catastrophic failure. After EM damage, a catastrophic failure process arises in which Eq. (3.9) is no longer in effect.

It is also noticed that there exists a relation between activation energy and pre-exponential constant A when TRACE technique is used to determined these two parameters (Hong and Crook 1985). This relation is expressed as

$$\ln A = \ln p_1 + p_2 E_a, \tag{3.16}$$

where p_1 and p_2 are two suitable constants. This relation is justified well in ref. (Hong and Crook 1985).

It is reported that though TRACE technique is useful for Al-Si strip to extract acceptable results, the same does not hold for multi-layer Al based interconnects (Finetti, Suni, Armigliato, Garulli and Scorzoni 1986). Resistance decrease due to the hillocks formation is explained as the possible factor for the technique to fail for multi-layer interconnects. Thus in principle, only resistance measurement near the cathode should give reliable data by TRACE technique.

3.7.2 *Standard Wafer-level Electromigration Accelerated Test (SWEAT)*

The Standard Wafer-Level Electromigration Acceleration Test (SWEAT) was proposed in 1985 by Root and Turner (Root and Turner 1985). It was developed as a fast method for obtaining a measure of metallization quality and providing control data to the semiconductor manufacturer.

The test structure used for SWEAT is a succession of narrow and wide regions linked by tapered zones as shown in Fig. 3.29. This particular structure geometry causes high current density divergences and high temperature gradients generated in the narrow regions. As a result, it accelerates electromigration in the narrow region. The wide regions act as heat sinks for the joule heat originating in the narrow regions.

SWEAT test approach (1994) is novel in that instead of controlling the independent variables, an attempt is made to control the dependent variable through the use of the Black equation given as follows.

$$t_{50} = \frac{A}{j^n} e^{\frac{E_a}{kT}}, \tag{3.17}$$

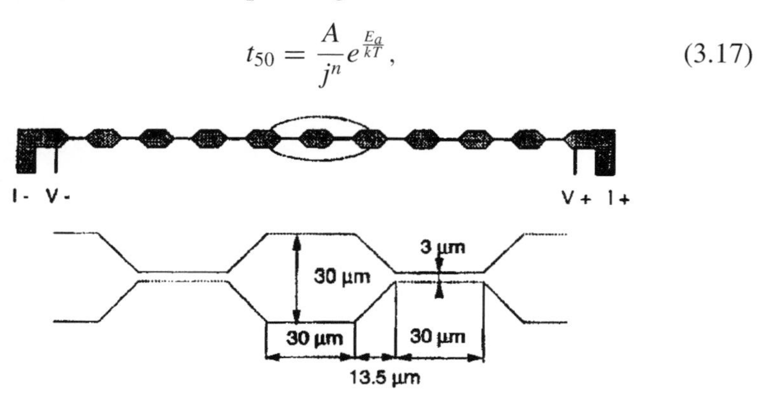

Fig. 3.29. SWEAT structure. (Giroux, Gounelle, Mortini, Ghibaudo, 'Wafer level electromigration tests on NIST and SWEAT structures', Proceeding of the International Conference on Microelectronic Test Structures, p. 229–232), Copyright © 1995 IEEE.

where

t_{50}: median time to failure
A: material constant based on microstructure and geometric properties of the test line
j: current density
n: current exponent index
E_a: activation energy
k: Boltzmann's constant
T: absolute temperature of test line

From the Black Equation of Eq. (3.17), the median time to failure of a metal line undergoing electromigration is a function of two independent variables, T and j. Thus, the execution of SWEAT has the following steps (Scorzoni, Neri, Caprile and Fantini 1991):

(a) The TCR (Temperature Coefficient of Resistivity) of the stripe and its resistance R_0, at a given temperature, are measured in order to calculate, during step (b), the actual temperature by measuring the current value of resistance. The detail of this methodology can be obtained from JEDEC (1995).
(b) The relationship between the electrical power p dissipated inside the stripe, and the corresponding increase of the temperature, is experimentally evaluated. The temperature is calculated by using the previously measured values of TCR and R_0.
(c) A minimum acceptable lifetime TTF_0 of the metallization system was defined. It is derived according to Eq. (3.17), at the device normal operating conditions including temperature T_o and current density j_o.
(d) The accelerating stress factor, S_F, is then defined (e.g. 10^8). Theoretically, it is defined by dividing the minimum acceptable lifetime TTF_0 with the required lifetime under test condition TTF, as shown in Eq. (3.18).

$$S_F = \frac{TTF_0}{TTF} = \left(\frac{j_0}{j}\right)^{-n} \exp\left[\frac{E_a}{k}\left(\frac{1}{T_0} - \frac{1}{T}\right)\right], \tag{3.18}$$

where

TTF_0: minimum acceptable lifetime
TTF: required lifetime under test condition
S_F: acceleration factor

E_a: activation energy for the diffusion process
j: current density
n: current density exponent (assume to be 2)

(e) The current density j is gradually increased and the power $P = IV$ dissipated inside the resistor is continuously measured, by monitoring the current I and voltage drop V by mean of a computer-controlled measurement system. In this way, by using the relationship between P and T, it is possible to know the actual temperature. The current ramp stops at value j_s, which corresponds to a temperature T_s, at which the TTF of the stripe is equal to $t_{\min}/S_F$. This required a priori knowledge of the parameters A, n, and E_a, presumably obtained through classic testing on packaged test structures.
(f) The value of j is continuously adjusted in order to satisfy $TTF = t_{\min}/S_F$, until the metal stripe failure.

SWEAT uses a feedback control loop to adjust the stress current applied to the metallization such that temperature and current density of the structure maintain the selected MTF within error band.

As electromigration continue to change the resistance of the structure, feedback-control algorithm acts to maintain the estimated MTF within the error band. Finally, the structure failed when its resistance exceeds the failure resistance. The failure criterion is based upon the heated structure resistance rather than the ambient temperature resistance. This is reflected, through the estimation of structure temperature, as rapid decrease in estimated MTF. The test is then terminated and the results are logged. Then, the measured failure time is recorded.

If the measured failure time under this test condition is greater than $TTF = t_{\min}/S_F$, the wafer under the test is accepted, otherwise, it is rejected.

The test is used only as a qualitative indicator of metal performance. This is done by observing the variance between the targeted t_{50} and the measured t_{50}.

3.7.3 *Wafer-level isothermal Joule-heated electromigration test (WIJET)*

Wafer-level Isothermal Joule-heated Electromigration Test (WIJET) is an accelerated electromigration test developed by Jones and Smith (Jones and Smith 1987). It is capable of determining the characteristics of electromigration. This test method allows user to obtain information such as the median-time-to-failure t_{50} under known conditions, the distribution of lifetimes under given conditions and the activation energy E_a. It also produces some information about the current-density dependence.

WIJET tries to emulate the standard isothermal PET tests by using Joule heating of the sample to raise the temperature instead of a heated ambient. The resistance is monitored and its value is kept constant by adjusting the current to give a constant self-heating and hence temperature. The detail testing procedure and failure criterion can be obtained from JEDEC Standard "EIA/JESD61" (1997).

Experiment was done by Jones *et al.* (Jones and Smith 1987) to compare WIJET with the conventional PET method, and promising results were obtained. The accuracy of WIJET with respect to the conventional method is attributed from various factors as follows:

(a) Line temperature is directly controlled so it is easier to avoid lattice diffusion.

During the test, the maximum temperature, T_m, must be defined. T_m is the maximum temperature before the dominant transport mechanism changes from grain-boundary diffusion to lattice diffusion. It is shown that this technique of testing can provide the maximum lifetime test acceleration within the constraint of low temperature or grain-boundary diffusion being the dominant transport mechanism. For a Al-Si test lines, T_m is approximately 345°C.

Maintaining the test below this temperature is critical in order to obtain accurate results. This is to make sure that while performing the test, it is within the constraint of grain-boundary diffusion being the dominated transport mechanism (Lloyd and Koch 1992).

(b) The capability to determine activation energy

Activation energy is one of the important characteristics to define electromigration. SWEAT does not facilitate the capability to determine this important parameter of electromigration. WIJET provides the capability of determining the activation energy.

Unlike SWEAT, WIJET is possible to obtain quantitative results of MTF (1997). This is possible as unlike SWEAT, where the median time to failure

(MTF) as predicted by the Black equation is the control parameter, WIJET uses temperature as its control parameter. Thus, by fixing temperature through adjusting current density or hot-chuck temperature during the test, median-time-to-failure of a test line can be obtained. Repeat this process with different test temperatures and we can obtain lifetimes at different test temperatures, and hence the activation energy.

(c) Absence of significance temperature gradient

WIJET uses interlevel dielectrics to achieve a higher maximum current density than SWEAT structure mentioned earlier. Long metal test lines over a flat oxide on a silicon wafer are used to provide a uniform test temperature along each line. As compared with SWEAT structure, its temperature gradient along the structure does not vary as much. Thus, it can achieve a higher level of accuracy when test results are extrapolated.

3.7.4 *Wafer level constant current electromigration test (Lee, Tibel and Sullivan 2000)*

The constant current test method simply applies a constant current of the same magnitude to all lines, without making any adjustments for individual geometric differences. The flow chart of the testing procedures is as depicted in Fig. 3.30. The beginning steps are the same as with the WIJET. The current was ramped to the target current in steps as with the WIJET. When the target current was reached at the end of the ramp, the Joule heating temperature was recorded as stabilized temperature.

3.7.5 *Breakdown energy of metal (BEM)*

The Breakdown Energy of Metals (BEM) approach was introduced by Hong and Crook (Hong and Crook 1985) in 1985. The control algorithm consists of a current ramp (staircase) applied to the line under test, with self-heating used to supply stress temperatures. Because the peak current is essentially unbounded, care must be taken to limit the peak current so that the temperature does not become so large as to reach unreasonable levels, such as the melting point of aluminum. Like the SWEAT technique, the temperature is monitored at each step in the staircase. The most intriguing aspect of this approach is the calculation of the so-called energy to failure,

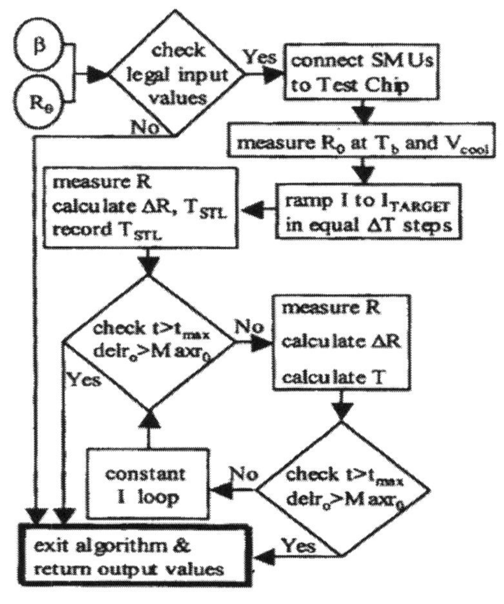

Fig. 3.30. Constant current algorithm flow chart. (Lee, Tibel and Sullivan, 'Comparison of Isothermal, Constant Current and SWEAT Wafer level EM testing methods'. Proc. IRW Final Report, p. 61–69), Copyright © 2000 IEEE.

E_f, computed over the current ramp as

$$E_f = \frac{1}{l}\int_0^{t_f} I^2(t)R(t)e^{-E_a/kT(t)}dt, \tag{3.19}$$

where

l: strip length
t_f: time to failure
R: strip resistance
I: applied current

The quantity E_f can be considered as the amount of energy per unit length required to force the EM failure of the test line. Figure 3.31 show that E_f values measured on a wafer obeys a lognormal distribution. The 50% point of the E_f distribution, i.e. the median energy to fail (MEF), is the parameter of merit of the technique and enables comparison of EM performances of different conductors under test. The following relationship

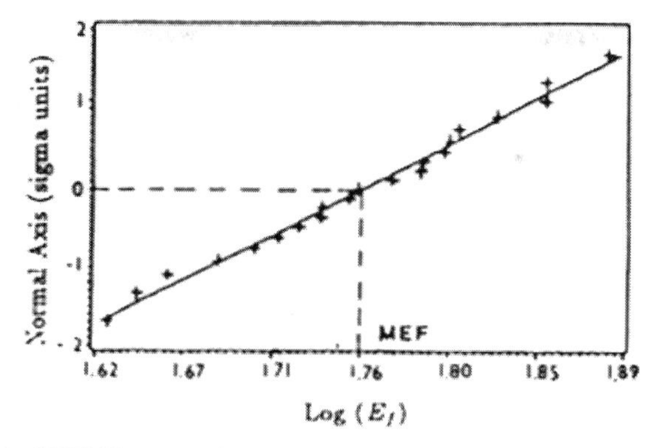

Fig. 3.31. Typical BEM lognormal distribution of failure energies on water; the normal axis centered on the distribution 50% point, The MEF value is also shown. (Hong and Crook, 'Breakdown Energy of Metal (BEM) — A new technique for monitoring metallization reliability of wafer level', Proc. of International Reliability Physics Symposium, p. 108–114), Copyright © 1985 IEEE.

between *MTF* and *MEF* is obtained (Hong and Crook 1985) as

$$MEF = MTF(RI^2/l)e^{-E_a/kT}. \tag{3.20}$$

There are two basic conditions that must be satisfied for any accelerated stress test in order for the test to provide meaningful input for production and quality processes. The first condition is that a component, operating under an increased stress, will have exactly the same failure mechanisms as seen when used at normal stress. The second condition is the requirement of an acceleration model that bridge the stress gap, so that data obtained under the high stress conditions can be extrapolated to the normal stress conditions (Tobias and Trindade 1986). However, these two basic conditions are generally not carefully monitored in the industry when reliability testing is being conducted. Let us examine the case of SWEAT as the high stress conditions in SWEAT can bring in unintended new failure mechanisms not normally present in normal operation, thereby adding uncertainties.

The pitfalls of SWEAT in its ability to detect process changes and the correlation between its test results and those of conventional electromigration tests are reported in references (Dion 1993; Giroux, Gounelle, Mortini and Ghibaudo 1995; Menon and Choudhury 1997; A. Scorzoni, Impronta, Munari and Fantini 1999; Foley, Molyneaux and Mathewson 1999). The pitfalls of SWEAT are investigated from the perspective of reliability statistics by Tan *et al.* (ref), follow the outlined given by Meeker (Meeker 1998; Meeker and Escobar 1998), as outlined by Meeker (Meeker 1998; Meeker and Escobar 1998).

3.7.6 *Pitfalls of SWEAT*

(a) Multiple (Unrecognized) Failure Mechanisms

Meeker (Meeker 1998; Meeker and Escobar 1998) pointed out that the high levels of accelerating variables could induce failure mechanisms that would not be observed at normal conditions. This would violate the first basic requirement of an accelerated stress testing. For the case of SWEAT, this is applicable as follows.

(i) Consequence of high temperature condition

One of the essential conditions for electromigration failure to occur is the occurrence of a divergence of vacancy flux somewhere in the metal line that allows voids or extrusions to form. Diffusivity is one of the main factors that affect the vacancy flux.

For a given thickness of the metal line, the vacancy diffusion paths will change from grain boundaries to through-lattice, as temperature changes. Lloyd finds that the through-lattice diffusion (also called bulk diffusion) can be neglected for temperatures at operating condition. The ratio of the number of vacancies diffusing through the lattice to those diffusing along the grain boundary increases as temperature increases. The ratio rises above 1 at approximately 500K for Al metal with average grain size 10 times less than the width of the metal line. If the width of the metal line narrows to the average grain size, then the ratio rises above 1 at an increased temperature, approximately 600K as shown in Fig. 3.9. This is further evidenced by the fact that at operating temperature failure sites are characterized by crack-like features, showing that electromigration takes place via the grain boundaries diffusion. However, at higher temperatures, more areas around the grain boundary are missing, indicating that the bulk diffusion becomes significant.

SWEAT subjects a metal line to a very high maximum temperature (>600K) during testing (Foley, Molyneaux and Mathewson 1999), making bulk diffusion prominent during testing. Hence, we have two co-existing diffusion mechanisms with different activation energies present, as shown in Fig. 3.32 (Dion 1993). Therefore, the test results cannot faithfully represent

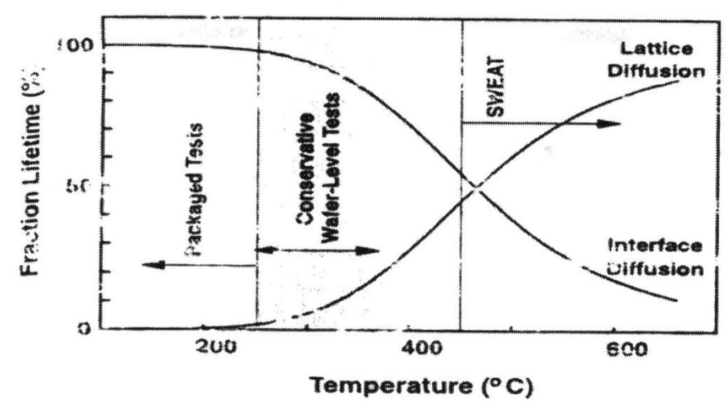

Fig. 3.32. The relative contribution to metal failure due to different failure mechanism. Reprinted from Microelectronics Reliability, 33(11/12), Dion M.J, 'On the status of wafer level metal integrity testing', p. 1807–1827, Copyright © 1993, with permission from Elsevier.

the performance of the metal line at normal stress conditions. This becomes worse especially when the comparison of test data is based on the Black equation with *a prior* knowledge of the activation energy obtained using conventional test conditions. Obviously, the activation energy would be much different under the SWEAT condition.

Furthermore, at high temperatures, the thermal expansion mismatch can be severe. The resulting stresses in the metal line could induce or retard the formation of hillock and whisker. The enhancement of hillock and whisker formation is the result of locally enhanced diffusivities of vacancies, resulting in severe flux divergence. The retardation in hillock and whisker formation is the result of self-healing due to the backflow of metal atoms (Nikawa 1981; Li, Zhang, Ji, Wang, Cheng and Gao 1992; Li, Bauer, Mahajan and Milnes 1992). The SWEAT procedure does not account for this self-healing.

With the bulk diffusion dominating, the electromigration process becomes microstructure insensitive (A. Scorzoni, Impronta, Munari and Fantini 1999). This may render SWEAT meaningless in predicting metal line electromigration performance under normal operating conditions.

(ii) Consequence of high current density and temperature gradient

A SWEAT structure produces a high temperature gradient. Under normal operating condition, the temperature gradient has an impact to the MTF of a metal line. Guo *et al.* (Weiling, Zhiguo, Tianyi, Yaohai, Changhua and Guangdi 1998) showed that for a negative temperature gradient, i.e., the electron flow in the direction of increasing temperature, the electromigration MTF is found to be approximately doubled as compared with that in a positive gradient. This occurs even at a temperature gradient as low as 280°C/mm while current densities are in the range of 1×10^6 to 3×10^6 A/cm^2. In SWEAT, the temperature gradient is approximately 5000°C/mm (Quintard, Dilhaire, Phan and Claeys 1999).

Also, the current density during SWEAT testing is very high, in the order of 1.0×10^7 A/cm^2. Under these conditions, Giroux *et al.* (Giroux, Gounelle, Vialle, Mortini and Ghibaudo 1994) showed that the ion flux divergence is much less dependent on temperature gradient. As temperature gradient is common for normal metal lines during device operation, the failure mechanism in SWEAT could be different from that during the normal operating condition.

(iii) Consequence of using the Black Equation

Foley (Foley 1998) makes a comparison between normally and highly accelerated electromigration testing on the 5 μm-wide NIST line. He concluded that for extrapolations of test results obtained from accelerated tests, it is safe to use the Black equation only in the case of normally accelerated stress condition. Here normally accelerated stress conditions are conditions with temperature below 240°C and current densities of $2.0 - 3.0 \times 10^6$ A/cm^2. Highly accelerated stress conditions are those with temperature of 250°C and current densities of $4.5 - 6.5 \times 10^6$ A/cm^2.

By using the Black equation, Foley (Foley 1998) found that the MTF calculated using the data from the normally accelerated condition is 500,327 hrs, and that for highly accelerated condition is 1,706,713 hrs. Thus, the use of high current density with the Black equation will result in an over-estimate of lifetime. This is so because the parameters in the Black equation are not the same under different stress current densities, and it is mainly due to the difference in the current density exponent n. Although a theoretical value of 2 has been demonstrated, the extracted value of n was found to be higher than the theoretical value. It indicates the possible presence of Joule heating-induced temperature gradient.

For normally accelerated condition, failure analysis showed that extended voids were coupled with hillocks downstream as shown in Fig. 3.33. For highly accelerated condition, clusters of voids were formed

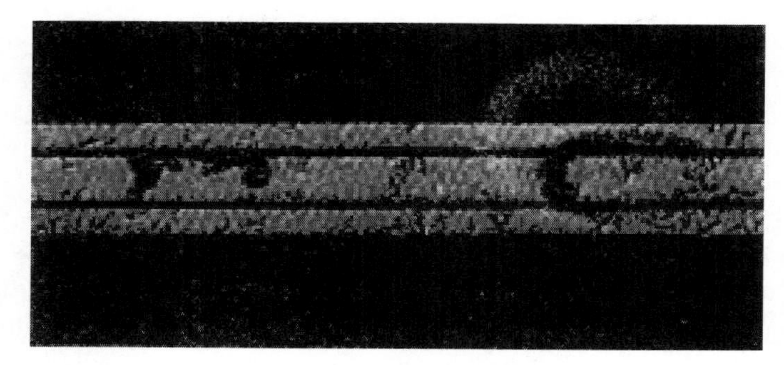

Fig. 3.33. Line-edge, extended void and hillock in a specimen at normally accelerated stress condition (240°C and 2 MA/cm^2). Reprinted from Microelectronics Reliability, 38, Foley, 'A comparison between normally and highly accelerated electromigration tests', pp. 1021–1027, Copyright © 1998, with permission from Elsevier.

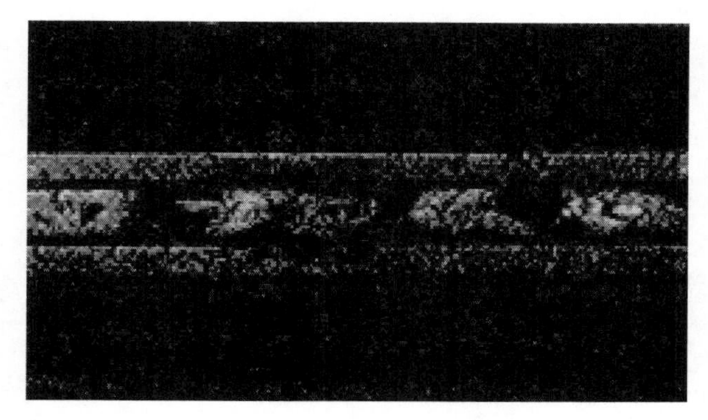

Fig. 3.34. Line-edge, extended void and hillock in a specimen at highly accelerated stress condition (250°C and 4.5 MA/cm^2). Reprinted from Microelectronics Reliability, 38, Foley, 'A comparison between normally and highly accelerated electromigration tests', pp. 1021–1027, Copyright © 1998, with permission from Elsevier.

along the line, as shown in Fig. 3.34. As stress condition worsen, the average size, the number and the degree of void clustering of line-edge voids increases.

The reason for an increase of the acceleration factor n is due to the clusters of voids formed along the line. This amount of void clustering increases the influence of temperature gradients at defects, and temperature gradient is one of the failure mechanisms that causes electromigration failure. It is this existence of an increasing temperature gradient that induces an increase of the acceleration factor, n.

(b) Failure to properly quantify uncertainty

As reliability testing is a statistical estimate, and there is always uncertainty in statistical estimates, and hence basing decisions only on point estimates can be misleading. Even if the statistical confidence intervals are estimated to quantify uncertainty, they are usually computed either from limited data and/or a large amount of extrapolation in temperature, or by using an assumed life-temperature relationship with imprecisely known activation energy. These would not provide a proper quantification of uncertainty (Meeker 1998; Meeker and Escobar 1998). In SWEAT, the temperature of the metal line is so much higher than the normal stress condition, thus a larger amount of extrapolation in temperature is needed. Also the activation energy of the life-temperature relationship is not known precisely due to the co-existence of more than one failure mechanism. The statistical confidence bounds thus calculated are inaccurate. This makes the usefulness of the test data in jeopardy.

The temperature estimation of the metal lines during SWEAT test could also introduce large variation of the test results. This makes estimated confidence interval too large to be meaningful. The large variation from temperature estimation is due to the fact that the mean temperature of the structure is evaluated using the temperature sensitivity of the resistance also known as TCR (Temperature Coefficient of Resistance). The practical use of the TCR depends on two factors. Firstly, the resistivity must be a linear function of the temperature. Secondly, the resistivity of a metallization ($\Delta R/R$) must be stable such that it will not suffer any irreversible changes (Schafft 1992; 1995). It is observed that departure from linearity begins somewhat below 300°C (Hatch 1984) for Al. In SWEAT, the temperature is much higher. When combined with the presence of large temperature gradient and the occurrence of irreversible change in resistivity during SWEAT, the use of TCR in estimating temperature results in serious error. The use of high current density can also make the amount of temperature rise due to Joule heating much more sensitive to the underneath dielectric thickness (Giacomo 1997). This will introduce another uncertainty in the results.

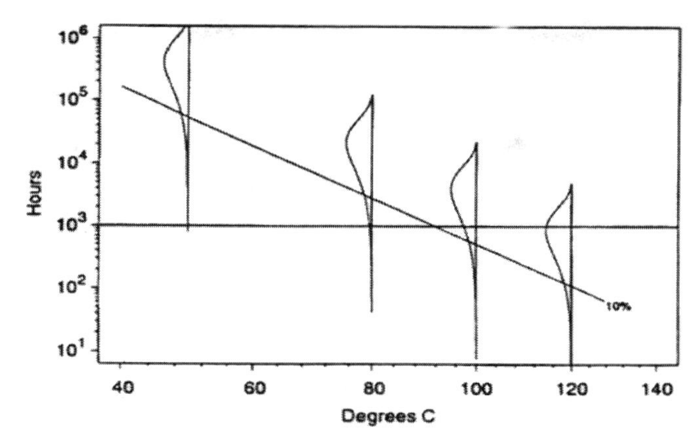

Fig. 3.35. Result of single failure mechanism. (Meeker, 'Pitfalls of accelerated testing', IEEE Trans. on Reliability, 47(2), p. 114–118), Copyright © 1998 IEEE.

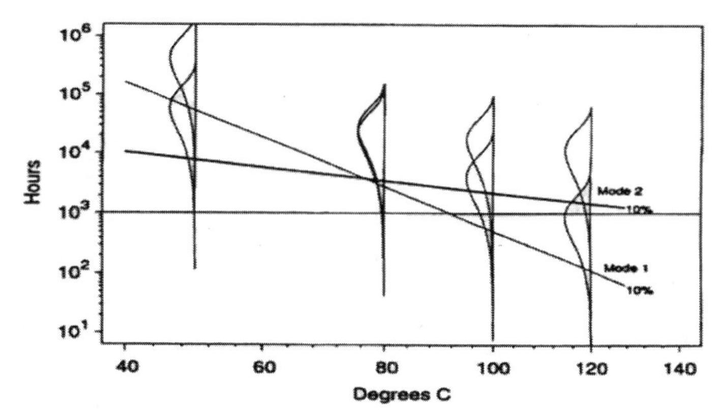

Fig. 3.36. Result of two failure mechanisms. (Meeker, 'Pitfalls of accelerated testing', IEEE Trans. on Reliability, 47(2), p. 114–118), Copyright © 1998 IEEE.

(c) Masked Failure Mechanism

Figure 3.35 shows a graph of what might illustrate the results of a typical accelerated life test if there was just a single failure mechanism and if increased temperature accelerated in a simple manner.

However, it is possible that for an accelerated test, while focusing on one-known failure mechanism, one might mask another as shown in Fig. 3.36. In SWEAT, as discussed in the earlier section, it introduces another failure mechanism called lattice diffusion at a very high temperature.

Lattice diffusion is negligible at normal operating condition (at much lower temperature), hence what may cause the metal line to fail at accelerated testing may not fail at an actual operating condition. As illustrated in Fig. 3.36, it is often the masked failure mechanism that is the first one to show up in the field. In such cases, the masked failure mechanism will dominates the reported field failures.

(d) Faulty comparison

In view of the pitfalls mentioned earlier, some may use SWEAT for comparing alternatives, instead of predicting the reliability of interconnections. This is based on the assumption as illustrated in Fig. 3.37. In Fig. 3.37, if vendor-1 is better than vendor-2 in an accelerated testing, than the same would be true in actual operating condition.

However, a possible scenario would be different failure mechanism occurred at operating and accelerated conditions. The early failures for vendor-2 might be masking the failure mechanism in the test results for vendor-1, as shown in Fig. 3.38. Thus, we cannot use accelerated life test to compare products that have different kinds of failure mechanisms.

As mentioned earlier, a SWEAT is designed to give only qualitative information of the reliability of the metal line by mean of comparison. However, it failed to account for different failure mechanisms occurred at different stress condition. Thus, situation such as Fig. 3.38 could easily occur during SWEAT testing.

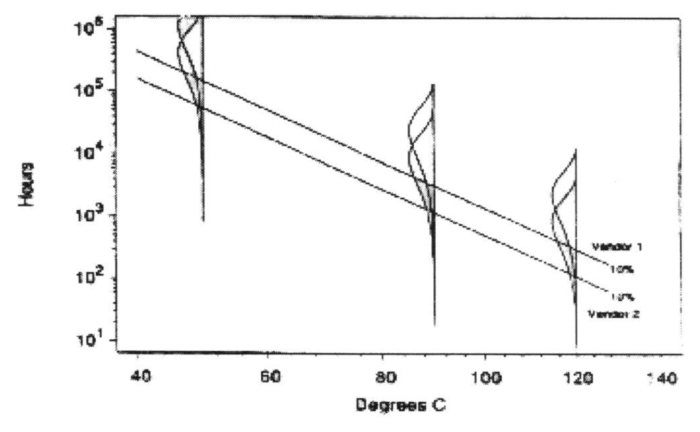

Fig. 3.37. Well-behaved comparison of 2 products. (Meeker, 'Pitfalls of accelerated testing', IEEE Trans. on Reliability, 47(2), p. 114–118), Copyright © 1998 IEEE.

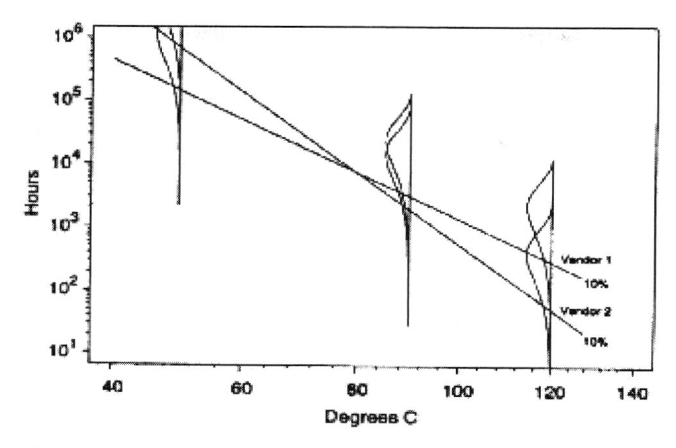

Fig. 3.38. Comparison with evidence of different failure mechanisms. (Meeker, 'Pitfalls of accelerated testing', IEEE Trans. on Reliability, 47(2), p. 114–118), Copyright © 1998 IEEE.

(e) Drawing Inaccurate Conclusions on the Basis of Specially Built Prototype Test Units

Unreliable conclusions can result from an accelerated life test if the test-units differ in important ways from actual production units.

SWEAT uses a specially built test structure that is designed to yield very large thermal gradients. This enables maximized currents to be forced through the narrow region. Since the current density is high, it results in shorter time to failure. Though the use of self-heating in wafer-level tests provides one of the advantages from a speed and simplicity standpoint, it also causes uncertainties in the accuracy of the test results obtained.

Figure 3.39 shows the temperature profile along a SWEAT structure. The temperature throughout the test structure is not uniform. At different regions of the structure, there are different temperatures. As a result, thermometry is uncertain. An added difficulty is that very short narrow line segments are inefficient at sampling line grain structure, defects and potential nucleation sites (Pierce and Brusius 1997).

Most of the existing electromigration knowledge was gathered presuming an uniform structure temperature. These variations in line temperature along the line significantly challenge any investigator attempting to understand results with existing model.

As suggested by Pierce (Pierce and Brusius 1997), the ideal structure will be the test structure most often used in classic packaged structure

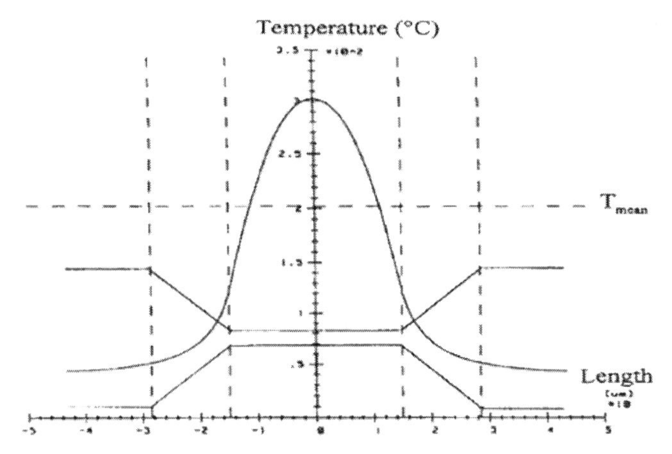

Fig. 3.39. Temperature profile along SWEAT structure. (Giroux, Gounelle, Mortini, Ghibaudo, 'Wafer level electromigration tests on NIST and SWEAT structures', Proceeding of the International Conference on Microelectronic Test Structures, p. 229–232), Copyright © 1995 IEEE.

tests, the NIST structure. Because the length to width ratio is so large, the temperature is very uniform until the end of lines are reached. Such a strip will provide an adequate sample of grain structures and minimize longitudinal thermal gradient.

Other pitfalls such as accelerating reliability test parameters can cause deceleration for the failure mechanism concerned (Meeker 1998; Meeker and Escobar 1998) is obviously presence in SWEAT test structure.

Therefore, one can see that the data produced from the SWEAT cannot be applied to the study of electromigration of metal lines. Two alternative wafer level electromigration tests are developed (Scorzoni, Impronta, Munari and Fantini 1999; Foley, Molyneaux and Mathewson 1999).

Foley *et al.* (Foley, Molyneaux and Mathewson 1999) suggested a test that uses an isothermal test method on SWEAT type structures. The advantage of this method is that an attempt is made to control the temperature directly. Then, the test is less likely to be susceptible to excessive temperature (which is kept below 300°C). The pitfall due to the use of a specially built prototype remains, however. If the test is for a process technology qualitative indicator, the pitfall is not significant. On the other hand, for bamboo line and multiple level metallization, the method is inadequate.

Scorzoni *et al.* (A. Scorzoni, Impronta, Munari and Fantini 1999) proposed a moderately accelerated electromigration test by keeping the maximum temperature to 250°C, and stress current density to a level such that the temperature rise due to Joule heating is limited to below 40°C. The test structure used is similar to NIST, but with a 'Babal tower' at the end regions. These end regions are designed so as to reduce the problem of the occurrence of severe ion flux divergence at the bamboo/polycrystalline junction. One can see that with such a test structure and test method, all the pitfalls mentioned for SWEAT will be eliminated.

3.7.7 *Potential pitfalls of constant current test method*

Although the constant current test method is straight forward, but the Joule heating temperature of a line can vary substantially depending on the homogeneity of the line and thermal conductivity of the underlying material. This could create a temperature gradient, especially when voids begin to form in some parts of the test line. Hence, the inaccuracy of the test results could be resulted.

Also, if the test current is used for increasing the temperature of the test line as is usually the case for FWLR, the current will tend to be high. From the damage map as depicted in Fig. 3.40 (Kraft, Mockl and Arzt 1995), the failure mechanism at the test condition will be substantially

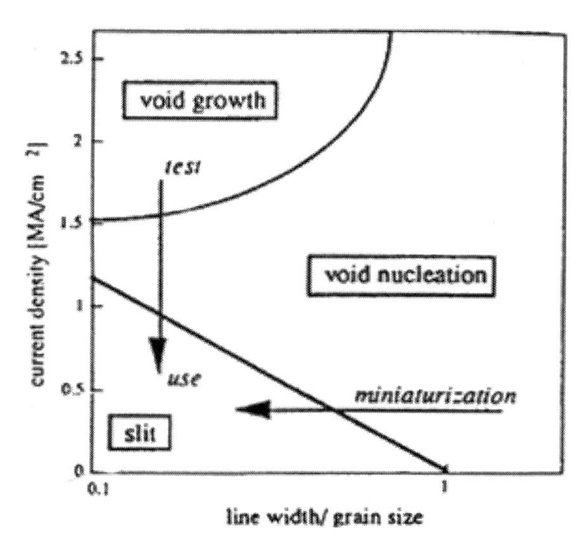

Fig. 3.40. Electromigration damage map for near-bamboo lines (Kraft, Mockl and Arzt 1995).

difference from that at the use condition, making extrapolation of the test results meaningless.

3.7.8 *Potential pitfall of breakdown energy method (BEM)*

When performing BEM, it is important to stress that in the expression for the failure energy, a number of assumptions have been made: (Hong and Crook 1985)

(a) The activation energy and the current density exponential (n) have been set equal to 0.5 eV and 2 respectively; However, the activation energy is technology and process dependent, and unless the activation energy is determined before the BEM test, the test results could be meaningless.
(b) Relation between MTF and the conductor line cross section has been assumed as linear; however, this relationship has yet to be verified.
(c) Strip temperature is assumed uniform. Hence, it depends on the homogeneity of the line and thermal conductivity of the underlying material. However, when voids begin to form in some parts of the test line, a temperature gradient is created. Hence, the inaccuracy of the test results could be resulted.

All these assumptions hindered the validity of the relationships reported. Thus, the MEF (Median energy to fail) can only be considered as a parameter describing the performance of the tested conductors with respect to the EM stress applied.

3.7.9 *Potential pitfalls of WIJET*

Although WIJET posses several advantages, it has several pitfalls as follows:

(a) Poor temperature control accuracy

A potential problem exists with WIJET and also other rapid tests is that the metal lines under test have resistance fluctuations induced by the degradation. Any change of the resistance at the ambient R_{ambient} and of α, the temperature coefficient of resistivity, may severely affect the temperature control.

In WIJET test, it depends on the line resistance for thermometry. Deviations in either the geometry or the temperature-dependent resistivity function from initial condition will result in thermometry error (Schafft 1992).

(b) Problems from high test current density

Current density used in WIJET is one parameter without any limitation. According to JESD61 (1997), the starting current density is set at 1×10^6 A/cm^2, and it can ramp up to more than 12×10^6 A/cm^2. At such a large current density, the failure mechanism will be much different from the use (circuit operation) condition as depicted in the damage map of Fig. 3.40.

3.7.10 *Summary*

From the above analysis, one can see that the most feasible FWLR is WIJET on NIST structure with 'Babal tower' at the end regions as suggested by Scorzoni *et al.* (A. Scorzoni, Impronta, Munari and Fantini 1999). To prevent the use of high current density, hot-chuck is used to supplement the heating of the test wafer, so that the current density need not be very high as it is no longer required to heat up the metal line to a large extent. This could ensure the failure mechanism at the test condition is the same as that under the use condition as depicted in the damage map of Fig. 3.40.

To ensure good temperature control, especially when large voids are formed during electromigration, WIJET is to be terminated at early failure, says at 0.5% change in resistance. Under such condition, the effect of voids on the temperature uniformity of the line will be minimum.

3.7.11 *Highly accelerated electromigration test*

Since many physical mechanisms are involved in an EM process occurring in metal thin film as discussed earlier, simply increasing the stress condition of EM test could render different physical mechanism governing the EM failures, thus making the prediction of EM performance through extrapolation of the test results invalid. On the other hand, manufacturing plant is constantly striving to shorten accelerated life testing without compromising the accuracy of the result, a highly accelerated electromigration test is recently proposed that proven to shorten 80% of the test time while ensuring the invariant of the physical mechanism (Tan, Li, Tan and Low 2006).

In this method, a combination of the simulation using Finite Element Modeling (FEM) of EM and statistical analysis of test data is employed.

Firstly, using the EM failure physics based FEM, they determined the accelerated stress conditions. Verification is then performed by conducting the actual reliability tests at the computed stress conditions followed by failure analysis. The test data is analyzed using statistical means such that any difference in mechanisms of EM can be detected, and thus a highly accelerated stress condition without any change in physical mechanism of EM can be derived.

The illustrated example in their work is an Al line-via structure with line thickness of 0.45 μm and via diameter of 0.288 μm and TiN barrier metal. To accelerate the EM test, either a higher current density or higher temperature is used. Using finite element analysis, the higher current density and temperature such that the physical mechanism remains the same are identified as shown in Table 3.8, and the void locations are found to be the same for all the three cases as obtained from the finite element analysis.

From Table 3.8, it can be seen that in all three cases, the dominant mechanism is stress gradient induced migration, which is consistent with the reports from other works (Tan and Roy 2007 and the references therein). The failure criterion is set as 10% of the relative resistance change.

Actual EM test are performed according to the conditions stated in Table 3.8, and with reliability data analysis software REDAS^{TM}, outlier points are first identified in the data and they are treated as censored data as they represent different failure mechanisms. Then Expectation and Maximization algorithm is used to examine the possibility of the presence of more than one mechanisms of EM, and all of the test data from the three

Table 3.8. Different stress conditions for highly accelerated EM test and their respective physical mechanisms. Here EWM, TM and SM are electron wind force induced migration, temperature gradient induced migration and stress gradient induced migration.

Test split	Test temperature (°C)	Current density (MA/cm^2)	% Contribution due to various driving force		
			EWM	TM	SM
Standard	200	6.9	9	1	90
Split 1	250	6.9	14	1	85
Split 2	200	12.3	241	1	75

stress conditions are found to have only one mechanism of EM which is believed to be the SM based on the FEM as shown in Table 3.8. Lognormal plot are then performed for the three set of data, and one found that the sigma for standard and split 1 are the same, indicating same physical mechanism, and the results are further confirmed by their TEM analysis.

On the other hand, the physical mechanism of EM for the split 2 is different from that of the standard as their slopes are different. This is believed to be due to the Blech back stress effect as the current density used in split 2 is very high. This Blech effect is not considered in the FEM developed in this work, and hence such a deviation is possible.

With these analysis, they determined that the highly accelerated stress condition chosen is 250°C at 6.9 MA/cm^2. As compared with the standard EM test used, the test time is reduced by more than 80%.

3.8 Practical Consideration in Electromigration Testing

There are several considerations that must be taken care of when performing electromigration test in order to ensure invariant of the failure mechanism at the stress and normal operating condition. If this condition cannot be fulfilled, the prediction of the electromigration lifetime from the accelerated testing will be in error (Elsayed 1996).

It has been found that many failure mechanisms for electromigration that occur at the accelerated conditions do not happen under normal operating conditions. On the other hand, there are failure mechanisms that occur at the normal operating conditions but are masked at the accelerated conditions. Therefore, careful examination of the matter in relation to the electromigration testing is necessary.

3.8.1 *Failure criteria used in EM testing*

A study on the recovery of electric resistance degraded by electromigration revealed that the failure criteria $\Delta R/R_0$ can invoke different defect mechanisms. The amount of resistance recovery is determined by the following phenomena (Ohfuji and Tsukada 1995):

(a) the transition from supersaturated condition to the saturated condition of vacancy
(b) mechanical stres relief

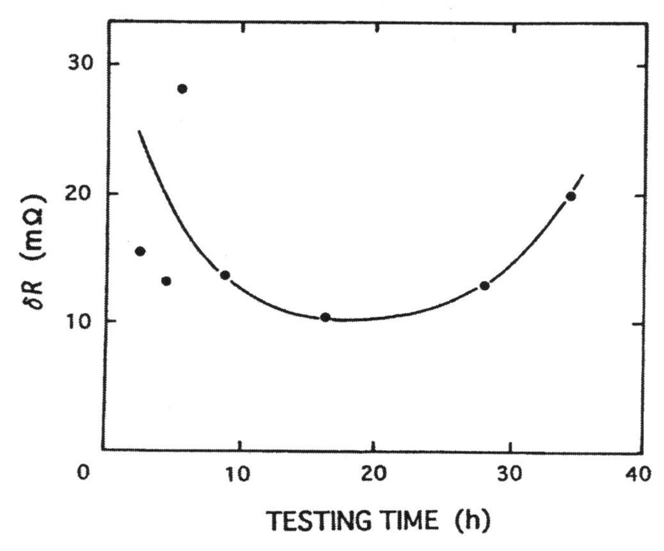

Fig. 3.41. Resistance recovery δR versus the time of EM testing to get to $\Delta R/R_{298} = 2\%$ for the samples EM tested at 298K. Reprinted with permission from Ohfuji and Tsukada, 'Recovery of electric resistance degraded by electromigration', Journal of Applied Physics, vol. 76(6), p. 3769. Copyright © 1995, American Institute of Physics.

(c) microstructural changes relieved by bulk diffusion
(d) dissociation of vacancy-impurity complexes

From the variation of resistance recovery as a function of the failure criteria (Fig. 3.41), one sees that the recovery of resistance is not proportional to the fractional variation of electrical resistance. This means that the resulting concentration of supersaturated condition of vacancy, mechanical stress, microstructural changes and concentration of vacancy-impurity complexes are different for different failure criteria as the applied stress is different for different test criteria. Even at a given failure criteria of 2%, the recovery of resistance depends on the testing time required to reach the 2% failure criteria, and the dependence is not monotony (Fig. 3.42). This could indicates that the reason for different lifetime is due to the different degree of stress dependence on the mechanical stress generation, concentration of supersatured vacancy, microstructural changes and concentration of vacancy-impurity complexes.

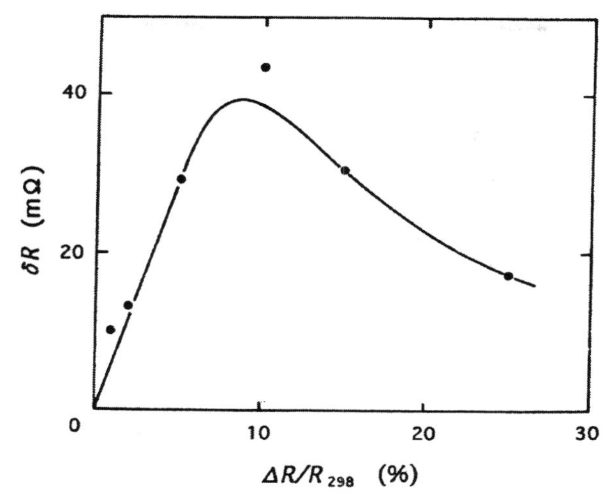

Fig. 3.42. Variation of resistance recovery δR as a function of the fractional variation of resistance during EM testing at 298K. Reprinted with permission from Ohfuji and Tsukada, 'Recovery of electric resistance degraded by electromigration', Journal of Applied Physics, vol. 76(6), p. 3769. Copyright © 1995, American Institute of Physics.

3.8.2 *Interpretation of the measured* $\Delta R/R_0$

EM test monitoring is usually done with $\Delta R/R_0$. However, interpretation of the measured $\Delta R/R_0$ is complicated by other processes, the test conditions, and the test structure itself. Especially, for layered structures, the conductive barrier layers allow continued current flow as the Al voids (J.E. Sanchez and Pham 1994).

Both the conductor resistivity and geometry determine the resistance and the detected $\Delta R/R_0$. Sanchez and Pham (J.E. Sanchez and Pham 1994) define intrinsic $\Delta R/R_0$ changes as due to changes in resistivity, while extrinsic changes are due to variations in geometry. Lattice defects normally dominate the resistivity effects which include the vacancies and solute concentration. For vacancy contributions, we have

$$\Delta\rho = a_v \Delta X_v, \tag{3.21}$$

where ΔX_v is the change in vacancy concentration and a_v is an experimentally determined coefficient which is $3\mu\Omega - \text{cm/at.\%}$ vacancies

(Venkatrkrishnan and Ficalora 1992). Calculation shows that the vacancies have a small effect on $\Delta R/R_0$.

Usual elemental additions to Al (e.g. Si, Cu etc.) substitutionally occupy Al lattice sites ("solutes") and often precipitate out as second phases. These solutes increase Al resistivity in proportional to their concentration. Precipitation processes deplete the Al lattice of the solutes, so that a drop in resistivity is associated with second phase formation. The solute-precipitation effect on resistivity can be expressed as (J.E. Sanchez and Pham 1994)

$$\Delta\rho_s = a_s \Delta X_s, \tag{3.22}$$

where ΔX_s is the change in solute concentration (in wt. %) and a_s is an experimentally determined coefficients. Calculation shows that the precipitation processes describe the $\Delta R/R_0$ drops typically found at the beginning of accelerated test in Al alloy interconnects.

The change in geometry of interconnect include the change in area and length. Bending of the Si substrate by external loading or thermal mismatch will strain interconnects on the substrate surface. It is found that for a strained Al conductor, its Poisson strain can be neglected (J.E. Sanchez and Pham 1994). Therefore, $\Delta R/R_0$ is linear with the strain, as is verified experimentally (J.E. Sanchez, Lloyd and Morris 1990).

Non-fatal voids in long single layer Al interconnects produce extremely small $\Delta R/R_0$ signals during electromigration testing prior to open circuit (Schreiber 1985). However, with layered structures, the barrier layers along with the Al allows for larger possible $\Delta R/R_0$ since the barrier layers provide a continuous path for current as the Al is through-voided as shown in Fig. 3.43. In this case, it can be found that (J.E. Sanchez and Pham 1994).

$$\left(\frac{\Delta R}{R_0}\right)_v = \frac{L_v}{L}\left[\left(\frac{\rho_B}{h_B}\right)\left(\frac{h_{Al}}{\rho_{Al}}\right) - 1\right]. \tag{3.23}$$

Thus, the magnitude of the signal is determined by the interconnect length L, and the thickness h and resistivity ρ of each layer in the interconnect stack. Based on the formula in Eq. (3.15), a plot of $\Delta R/R_0$ versus void length is shown in Fig. 3.44. The plot shows that improbably large voids are required to produce the $\Delta R/R_0 = 20\%$. This suggests that the large $\Delta R/R_0$ found is due to localized heating at the voided site as the large current are shunted into the barrier layers.

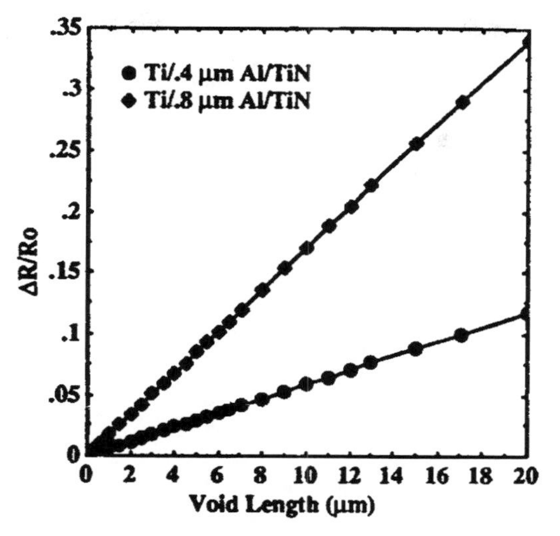

Fig. 3.43. Section of Al/barrier interconnect, of total length L, showing a void of length L_v in the Al. J.E. Sanchez Jr., and V. Pham, 'Interpretation of resistance changes during interconnect reliability testing', Materials Reliability in Microelectronics, Mater. Res. Soc. Symp. Proc. vol (338), Warrendale, PA, 1994, 459–464.

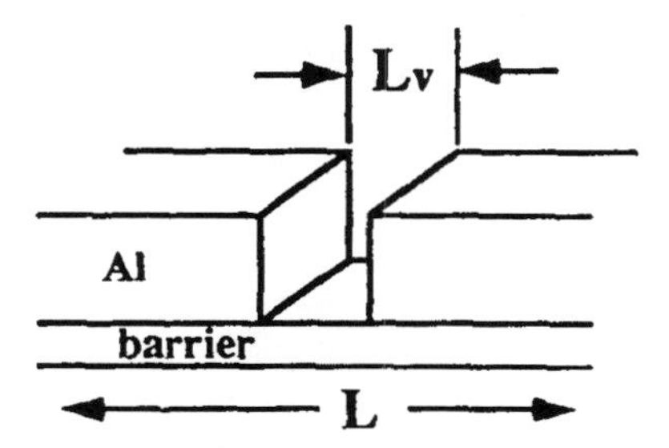

Fig. 3.44. Calculated $(\Delta R/R_0)_v$ for Ti/Al/TiN layered interconnects, assuming no local heating at void (R measured at low current). J.E. Sanchez Jr., and V. Pham, 'Interpretation of resistance changes during interconnect reliability testing', Materials Reliability in Microelectronics, Mater. Res. Soc. Symp. Proc. vol (338), Warrendale, PA, 1994, 459–464.

The resistance of a test stripe with a rectangular void ($l_v \times W$) is given as below (Filippi, Biery and Wachnik 1995).

$$\begin{aligned} \Delta R &= \frac{\rho_r l_v}{W t_r} - \frac{l_v}{W}\left(\frac{\rho_c \rho_r}{\rho_c t_c + 2\rho_c t_r}\right) \\ \frac{\Delta R}{R_o} &= \frac{l_v}{L}\left(\frac{\rho_r t_c}{\rho_c t_r} + 1\right) \end{aligned} \quad . \tag{3.24}$$

It should be noted that for a void size of ($l_v \times W$) the absolute resistance change is independent of the length of the structure, while the fractional resistance change is independent of the width of the structure (Filippi, Biery and Wachnik 1995). Also, only if the TCR for the Al and the barrier layer are similar, the fractional resistance change is independent of temperature.

3.8.3 *Actual temperature of test strips*

The temperature of the test line during an electromigration stress test is usually determined from a measurement of the resistance of the test structure and the use of TCR for the metalization. For the standard NIST structure, the resistance measured will include the resistance of the cooler end-contacts sections between the voltage tap and the test line. Thus an uncorrected resistance measurement will under-estimate the temperature of the test line. The degree of the under-estimation will depend on the magnitude of the joule heating and on the design of the test structure (Schafft 1987).

A difficulty with the observations of double layer metallizations is the temperature rise at those spots where the aluminum is depleted. Under user conditions, this rise in temperature will be much less, and therefore the damage might appear differently. This influences accurate reliability predictions when using highly accelerated stress data (Pinto 1991).

In actual circuits under real-life operating conditions, severe Joule heating is not a factor. The failure mode at high current density, and therefore, significant Joule heating becomes that of a temperature gradient-induced flux divergence and not the structurally induced flux divergence that is typical of field failure (Shatzkes and Lloyd 1986).

The ratio of number of diffusing vacancies into the lattice to those diffusing along the grain boundary to the surface depends on the line temperature as shown in Fig. 3.9 earlier. These two mechanisms have different Ea, which could affect the T_{50} extrapolation. This also raise a question on the suitability of standard accelerated tests, i.e. the results of high temperature tests may not be valid at use temperatures (Lloyd and Koch 1992).

Electromigration mass flux divergences can create vacancy supersaturations in the grain boundary which will flow to and annihilate at the surface via the boundary or via the adjacent lattice. How they flow, which is temperature dependent, determines the damage morphology (Lloyd and Koch 1992); and this has important implication in the validity of accelerated testing.

3.8.4 *Test structure used*

As mentioned in the earlier section on test structure, test structure used in EM testing is important so that the actual failure mechanisms in EM can be observed.

3.8.5 *Current density used*

Figure 3.40 shown earlier indicated three basic failure mechanisms found in electromigration, namely void growth, void nucleation and slit-like void formation (Kraft, Mockl and Arzt 1995). Flux divergence can either cause an existing void to grow by having metal atoms diffuse away from the void surrounding, or void nucleation where voids diffuse to meet each other at the triple points and nucleate. When the metal line becomes bamboo structure, void shape change occurs without any change in volume. This is due to the self-induced current density and temperature gradients (refer to Section 3.4). This phenomenon usually occurs at lower current density, rendering different failure mechanisms in metal line failing EM at accelerated testing and actual use.

Also, the addition of Cu which slows down the formation and/or growth of wedge type voids and enhances the lifetime under testing conditions can enhance the formation of slit-like voids, making the actual MTF lower than was predicted from the testing conditions (Kraft, Mockl and Arzt 1995). This is because in order to avoid the formation of slit voids, a high surface energy would be favourable. The addition of Cu and Si might lower the surface energy by segregation to the void surface and promote the formation of a slit.

3.8.6 *Short length effect*

The resistance as a function of time usually exhibits three different stages. A typical example of the behavior is as shown in Fig. 3.45. The first stage is a decrease in the resistance. This is followed in stage 2 by an increase in the resistance. In stage 3 additional resistance increase is either suppressed or continued at a slower rate than in stage 2. This stage 3 behavior is not always observed (Filippi, Biery and Wachnik 1995).

Stage 1 behavior is caused by Cu precipitation as discussed earlier. Stage 2 is caused by the depletion of Al due to electromigration. The increase

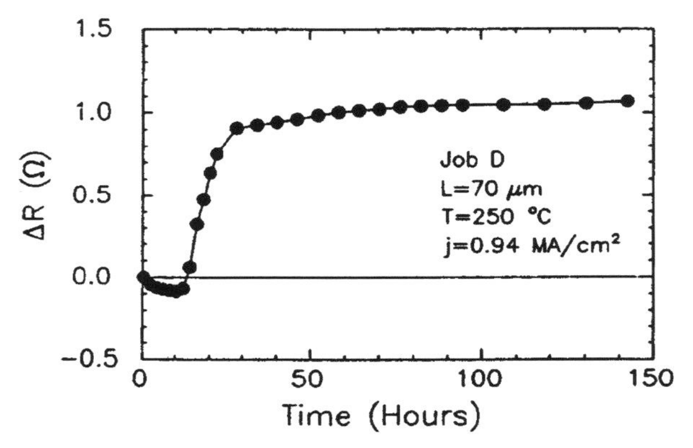

Fig. 3.45. Resistance shift vs. time for a 70 μm long sample tested at 9.4×10^5 A/cm^2 and 250°C. Reprinted with permission from Filippi, Biery and Wachnik, 'The electromigration short-length effect in Ti-AlCu-Ti metalization with tungsten studs', Journal of Applied Physics, vol. 78(6), p. 3756 Copyright © 1995, American Institute of Physics.

in resistance with time is evidence of continual EM-induced void growth. Stage 3 is possible because of the W studs. The studs act as diffusion barriers, which allow for the accumulation of Al atoms at the anode end of the chain and the resulting diffusional backflow (Filippi, Biery and Wachnik 1995). This is the evidence of short-length effect (also called the Blech effect) for EM. Another possibility of stage 3 is the presence of extrusions.

Another evidence of short-length effect is that longer lines increase in resistance at a faster rate than shorter lines, and the increase in resistance for longer lines is higher than that for shorter lines (Filippi, Biery and Wachnik 1995). Korhonen *et al.* (Korhonen, Borgesen, Tu and Li 1993) showed that in a finite line of length L, the time required to reach a given stress level at a particular reduced length (x/L) is proportional to L^2. The variable x is the distance from the end of the blocking boundary (W stud). This is discussed in the earlier section on Via-separation.

Further evidence of the electromigration short-length effect is the observed dependence of the lognormal sigma of EM lifetime with the current density and the maximum allowed resistance change increase (which is indirectly related to the failure criteria $\Delta R/R_0$). This is because if the short-length effect is present, the change in resistance vs. time will be as shown in Fig. 3.45, and for the failure criteria set near the saturation of ΔR, a small change in the value of the saturation ΔR due to change in current density or change in the maximum allowed resistance change, the time to failure will change significantly, renders a large lognormal sigma observed.

The presence of j_c below which no electromigration will occur is again due to the short length effect. The value of j_c is found to be a function of the stripe length and the level of resistance increase as a result of the short length effect. This can be explained by the fact that the stress gradient generated across the W-studs is not always constant over the entire length of the conductor. For a given time, the stress gradient is smaller for long lines as compared with short lines, hence the threshold product must be larger for shorter lines (Filippi, Biery and Wachnik 1995). Also, it is found that j_c is independent of temperature in the range 175–250°C (Filippi, Biery and Wachnik 1995).

3.8.7 *Failure model used in EM accelerated testing (deviation from Black equation)*

The empirical equation for EM is the Black Equation given by

$$t_{50} = aj^{-n} \exp(E_a/kT), \tag{3.25}$$

where a and n are constants, j is the current density, and other have their usual meaning. However, as mentioned in Chapter 2, direct application of Eq. (3.25) is not correct and several modifications have been made.

The value of n in Eq. (3.25) is found from experiments to vary from 1 to 16, especially when there is significant Joule heating which is not properly taken into account. It is generally agreed that n should have a value of 2. The derivation of diffusion equation where both Fickian diffusion and mass transport are taken into account prove that n should be 2 for EM (Shatzkes and Lloyd 1986). In other words, model based on void nucleation and growth by vacancy accumulation indicate that $n = 2$ occurs when void nucleation, or the attainment of a critical tensile stress for void formation is the limiting process (A.S. Oates 1995). The experimental evidence for this mechanism is that the number of voids increases with current density. This has been observed in both multilayer (Hinode, Furusawa and Homma 1993) and single layer Al films (Kraft, Sanchez and Arzt 1992). With this derivation, the time to failure for EM is given as follow (Shatzkes and Lloyd 1986).

$$\begin{aligned} t_f &= BT^2 j^{-2} \exp(E_a/kT) \\ B &= (2C_f/D_0)(k/Z^* e\rho) \end{aligned}, \tag{3.26}$$

where D_0 is the grain-boundary self-diffusivity, E_a is the activation energy for vacancy diffusion through grain-boundary, C_f is the critical value of vacancy concentration where failure begins to occur (therefore, whisker and extrusion are not considered). The time to failure equation differ from the Black equation primarily by the T^2 factor. Although both equations for the time to failure fit equally well to the experimental data, the activation energies derived are different, with the Black equation giving a lower value of E_a, and hence an under-estimate of the EM lifetime (Shatzkes and Lloyd 1986).

When metal line width go below 1 μm, the dominant contribution of voiding is from void growth (A.S. Oates 1995). In this case, the number of void does not increase with current density as shown experimentally and the rate of formation of flux divergence is proportional to the flux atoms, which is itself proportional to the current density (Huntington and Grone 1961), and a value of $n = 1$ is therefore expected.

At relative small current densities, it is found that the Black's empirical relation can no longer hold for the case where W-stud is presence. In fact, the lifetime data was found to obey a modified version of Black's equation expressed as (Filippi, Biery and Wachnik 1995):

$$t_{50} = B(j - j_c)^{-n} \exp(E_a/kT), \tag{3.27}$$

and j_c is called the critical current density.

Experimentally, n is found to be closer to 1 rather than 2. This is because the EM failure for interconnect with W-stud presence is not limited by void nucleation. It is due to the Al drift away from the W-stud. Blech (Blech 1976) suggested that the rate of Al depletion from the edge of a finite conductor is proportional to $(j - j_c)$. If the time to form a void of sufficient size t_f is inversely proportional to the drift velocity, then one would expect $t_f \propto 1/(j - j_c)$ (Kirchheim and Kaeber 1991; Lloyd 1991). However, no physical derivation is performed.

Therefore, in view of the various factors that will affect the EM failure mechanisms, practical EM testing need to be carried out with great care. All the above-mentioned issues require further research effort.

3.9 Failure Modes in Electromigration

There are basically two types of failure modes observed in EM failed units as described as follows:

3.9.1 *Open/resistance increase*

The open failure is due to the depletion of aluminum atoms which are being replaced by voids. Hence, a gradually increase in line resistance can be observed during the EM test. The increase in resistance could, however, be due to different mechanisms (Baerg 1997) as follows.

(a) For test structure with wide line, and terminated by two bond pads, temperature gradient might exist since the temperature of bond pad is clamped to the ambient temperature through the bonding wire. This temperature gradient can result in void formation.
(b) For layered structure, if the shunt layers are adjacent to the plugs, the voids will be formed near the upstream boundary consistently (Fig. 3.46). The resistance vs. stress time will be as shown in Fig. 3.47.
(c) If the shunt layers is not adjacent to the blocking boundary (Fig. 3.48), then even a void in the Al(Cu) next to the plug can cause open circuit failure. The resistance vs. stress time plot will be of erratic nature as shown in Fig. 3.49.
(d) Short lines (less than 100 μm) terminated by blocking boundaries are known to fail more slowly than long lines because of the compressive stress buildup as metal accumulates downstream (Blech 1976). If the stress current is below a certain threshold, a steady state is achieved in which the resistance does not change over time due to the short length effect. However, if the resistance has a jump as shown in Fig. 3.50, this

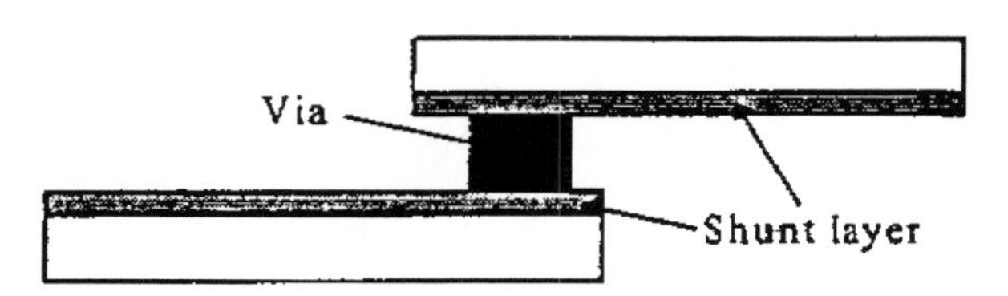

Fig. 3.46. Al interconnect with tungsten plug vias and refractory shunt layers adjacent to plugs. (Baerg, 'Recent problems in electromigration testing', Proceeding of the Annual International Reliability Physics Symposium, p. 211–215), Copyright © 1997, IEEE.

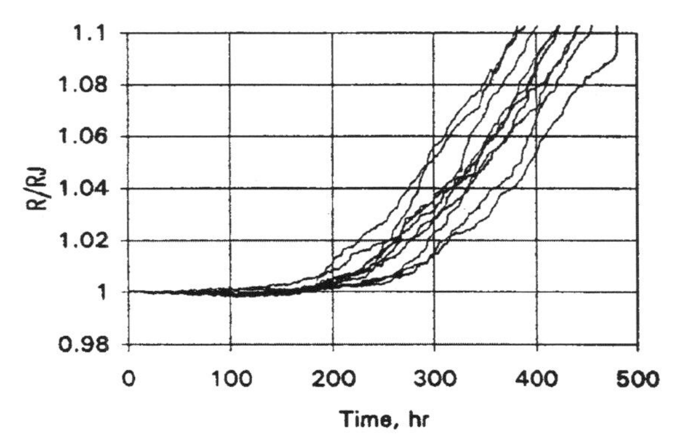

Fig. 3.47. Typical resistance change versus time during EM test for structure with refractory shunt layers adjacent to plugs. (Baerg, 'Recent problems in electromigration testing', Proceeding of the Annual International Reliability Physics Symposium, p. 211-215), Copyright © 1997, IEEE.

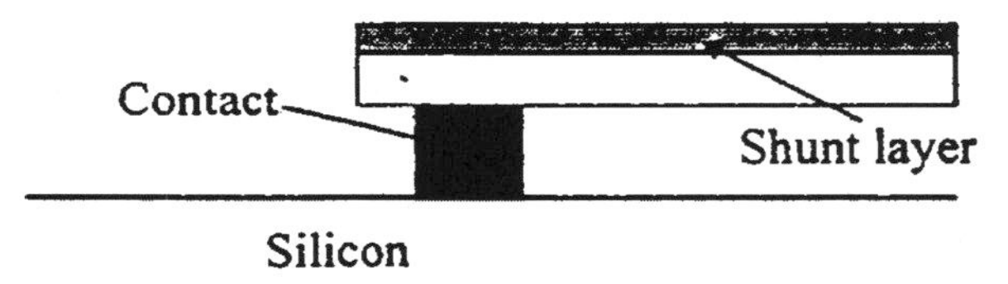

Fig. 3.48. Al interconnect with shunt layer not adjacent to the blocking boundary. (Baerg, 'Recent problems in electromigration testing', Proceeding of the Annual International Reliability Physics Symposium, p. 211–215), Copyright © 1997, IEEE.

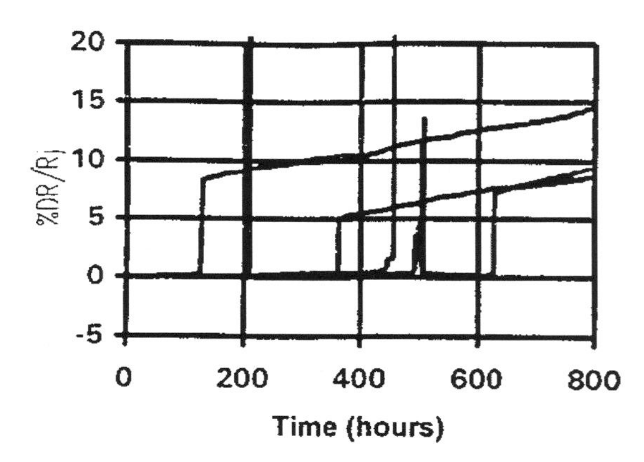

Fig. 3.49. Erratic behavior of resistance change for interconnect with shunt layer not adjacent to the blocking boundary. (Baerg, 'Recent problems in electromigration testing', Proceeding of the Annual International Reliability Physics Symposium, p. 211–215), Copyright © 1997, IEEE.

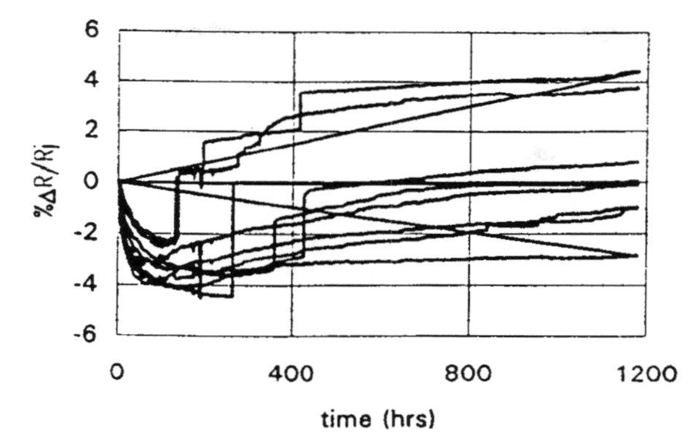

Fig. 3.50. Resistance versus time for Al interconnect with short length effect and delamination failure. (Baerg, 'Recent problems in electromigration testing', Proceeding of the Annual International Reliability Physics Symposium, p. 211–215), Copyright © 1997, IEEE.

is due to the delamination of the upper shunt layer from the W-plug (Baerg 1997).

Therefore, from the signature of the resistance change vs. stress time plot, various failure mechanisms can be identified.

3.9.2 *Short*

Hillock formation as a result of Al atoms accumulation can result in short circuit with the metal layer on top or below or side way. When there is capping layer on the top surface of metallization, hillock formation will be suppressed, and protrusions from metal sidewalls will occur. The result is the shorting between adjacent metal lines (Hu, Small and Ho 1993).

3.10 Test Data Analysis

While test structure and test procedures are important in ensuring the credibility of EM test data, correct test data analysis is also very important especially when extrapolation is need to assess the interconnect EM under normal operating condition.

Typically, the data analysis of EM test data is performed by placing the time to failure data points together with their respective estimated CDF obtained using Median Rank method (Elsayed 1996) on a lognormal probability graph paper or using reliability software for analysis to determine the failure distribution parameters, and hence the t_{50} or MTF of the interconnect.

Table 3.9. Time to failure of 20 samples.

1.184	3.34*	24.6	697.23
1.58*	5.62*	56.99	1349.01
1.783	6.32*	208.53	3715.83
2.239	7.23	276.77	4290.30
2.43*	24.14	335.86	6911.43

However, in reality, as EM test is performed at accelerated test conditions, outlier failure points and bimodal distribution due to the presence of more than one failure mechanism are often observed. The ignorance of their presence can render the analysis results and conclusion invalid.

As an example, Table 3.9 shows 15 time-to-failure (TTF) data simulated from a lognormal distribution with $\mu = 100$ days, $\sigma = 20$ days using random number generator. These TTF will resembles the actual test data if the interconnect EM failure distribution has a population with the same distribution parameters as given above. If 5 random failure points (marked as *) are added as outlier points, i.e. they are not from the same distribution, the lognormal probability plot will be as shown in Fig. 3.51. Figure 3.52 shows the probability plot if the 5 outlier points are not present. Comparing Figs. 3.51 and 3.52, the 5 outlier points are not obvious at all as the actual TTFs are distributed randomly as well (Tan and Roy 2007), and hence by merely looking at the probability plot, identification of outlier points may not be effective.

The t_{50} computed from the 20 TTF data points without taking the outlier points as censored points is 44.79 days which is far from the population value (100 days). If the 5 outlier points are identified and treated as censored points, the t_{50} is 135.22 days, a value closer to the population. Thus, different conclusions could be derived from the two t_{50} values, and one can see the importance of outlier point identification. Here an outlier point is referred to an experimental observation that lies outside the overall pattern of the underlying distribution describing majority of the experimental data points.

For the case of bimodal distribution, Fig. 3.53 shows a typical bimodal distribution for copper electromigration which is obtained from the work of Vairagar *et al.* (Vairagar, Mhaisalkar and Krishnamoorthy 2004). The study of Cu EM will be discussed in the next chapter.

To determine the two distributions within the bimodal distribution, an intuitive and common method is to fit two straight lines through the different set of points as shown in Fig. 3.53. The distribution parameters

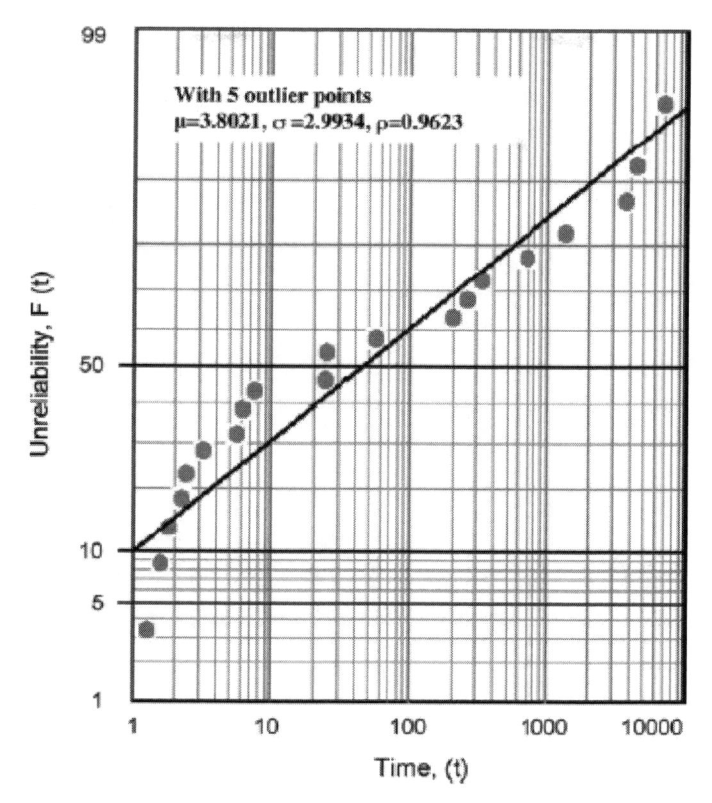

Fig. 3.51. Lognormal plot of the time to failure data shown in Table 3.9 with 5 outlier points included (Tan and Roy 2007). Reprinted from Materials Science and Engineering Review, 58, Tan and Roy, 'Electromigration in ULSI interconnects', pp. 1–75, Copyright © 2007, with permission from Elsevier.

can then be obtained from the best straight line through each set of points using the principle of probability plotting, and the proportion value for each component was obtained by finding the ratio of the approximate number of data points in each component. The mixture model thus obtained is given as:

$$pdf = 0.125 \times \Phi\left(\frac{\ln t - 5.70}{1.83}\right) + 0.875 \times \Phi\left(\frac{\ln t - 5.07}{0.62}\right). \quad (3.28)$$

However, the method cannot be applied when there is co-existence of two failure mechanisms in a failed unit. Also, as the data is a multi-censored data, the use of the simple median rank for the estimation of the CDF is not correct, and modified failure order points must be calculated. For example,

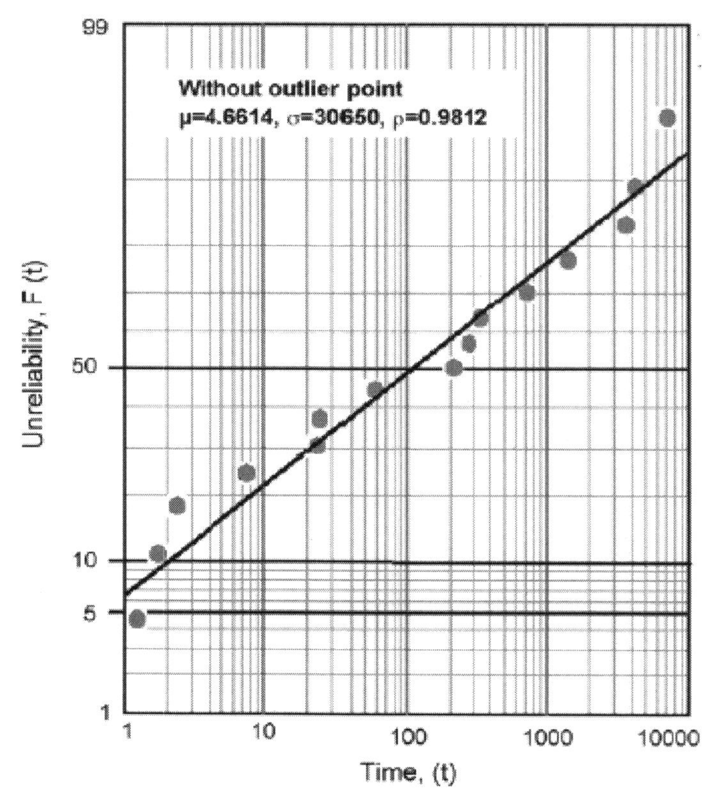

Fig. 3.52. Lognormal Plot of the time to failure data shown in Table 3.9 without the 5 outlier points. Reprinted from Materials Science and Engineering Review, 58, Tan and Roy, 'Electromigration in ULSI interconnects', pp. 1–75, Copyright © 2007, with permission from Elsevier.

the estimated CDF for the 4th data point should not be $(4 - 0.3)/(n + 0.4)$ where n is the total sample size, since the 4th data point is not the 4th failure according to failure mechanism 2, it is the 1st failure (For the detail of probability plotting and the estimation of CDF, please refer to the book by Elsayed). Likewise for the subsequent failure points that belong to failure mechanism 2. Furthermore, for failure points that fall on the intersection of the two straight lines, one cannot be sure that if these failures belong to failure mechanism 1 or 2. One way to resolve such a problem is to apply multiple censored data analysis (Tan, Anand, Zhang, Krishnamoorthy and Mhaisalkar 2005) if we can know the failure mechanism that each failure

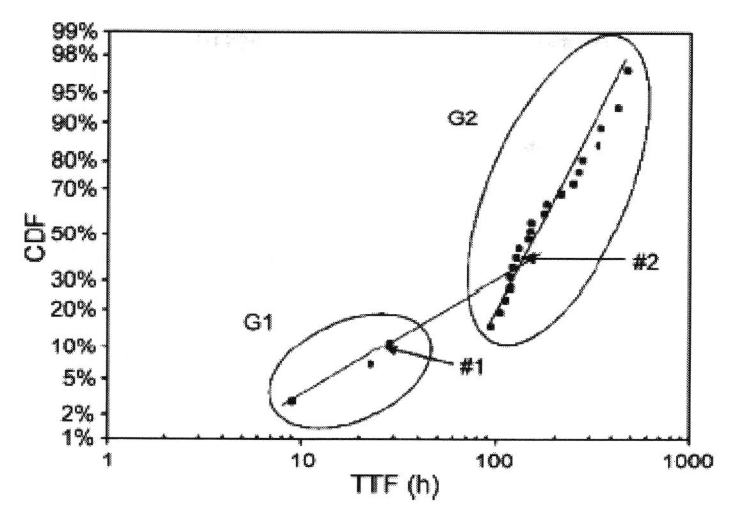

Fig. 3.53. Lognormal plot of the time to failure. Reprinted from Microelectronics Reliability, 44(5), Vairagar A., Mhaisalkar S.G., and Krishnamoorthy A., 'Electromigration behavior of dual-damascene Cu interconnect structures — structure, width and length dependences', p. 747–754, Copyright © 2004, with permission from Elsevier.

belong to. But this is practically impossible because failure analysis on all the failed units must be performed.

Therefore, the conventional method can only be used to show the existence of multiple failure mechanisms and the most probable number of failure mechanisms. Tan *et al.* (Zhang, Tan, Tan, Sim and Zhang 2004) have developed an analysis method that can help to identify the failure mechanism that each failure unit belong to using statistical method without failure analysis, and the method can also identify failure that contain co-existence of failure mechanism. With the application of such method, the actual mixture distribution for the failure data shown in Fig. 3.53 should be

$$pdf = 0.12 \times \Phi\left(\frac{\ln t - 2.9}{0.5}\right) + 0.88 \times \Phi\left(\frac{\ln t - 5.2}{0.48}\right), \tag{3.29}$$

which is significantly different from the previously determined pdf. Table 3.10 shows the Data Identification Matrix and the corresponding verification through failure analysis. The identified line failure and via failure micrographs are shown in Figs. 3.54 and 3.55, respectively.

In order to further verify the accuracy of the EM algorithm in this application, the method is also applied to another test results from the

Table 3.10. Date Identification Matrix from the EM algorithm for the test data obtained from 50 μm long M2 test structure.

Failure no.	TTF	Data identification matrix		FIB analysis	Failure no.	TTF	Data identification matrix		FIB analysis
		Probability due to failure mechanism #1	Probability due to failure mechanism #2				Probability due to failure mechanism #1	Probability due to failure mechanism #2	
1	9.04	1	4.23E–08	—	13	148.74	2.31E–05	0.99998	—
2	23.06	0.99934	0.000664	Line EM	14	150.08	2.13E–05	0.99998	—
3	28.55	0.99436	0.00564	Line EM	15	175.45	5.09E–06	0.99999	—
4	94.18	0.0016327	0.99837	—	16	179.66	4.10E–06	1	—
5	105.05	0.00058541	0.99941	—	17	213.33	8.62E–07	1	—
6	112.01	0.00032125	0.99968	—	18	247.21	2.29E–07	1	—
7	118.24	0.00019389	0.99981	—	19	263.12	1.31E–07	1	—
8	118.37	0.00019192	0.99981	—	20	274.24	9.04E–08	1	VIA EM
9	120.97	0.00015677	0.99984	—	21	331.73	1.67E–08	1	—
10	126.21	0.0001057	0.99989	VIA EM	22	345.25	1.17E–08	1	—
11	130.23	7.90E–05	0.99992	—	23	425.63	1.88E–09	1	—
12	145.35	2.86E–05	0.99997	—	24	474.34	7.33E–10	1	—

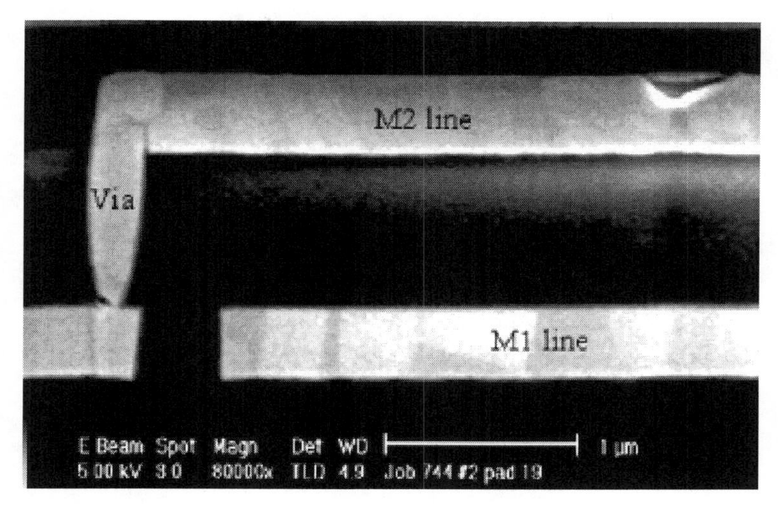

Fig. 3.54. FIB micrograph for line related EM failure correspond to the failure denoted as #1 in Fig. 3.53. Reprinted from Microelectronics Reliability, 44, Zhang G. Tan CM, Tan KT, Sim KY and Zhang WY, 'Reliability improvement in Al metalization: A combination of statistical prediction and failure analytical methodology', pp. 1843–1848, Copyright © 2004, with permission from Elsevier.

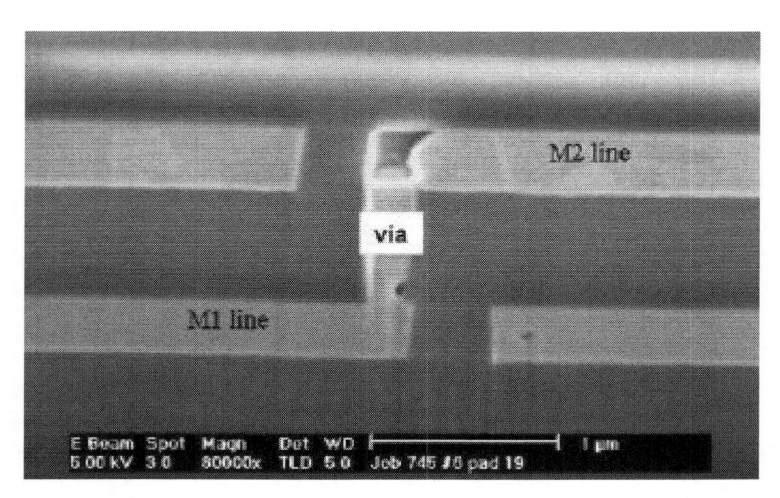

Fig. 3.55. FIB micrograph for line related EM failure correspond to the failure denoted as #2 in Fig. 3.53. Reprinted from Microelectronics Reliability, 44, Zhang G. Tan CM, Tan KT, Sim KY and Zhang WY, 'Reliability improvement in Al metalization: A combination of statistical prediction and failure analytical methodology', pp. 1843–1848, Copyright © 2004, with permission from Elsevier.

Table 3.11. Data Identification Matrix from the EM algorithm for the test data obtained from 100 μm long M2 test structure.

		Data identification matrix					Data identification matrix		
Failure no.	TTF	Probability due to failure mechanism #1	Probability due to failure mechanism #2	FIB analysis	Failure no.	TTF	Probability due to failure mechanism #1	Probability due to failure mechanism #2	FIB analysis
1	19.33	0.99994	5.57E–05	—	16	134.23	8.43E–05	0.99992	—
2	45.50	0.6298	0.3702	Mixed	17	134.91	8.43E–05	0.99992	—
3	82.11	0.005935	0.99407	—	18	147.56	4.03E–05	0.99996	—
4	82.11	0.005935	0.99407	—	19	150:29	3.42E–05	0.99997	
5	83.04	0.005422	0.99453	—	20	156.05	2.48E–05	0.99998	VIA EM
6	83.73	0.004954	0.99505	—	21	156.35	2.48E–05	0.99998	—
7	86.07	0.003782	0.99622	VIA EM	22	159.70	2.11E–05	0.99998	—
8	94.33	0.001695	0.9983	—	23	171.49	1.21E–05	0.99999	—
9	103.36	0.000769	0.99923	—	24	193.64	4.39E–06	1	—
10	111.44	0.00042	0.99958	—	25	218.37	1.77E–06	1	—
11	120.01	0.000212	0.99979	—	26	220.17	1.77E–06	1	—
12	122.00	0.000195	0.99981	—	27	230.66	1.22E–06	1	—
13	122.44	0.000179	0.99982	—	28	232.14	1.13E–06	1	—
14	123.77	0.000118	0.99988	—	29	276.51	3.30E–07	1	—
15	132.76	9.16E–05	0.99991	—	30	281.89	2.87E–07	1	—

same work but with M2 length equal to $100\,\mu$m. Table 3.11 shows the corresponding data identification matrix and the FIB analysis results. In this case, the second failure unit showed co-existence of two failure mechanisms from the analysis, and it is verified by the corresponding FIB analysis as shown in Fig. 3.56.

The method that Tan *et al.* (Zhang, Tan, Tan, Sim and Zhang 2004) developed is basing on the statistical methods for outlier point identification and missing data analysis. An outlier is an observation that appears to deviate markedly from other members of the sample in which it occurs (Grubbs 1969). Outlier detection has been used for centuries to detect and remove (in some cases) anomalous observations from data, and principled and systematic techniques have now been developed for the outlier detection such as the Chauvenet's criterion, Grubbs' test for outliers, Peirce's criterion, and Dixon's Q test (Rousseeuw and Leroy 1996; Hodge and Austin 2004; Peirce 1986).

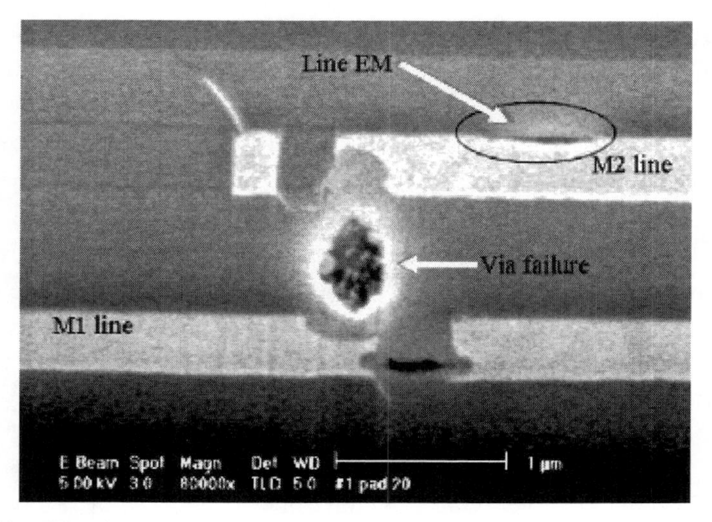

Fig. 3.56. FIB micrograph for the failure correspond to the second failure in Table 3.11 which indicated a mixed failure mechanism. Reprinted from Microelectronics Reliability, 44, Zhang G. Tan CM, Tan KT, Sim KY and Zhang WY, 'Reliability improvement in Al metalization: A combination of statistical prediction and failure analytical methodology', pp. 1843–1848, Copyright © 2004, with permission from Elsevier.

For the missing data analysis, Expectation and Maximization (EM) algorithm is employed. In statistics, an **expectation-maximization (EM) algorithm** is a method for finding maximum likelihood estimates of parameters in statistical models, where the model depends on unobserved latent variables such as the number of different underlying physical mechanisms and the probability of each failed unit belonging to these different mechanisms. EM is an iterative method which alternates between performing an expectation (E) step, which computes the expectation of the log-likelihood evaluated using the current estimate for the latent variables, and a maximization (M) step, which computes parameters maximizing the expected log-likelihood found on the *E* step. These parameter-estimates are then used to determine the distribution of the latent variables in the next *E* step. The EM algorithm was explained and given its name in a classic 1977 paper by Dempster, Laird, and Rubin (1977). The advantage of the above mentioned statistical methods is that one does not need to have a large data set for the analysis.

Figure 3.57 shows the CDF plots for the two lognormal components in the mixture distribution given by Eq. (3.29), as determined from the method developed by Tan *et al.* (Zhang, Tan, Tan, Sim and Zhang 2004) with their proportions as weighted factors for them respectively. One can see that while the proportion for the via-related EM is high, indicating that it is the dominant failure mechanism, the line-related EM is the early failure under the EM stress condition. In general, for the test data like Table 3.10, via related failure will usually receive higher attention from the reliability test data analysis, as it is dominant. However, in device application, line related failure should be of more concern as it determines the time to first failure.

One also needs to be cautious that since different failure mechanisms have different relationship with the accelerated stress condition, early failure mechanism observed under stress condition might no longer be the early failure as depicted schematically in Fig. 3.57. This is indeed observed in the work by Zhang *et al.* (Gan, Tan and Zhang 2003).

From the above discussion, the mixture distribution and outlier points analysis of EM test data are important so that correct extrapolation can be made, and hence the correct early failure mechanism can be identified and addressed.

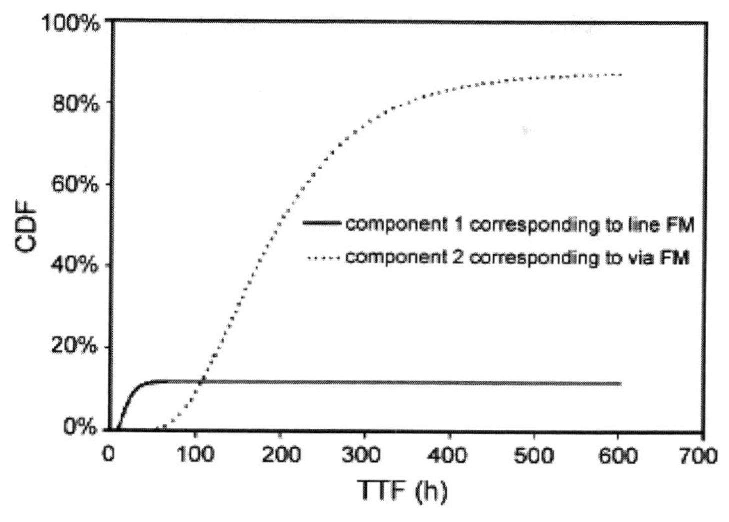

Fig. 3.57. CDF plot for the two lognormal components identified in Table 3.10. Reprinted from Microelectronics Reliability, 44, Zhang G. Tan CM, Tan KT, Sim KY and Zhang WY, 'Reliability improvement in Al metalization: A combination of statistical prediction and failure analytical methodology', pp. 1843–1848, Copyright © 2004, with permission from Elsevier.

3.11 Failure Analysis on EM Failures

After EM test, besides the test data analysis, failure analysis on the failed units are also important in order to verify that the failures are indeed electromigration related and identified the failure mechanisms.

The physical failure analysis (FA) is the postmortem of sample which undergone EM stressing. The FA is performed on stressed sample by various analytical tools such as optical microscope, FIB-SEM, TEM etc. to investigate the root cause, failure location, fast diffusion path etc. Broadly, the FA can be divided into two groups, namely the destructive and nondestructive FA.

To begin the failure analysis, non-destructive FA is to be performed. Usually, optical microscope is used first, especially when the sample line width is in the order of few microns. In some cases, the top passivation layer/surrounding materials may need to be removed chemically in order to observed the failure under optical microscope. For smaller line width, Scanning electron microscope will be necessary.

After the microscopy examination and the failure is identified to be interconnect related and not corrosion induced, the next step is to locate the failure sites non-destructively. The common method employed is OBIRCH (optical beam induced resistance change). However, there are also other methods such as EBICH (electron beam induced current change) and magnetic imaging.

The OBIRCH method employs two main processes simultaneously, namely the laser-beam heating and resistance-change detection. The resistance change caused by laser-beam heating depends on the temperature increase. When a laser beam is irradiated, the generated heat is transmitted freely across areas that are free from defects or voids. However, heat transmission is impeded at defects such as voids. This creates nonuniform temperature increase which results in different resistance change pattern. The resolution of the technique is improved from 400 nm to 50 nm by using near-field optical probe instead of laser source (Nikawa, Saiki, Inoue and Ohtsu 1999). A successful application of the technique in the study of FA on EM stressed sample is reported by Hoshino (Hoshini 2000). It is shown that the technique is reliable for detecting the voided zone. 3D object reconstruction by electron tomography is the promising future method for nondestructive FA (Riege, Prybyla and Hunt 1996).

Figure 3.58 shows a typical micrograph of OBIRCH image where the red spot represents the suspect site of void. Figure 3.59 shows the corresponding FIB cross-section of the interconnect confirming the void site (2009).

Figure 3.60 compares the images using near-field and conventional OBIRCH, showing the improved resolution. Also, with the near-field probe, the optical probe induced resistance change caused by heating can be observed using a metallized probe without interference from a photocurrent created by the electron-hole pair generation, reducing the noise in the resistance change signal.

To further improve the image resolution, and to reconstruct the void in 3D, Gan *et al.* (Gan, Tan and Zhang 2003) used electron beam instead of optical beam as heat source for void location identification, and improve the

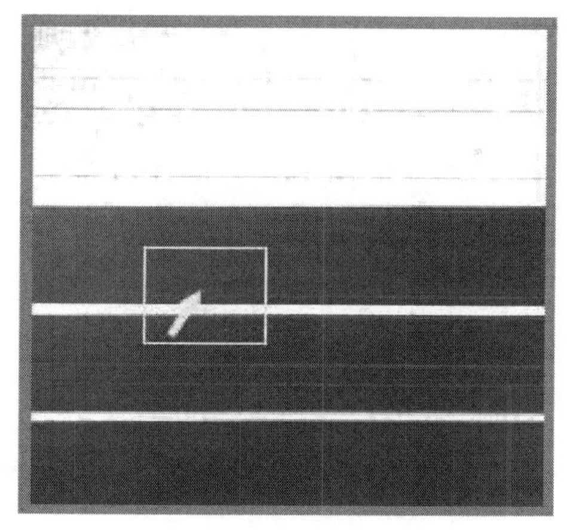

Fig. 3.58. Example of OBIRCH signal on micrograph showing the possible void location (2009).

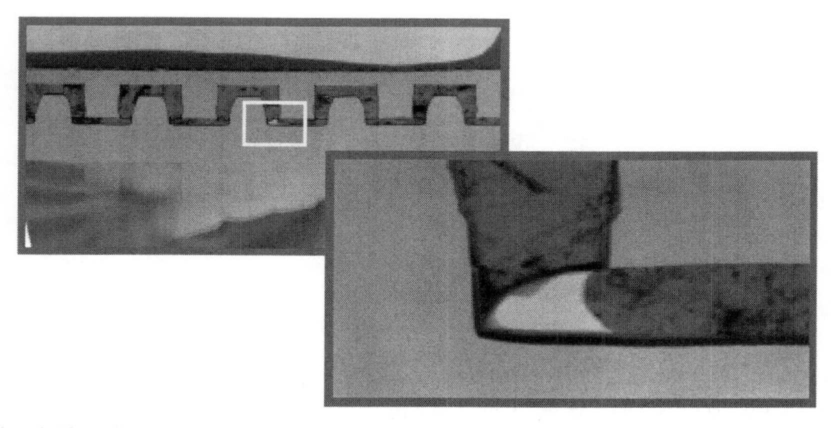

Fig. 3.59. Corresponding FIB/TEM cross-sectional view confirming the location of void (2009).

resolution further. They also developed an algorithm where the identified void volume can be reconstructed in 3D.

The use of magnetic imaging to observe current distribution in an interconnect and hence observe the time evolution of voids can also provide a powerful tool for void identification as well as a good understanding of the void evolution under passivation and non-destructively. Rous *et al.* (Rous,

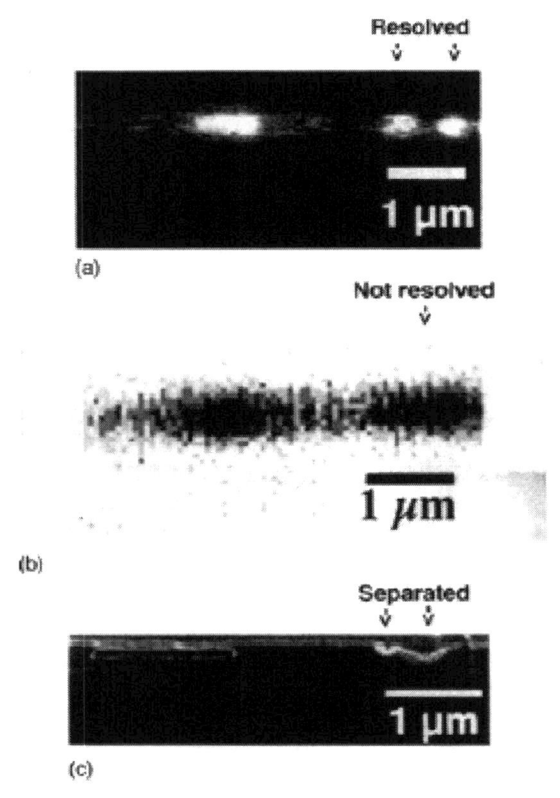

Fig. 3.60. (a) Near-field OBIRCH image; (b) Conventional OBIRCH image and (c) cross-sectional scanning ion microscopy image. Reprinted with permission from Nikawa, Saiki, Inoue and Ohtsu, 'Imaging of current paths and defects in Al and TiSi interconnects on very-large-scale integrated-circuit chips using near-field optical probe stimulation and resulting resistance change', Applied Physics Lett., vol. 74, pp. 1048–1050. Copyright © 1999, American Institute of Physics.

Yongsunthon, Stanishevsky and Williams 2004) used the principle of magnetic force microscopy to perform direct imaging of current distributions with submicron resolution. Magnetic force microscopy is used to measure the curvature of the magnetic field generated by a current carrying structure. They performed experiments on 11.5 μm wide and 130 nm high Cr/Au metal line and observe the current distribution as shown in Fig. 3.61. The results were compared with the finite element calculation as shown in Fig. 3.62 and just about perfect agreement was obtained.

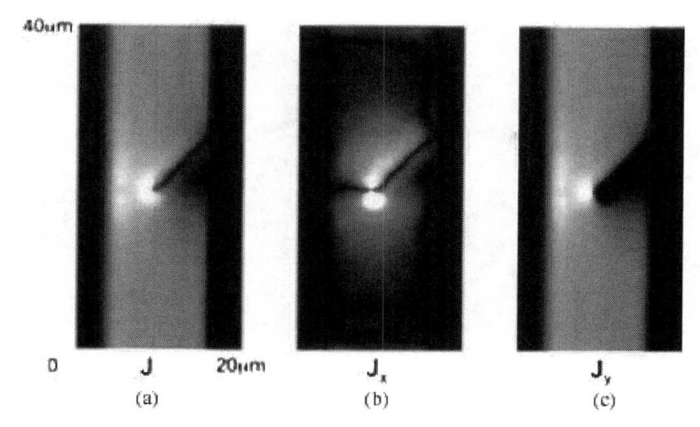

Fig. 3.61. Current densities determined by inversion of MFM phase image. From left to right, the panels show: the total current density, the component of the current density perpendicular to the lines edges, and the component of the current density parallel to the line edges. Current values are normalized to a value $J_{\text{total}} = 1$ far from the defects. The color scale ranges from 0 to 4 for the total and parallel current density and from 20.4 to 0.4 for the perpendicular current density. Reprinted with permission from Rous, Yongsunthon, Stanishevsky and Williams, 'Real-space imaging of current distributions at submicron scale using magnetic force microscopy: Inversion methodology', Journal of Applied Physics, vol. 95(5), pp. 2477–2486. Copyright © 2004, American Institute of Physics.

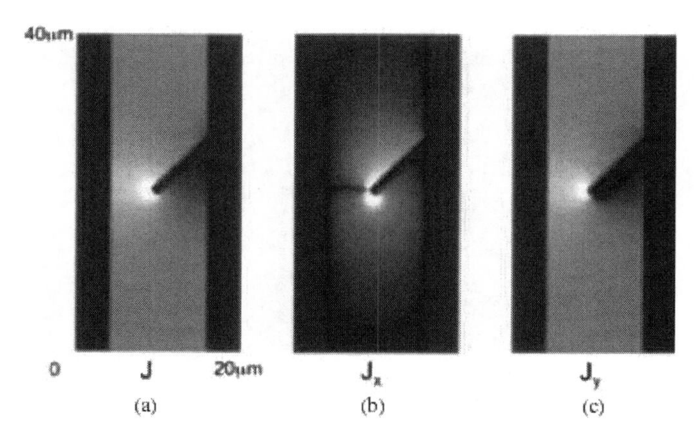

Fig. 3.62. Current densities determined by a finite-element calculation for a model of the sample shown in Fig. 3.61. The format of this figure and the color scales are the same as for the inverted images shown in Fig. 3.61. Reprinted with permission from Rous, Yongsunthon, Stanishevsky and Williams, 'Real-space imaging of current distributions at submicron scale using magnetic force microscopy: inversion methodology', Journal of Applied Physics, vol. 95(5), pp. 2477–2486. Copyright © 2004, American Institute of Physics.

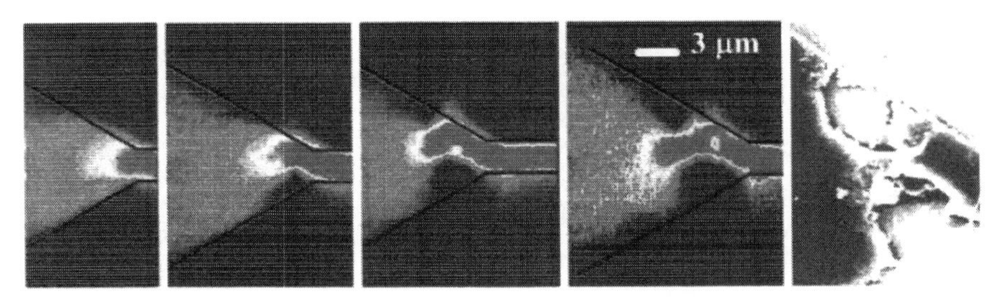

Fig. 3.63. (a)–(d) Time evolution of the current density flowing through the cathode of a passivated 3 μm conductor undergoing electromigration. Solid black lines denote the initial edges of the conductor. (e) To the far right is a postmortem SEM micrograph of the stripped sample confirming the failure mechanism. Reprinted with permission from Schrag and Xiao, 'Submicron electrical current density imaging of embedded microstructures', Applied Physics Lett., vol. 82(19), pp. 3272–3274. Copyright © 2003, American Institute of Physics.

Schrag and Xiao (Schrag and Xiao 2003) also developed a scanning magnetic microscopy technique for non-invasively imaging submicron magnetic fields from embedded microscopic electrical circuits. The technique operates at ambient temperature and can image buried layers with high resolution by using a nano-scale magnetoresistance sensor. With this technique, the time evolution of the current density flowing through the cathode of a 3 μm Al conductor undergoing electromigration can be observed as shown in Fig. 3.63.

Another method to locate the void in interconnect is to use optical microscopy imaging proposed by Li *et al.* (Li, Dasika, Heskett and Tang 2007). They found that there is a relationship between the reflectivity and conductivity of metal given as follows (Li, Dasika, Heskett and Tang 2007).

$$R = 1 - \frac{v_1}{\sigma_0}, \tag{3.30}$$

where R is the reflectivity, v_1 is a characteristic frequency which is around $10^{15}\ \text{s}^{-1}$ for metal, σ_0 is the d.c. conductivity of the metal.

Experimental results indeed showed the correlation as shown in Fig. 3.64. Thus, as the optical beam is scanned across the interconnect, segment of the interconnect with void will be revealed due to its increased resistance.

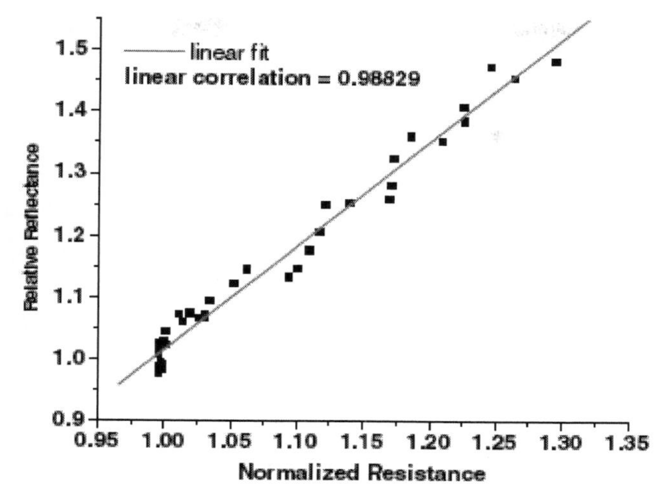

Fig. 3.64. Relative reflectance versus normalized resistance with a linear correlation of 0.99 for a 10 μm $\times$ 280 μm $\times$ 0.85 μm Al (0.5% Cu) interconnect during a 5 MA/cm^2, 200°C accelerated test. Li, Dasika, Heskett and Tang, 'Optical microscopy imaging method for detection of electromigration: Theory and experiment', Phys. Stat. Solidi A, 2007, 204, (5), pp. 1589–1595. Copyright Wiley-VCH Verlag GmbH & Co. KGaA. Reproduced with permission.

Once the failure sites are located, FIB cut will be performed to examine the sites more clearly with the help of SEM.

Transmission Electron Microscope (TEM) is also used in FA, especially for sample of line-width below 0.2 μm and when better resolution than SEM is desired. With the continuous decrease in the line-width of EM samples, TEM is expected to become essential for FA on EM stressed samples. TEM micrograph of void is shown in Fig. 3.59. TEM is also very useful to examine the change in the interconnect micro-structure after electromigration, which will shed light on the failure mechanism underlying EM.

X-ray is another technique by which FA can be performed, however it is not found to be suitable for submicron interconnects.

To identify the failure mechanisms underlying EM process, besides TEM, a common method is to compute the activation energy from the time to failure data using the Black's equation. However, the activation energy computed using the Black's equation is a lumped parameter of different physical processes, and the initiating physical process or the dominating physical process cannot be identified from the Black's equation. One method of finding the underlying physical processes is the use of noise measurement. Emelianov *et al.* (Emelianov, Ganesan, Puzic, Schulz, Eizenberg, Habermeier and Stoll 2003) used the $1/f$ noise measurement technique to determine the distribution of the activation energies of the processes involved from a single sample at progressive electromigration damaging. However, one needs to note that the energy distribution of the activation energy of the $1/f$ noise sources has little if anything to do with the activations of electromigration activation energies. One is electronics or trapping of electrons and holes by electronic traps while the other is the atomic or the trapping of atoms by the atomic traps on the surface of the voids. However, there could be a correlation between the atomic traps and the electronic traps, which is easily derived theoretically based on simple or more complicated models. This would be a good subject for future experiments.

Tan *et al.* (Tan and Lim 2002) performed time frequency analysis using Wigner-Ville distribution on the noise data from Al interconnection undergoing EM test. Using the method, they are able to show the occurrence of different physical processes during the EM progression as shown in Fig. 3.65.

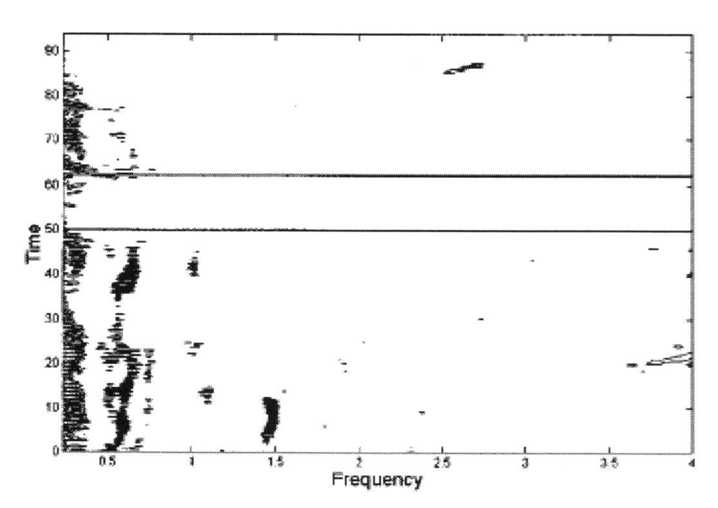

Fig. 3.65. Time-frequency plot of the noise signal as Al interconnect undergoing EM testing, showing the presence of different frequency components at different time, signify the different underlying physical processes governing the EM. (Tan and Lim, 'Application of Wigner-Ville Distribution in Electromigration Noise Analysis', IEEE Trans. Devices and Materials Reliability, 2(2), p. 30–35), Copyright © 2002 IEEE.

3.12 Conclusion

In this chapter, the physics of EM in Al are linked to the various EM tests results reported, and the effect of interconnection processes and design on its EM performances are explained and summarized. The practical considerations of EM test in order to accurately assess an interconnect EM reliability are discussed in detail which include the test structure design, effect of test conditions, test data analysis and the various pitfalls for the different EM tests. Failure analysis methodologies of EM failure are also presented.

From the various experimental results of Al interconnections, the following may be noted in order to enhance the EM lifetime of the interconnections:

(a) Adding Cu impurity into Al interconnection, but no pre-annealing
(b) Ensure the interconnect grain texture is [111] rich
(c) Minimize the step height and width transition
(d) Avoid 90° turn in interconnection
(e) Use reservoir in interconnection
(f) Lower the interconnect temperature due to Joule heating by reducing the current density
(g) Via layout and separation should be properly design, making use of the Blech effect
(h) Use thinner passivation layer
(i) Use barrier layer with oxygen implant or vacuum break

From the discussion in this chapter, one can see the complexity of EM and also the vast amount of knowledge gained from the many works done in Al EM. This will provide us a basis for the understanding of EM in Cu interconnect which is becoming a main stream of interconnect in ULSI today and will be discussed in the next Chapter.

References

(1994). A Procedure for Executing SWEAT. *JEDEC Publication JEP 119.*

(1995). Standard method for measuring and using the temperature coefficient of resistance to determin the temperature of a meallization line, JEDEC Solid State Technology Association.

(1997). Isothermal Electromigration Test Procedure. *EIA/JEDEC Standard 61.*

(2009). *Toshiba Semiconductor Reliability Handbook*, http://toshiba-sdcard.com.cn/eng/shared/reliability_pdf/bde0128e_4_4_4.pdf.

Oates, A. S. (1995). Current density dependence of electromigration failure of submicron width multilayer Al alloy conductors. *Applied Physics Lett.* **66**(12): 1475.

Scorzoni, A., Impronta, M., Munari, I. D. and Fantini, F. (1999). Proposal for a standard procedure for moderately accelerated electromigration tests on metal lines. *Microelectronics Reliability* **39**(5): 615–626.

Agarwala, B. N., Attardo, M. J. and Ingraham, A. P. (1970). Dependence of electromigration induced failure time on length and width of aluminum thin film conductors. *Journal of Applied Physics* **41**: 3954.

Agarwala, B. N., Digiacomo, G. and Joseph, R. R. (1976). Electromigration damage in aluminium-copper films. *Thin Solid Films* **34**: 165–169.

Agarwala, B. N., Patnaik, B. and Schmitzel, R. (1972). *Journal of Applied Physics* **43**: 1487.

Ainslie, N. G., *et al.* (1972). *Applied Physics Lett.* **20**: 173.

Amazawa, T. and Arita Y. (1991). A 0.25 μm via plug process using selective CVD aluminum for multilevel interconnection. *Proc. of IEEE IEDM*: 265–268.

Atakov, E. M. (1990). Electromigration resistance of TiN-layered Ti-doped Al interconnects. *Proc. of the IEEE VMIC*: 360–362.

Attardo, M. J. and Rosenberg R. (1970). Electromigration damage in aluminum film conductors. *J. Appl. Phys.* **41**: 2381–2386.

Avinum, M., Barel, N., Kaplan, W. D., Eizenberg, M., Naik, M., Guo, T., Chen, L. Y., Mosely, R., Littau, K., Zhou, S. and Chen, L. (1998). Nucleation and growth of CVD Al on different types of TiN. *Thin Solid Films* **320**: 67–72.

Baerg, W. and Wu K. (March 1991). Using metal grain size distributions to predict electromigration performance. *Solid State Technology*: 35–37.

Berenbaum, L. (1971). *Journal of Applied Physics* **42**: 880.

Black, J. R. (1969). Electromigration — A brief survey and some recent results. *IEEE Trans. Electron. Devices* **16**(4): 338–347.

Blech, I. A. (1976). Electromigration in thin aluminum films on titanium nitride. *Journal of Applied Physics* **47**: 1203–1208.

Blech, I. A. and Herring, C. (1976). Stress generation by electromigration. *Appl. Phys. Lett.* **29**: 131–133.

Blech, I. A. and Meieran, E. S. (1969). Electromigration in thin Al films. *Journal of Applied Physics* **40**(2): 485–491.

Brown, D. D., Korhonen, M. A., Borgesen, P. and Li, C. Y. (1994). Predicting and comparing electromigration failure for different test structures. *Material Research Society Symposium*, **338**: 435.

Clement, J. J. (1997). Reliability analysis for encapsulated interconnect lines under DC and pulsed DC current using a continuum electromigration transport model. *J. Appl. Phys.* **82**: 5991–6000.

Clement, J. J. and Lloyd, J. R. (1992). Numerical investigations of the electromigration boundary value problem. *Journal of Applied Physics* **71**: 1729–1731.

Clement, J. J. and Thompson, C. V. (1995). Modeling electromigration-induced stress evolution in confined metal lines. *Journal of Applied Physics* **78**: 900.

Dempster, A. P., Laird, N. M. and Rubin, D. B. (1977). Maximum likelihood from incomplete data via the EM algorithm. *Journal of the Royal Statistical Society. Series B (Methodological)* **39**(1): 1–38.

Dioin, M. J. (2000). Electromigration lifetime enhancement for lines with multiple branches. *IEEE International Reliability Physics Symposium*: 324–332.

Dion, M. J. (1993). On the status of wafer level metal integrity testing. *Microelectronics Reliability* **33**(11/12): 1807–1827.

Dion, M. J. (2001). Reservoir modeling for electromigration improvement of metal systems with refractory barriers. *IEEE International Reliability Physics Symposium*: 327–333.

Dirks, A. G., Tien, T. and Towner, J. M. (1986). AI-Ti and AI-Ti-Si thin alloy films. *Journal of Applied Physics* **59**(6): 2010–2014.

Dixit, G. A., Paranjpe, A., Hong, Q. Z., Ting, L. M., Luttmer, J. D., Havemann, R. H., Pual, D., Morrison, A., Littau, K., Eizenberg, M. and Sinha, A. K. (1995). A novel 0.25 μm via plug process using low temperature CVD Al/TiN. *Proc. of IEEE IEDM*: 1001–1003.

Doan, J. C., Lee, S., Lee, S.-H., Flinn, P. A. and Bravman, J. C. (2001). Effects of dielectric materials on electromigration failure. *Journal of Applied Physics* **89**(12): 7797–7808.

Dorgelo, A. M., Vroemen, J. A. M. W. and Wolters, R. A. M. (2001). An additional effect of texture on the electromigration behavior of aluminum. *Microelectronic Engineering* **55**: 337–340.

Elsayed, E. A. (1996). *Reliability Engineering*, Addison Wesley Longman, Inc.

Emelianov, V., Ganesan, G., Puzic, A., Schulz, S., Eizenberg, M., Habermeier, H. U. and Stoll, H. (2003). Investigation of electromigration in Copper Interconnects by noise measurement. *Proceeding of SPIE*. M. B. Weissman, N. E. Israeloff and A. S. Kogan, **5112**: 271–281.

Endicott, G. L., Bouldin, D. P. and Miller, L. A. (1992). On the modeling and scaling of Tungsten-stud-related electromigration. *Proc. of 9th International VLSI multilevel interconnect conference*: 434.

Filippi, R. G., Biery, G. A. and Wachnik, R. A. (1995). The electromigration short-length effect in Ti-AlCu-Ti metalization with tungsten studs. *Journal of Applied Physics* **78**(6): 3756.

Filippi, R. G., Wachnik, R. A., Eng, C. P., Chidambarrao, D., Wang, P. C. and White, J. F. (2002). The effect of current density, strip length, stripe width, and temperature on resistance saturation during electromigration testing. *Journal of Applied Physics* **91**(9): 5878–5795.

Finetti, M., Ronkainen, H., Blomberg, M. and Suni, I. (1986). *MRS 811*: 811.

Finetti, M., Scorzoni, A., Armigliato, A., Garulli, A. and Suni, I. (1988). *Solid State Devices*. G. Soncini and P. U. Calzolari, North-Holland: 361–364.

Finetti, M., Suni, I., Armigliato, A., Garulli, A. and Scorzoni, A. (1987). Correlation between resistance behavior and mass transport in Al-Si/Ti multilayer interconnects. *Journal of Vacuum Science and Technology* **A5**: 2854–2858.

Finetti, M., Suni, I., Santi, G. D., Bacci, L. and Caprile, C. (1986). *MRS 71*: 297.

Fischer, F. and Neppl, F. (1984). Sputtered Ti-doped AI-Si for enhanced interconnect reliability. *22nd IEEE International Reliability Physics Symp.*: 190–192.

Foley, S. (1998). A comparison between normally and highly accelerated electromigration tests. *Microelectronics Reliability* **38**: 1021–1027.

Foley, S., Molyneaux, J. and Mathewson, A. (1999a). An Evaluation of fast wafer level test methods for interconnect reliability control. *Microelectronics Reliability* **39**: 1707–1714.

Foley, S., Molyneaux, J. and Mathewson, A. (1999b). Evaluation of test methods and associated test structures for interconnect reliability control. *Proc. of IEEE International Conference on Microelectronic Test Structures*: 161–172.

Foley, S., Ryan, A., Martin, D. and Mathewson, A. (1998). *Microelectronics Reliability* **38**: 107.

Fujii, M., Koyama, K. and Aoyama, K. (1996). Reservoir length dependence of EM lifetime for Tungsten via chain under low current stress. *Proc. of 13th Int. VLSI Multilevel interconnection Conf.*: 312–317.

Fujii, T., Okuyama, K., Moribe, S., Torii, Y., Katto, H. and Agatsuma, T. (June 1989). Comparison of electromigration phenomenon between aluminum interconnection of various multilayered materials. *IEEE Proceedings of VMIC*: 477–483.

Gan, Z., Tan, C. M. and Zhang, G. (2003). Nondestructive void size determination in copper metallization under passivation. *IEEE Transactions on Device and Materials Reliability* **3**(3): 69–78.

Giacomo, G. D. (1997). *Reliability of Electronic Packages and Semiconductor Devices*, McGraw-Hill USA.

Giroux, F., Gounelle, C., Vialle, N., Mortini, P. and Ghibaudo, G. (1994). Current and temperature distribution impact on electromigration failure location in SWEAT structure. *Proceedings of the 1994 International Conference on Microelectronic Test Structures*: 214–217.

Giroux, F., Gounelle, C., Mortini, P. and Ghibaudo, G. (1995). Wafer level electromigation tests on NIST and SWEAT Structures. *IEEE International Conference on Microelectronic Test Structures*: 229–232.

Gleixner, R. J. and Nix, W. D. (1996). An analysis of void nucleation in passivated interconnect lines due to vacancy condensation and interface contamination. *Materials Reliability in Microelectronics VI. Symposium.* San Francisco, CA, USA: 475–480.

Gonzalez, J. L. and Rubio, A. (1997). Shape effect on electromigration in VLSI interconnects. *Microelectronics Reliability* **37**(7): 1073.

Grubbs, F. E. (1969). Procedures for detecting outlying observations in samples. *Technometrics* **11**: 1–21.

Hatch, J. E. (1984). Aluminum properties and physical metallurgy. Chapter 1, *American Society for Metals.*

Hinode, K., Furusawa, T. and Homma, Y. (1993). Dependence of electromigration damage on current density. *Journal of Applied Physics* **74**(1): 201.

Hinode, K. and Homma, Y. (1990). Improvement of electromigration resistance of layered aluminum conductors. *Proc. of International Reliability Physics Symposium*: 25.

Ho, P. S. (1970). *Journal of Applied Physics* **41**: 64.

Hodge, V. J. and Austin, J. (2004). A survey of outlier detection methodologies. *Artificial Intelligence Review* **22**: 85–126, Kluwer Academic Publishers.

Hong, C. C. and Crook, D. L. (1985). Breakdown Energy of Metal (BEM) — A new technique for monitoring metallization reliability of wafer level. *Proc. of the 23rd IEEE International Reliability Physics Symposium*: 108–114.

Hoshini, K. (2000). Early increase in resistance during electromigration in AlCu-plugged via structure. *Japanese Journal of Applied Physics* **39**: 994–998.

Hosoda, T., Yagi, H. and Tsuchikawa, H. (1989). Effects for copper and titanium addition to aluminum interconnects on electro- and stressmigration open circuit failures. *27th IEEE International Reliability Physics Symp.*: 202–206.

Hu, C. K. and Small, M. B. (1993). Electromigration failure of bamboo structured lines by interfacial transport. *Proc. of 10th International VLSI Multilevel Interconnection Conference*: 265.

Hu, C. K., Small, M. B. and Ho, P. S. (1993). Electromigration in Al(Cu) two-level structures: effect of Cu and Kinetics of damage formation. *Journal of Applied Physics* **74**(2): 969.

Hummel, R. E., DeHoff, R. T. and Geier, H. J. (1976). Activation energy for electrotransport in thin aluminum films by resistance measurements. *Journal of Phys. Chem. Solids* **37**: 73–80.

Huntington, H. B. and Grone, A. R. (1961). Current-induced marker motion in gold wires. *J. Phys. Chem. Solids.* **20**: 76–87.

J.E Sanchez, J. (1986). *Proc. of Tungsten and Other Refractor Metals for VLSI Applications*. E. Broadbent: 395.

J.E Sanchez, J. and Morris, J. J. W. (1991). Microstructual analysis of electromigration-induced voids and hillocks. *Material Research Society Symposium* **225**: 53.

J.E Sanchez, J., Lloyd, J. R. and Morris, J. J. W. (1990). *Journal of Electronic Materials* **19**: 1213.

J.E. Sanchez, J. and Pham, V. (1994). Interpretation of resistance changes during interconnect reliability testing. *Material Research Society Symposium* **338**: 459.

Jones, R. E. and Smith, L. D. (1987). A new wafer level isothermal Joule heated electromigration test for rapid testing of integreated circuit interconnect. *Journal of Applied Physics* **61**: 4670–4678.

Kawasaki, H., Gall, M., Jawarani, D., Hernandez, R. and Capasso, C. (1998). Electromigration failure model: Its application to W plug and Al-filled vias. *Thin Solid Films* **320**: 45–51.

Kirchheim, R. and Kaeber, U. (1991). *Journal of Applied Physics* **70**: 172.

Kisselgof, L., Elliott, L. J., Maziarz, J. J. and Lloyd, J. R. (1991). Electromigration lifetime and step coverage in Al/Cu/Si thin film conductors. *Material Research Society Symposium* **225**: 107.

Knorr, D. B., Rodbell, K. P. and Tracy, D. P. (1991). Texture and microstructure effects on electromigration behavior of Aluminum metallization. *Material Research Society Symposium* **225**: 21.

Koizumi, H. and Hiraoka, K. (1995). The blocking barrier effect on aluminum electromigration due to titanium layers in multilayered interconnects of LSI's. *IEEE Electron Devices Lett.* **16**(7): 298.

Korhonen, M. A., Borgesen, P., Tu, K. N. and Li, C.-Y. (1993). Stress evolution due to electromigration in confined metal lines. *Journal of Applied Physics* **73**: 3790–3799.

Kraft, O., Sanchez, J. J. E. and Arzt, E. (1992). *Material Research Society Symposium* **265**: 119.

Kraft, O., Mockl, U. E. and Arzt, E. (1995). Shape changes of voids in bamboo lines: A new electromigration failure mechanism. *International Quality and Reliability Engineering* **11**: 279.

Kwok, T., Tan, C., Moy, D., Estabil, J. J., Rathore, H. S. and Basavaiah, S. (1990). Electromigration in a two-level Al-CU interconnection with W studs. *Proc. of the VMIC Conference*, 106–112.

Le, H. A., Ting, L., Tso, N. C. and Kim, C. U. (2002). Analysis of the reservoir length and its effect on electromigration lifetime. *Journal of Material Research* **17**: 167–171.

Learn, A. J. and Shephered, W. H. (1971). Reduction of electromigration-induced failure in aluminum metallization through anodization. *Proc. of IEEE International Reliability Physics Symposium*, 129–134.

Lee, S.-H., Bravman, J. C., Doan, J. C., Lee, S. and Flinn, P. A. (2002). Stress induced and electromigration voiding in aluminum interconnects passivated with silicon nitride. *Journal of Applied Physics* **91**(6): 3653–3657.

Lee, T. C., Tibel, D. and Sullivan, T. D. (2000). Comparison of isothermal, constant current and SWEAT wafer level EM testing methods. *IRW Final Report*: 61–69.

Li, L. H., Dasika, V., Heskett, D. and Tang, W. H. (2007). Optical microscopy imaging method for detection of electromigration: Theory and experiment. *Phys. Stat. Solidi A* **204**(5): 1589–1595.

Li, X. X., Zhang, W., Ji, Y., Wang, Z., Cheng, Y. H. and Gao, G. B. (1992a). Increase in electromigration reistance by enhancing backflow effect. *Proc. of IEEE International Reliability Physics Symposium*, 211–216.

Li, Z., Bauer, C. L., Mahajan, S. and Milnes, A. G. (1992). Degradation and subsequent healing by electromigration in Al-1 wt% Si thin films. *Journal of Applied Physics* **72**(5): 1821.

Lloyd, J. R. (1982). *Thin Solid Films* **91**: 175.

Lloyd, J. R. (1991). Electromigration failure. *Journal of Applied Physics* **69**(11): 7601–7604.

Lloyd, J. R. (1991). Mechanical stress and electromigration failure. *Material Research Society Symposium* **225**: 47.

Lloyd, J. R. (1994). Reliability modelling for electromigration failure. *Quality and Reliability Engineering International* **10**: 303.

Lloyd, J. R. (1999). Electromigration and mechanical stress. *Microelectronic Engineering* **49**(1–2): 51–64.

Lloyd, J. R. and Koch, R. H. (1988). Study of electromigration-induced resistance and resistance decay in Al thin-film conductors. *Applied Physics Lett.* **52**: 194–196.

Lloyd, J. R. and Koch, R. H. (1992). Electromigration-induced vacancy behavior in unpassivated thin films. *Journal of Applied Physics* **76**(6): 3231.

Lloyd, J. R. and Smith, P. M. (1983). *J. Vac. Sci. Technol.* **A1**: 455.

Madden, E., Maiz, J., Flinn, P. and Madden, M. (1991). *Applied Physics Lett.* **59**: 129.

Maiz, J. A. and Segura, I. (1988). A resistance change methodology for the study of electromigration in Al-Si interconnects. *Proc. of IEEE IRPS*, 209–215.

Mazumder, M. K., Yamamoto, S., Maeda, H., Komori, J. and Mashiko, Y. (2001). Mechanism of pre-annealing effect on electromigration immunity of Al-Cu line. *Microelectronics Reliability* **41**: 1259–1264.

Meeker, W. Q. (1998). Pitfalls of accelerated testing. *IEEE Trans. on Reliability* **47**(2): 114–118.

Meeker, W. Q. and Escobar, L. A. (1998). *Statistical Methods for Reliability Data*, John Wiley & Sons Inc., USA.

Menon, S. S. and Choudhury, R. K. (1997). A candid comparison of the SWEAT technique and the conventional test procedure for electromigration study in sub-half micron ULSI interconnects. *International Integrated Reliability Workshop Final Report*, 75–79.

Munari, I. D., Vanzi, M., Scorzoni, A. and Fantini, F. (1995). On the ASTM electromigration test structure applied to Al-1%Si/TiN/Ti bamboo metal lines. *International Quality and Reliability Engineering* **11**: 33.

Nemoto, T. and Nggami, T. (1994). Segregation of Cu to grain boundaries by aging treatment and its effect on EM resistance for AlCu/TiN lines. *Proc. of International Reliability Physics Symposium*, 207.

Nikawa, J. (1981). Monte Carlo Calculations based on the generalized electromigration failure model. *Proc. of IEEE International Reliability Physics symposium*, 175–181.

Nikawa, K., Saiki, T., Inoue, S. and Ohtsu, M. (1999). Imaging of current paths and defects in Al and TiSi interconnects on very-large-scale integrated-circuit chips using near-field optical probe stimulation and resulting resistance change. *Applied Physics Lett.* **74**: 1048–1050.

Nishimura, H., Okuda, Y., Ueda, T., Hirata, M. and Yano, K. (1990). Effect of stress in passivation layer on electromigration lifetime for Vias. *IEEE Symposium on VLSI Technology*, 31–32.

O'Connor, P. D. T. (1995). *Practical Reliability Engineering*, John Wiley.

Oates, A. S., Nkansah, F. and Chittipeddi, S. (1992). Electromigration-induced drift failure of via contacts in multilevel metallization. *Journal of Applied Physics* **72**(6): 2227.

Ohfuji, S. I. and Tsukada, M. (1995). Recovery of electric resistance degraded by electromigration. *Journal of Applied Physics* **76**(6): 3769.

Ohring, M. (1971). *Journal of Applied Physics* **42**: 2653.

Ondrusek, J. C., Nishimura, A., Hoang, H. H., Sugiura, T., Blumenthal, R., Kitagawa, H. and McPherson, J. W. (1988). Effective kinetic variations with stress duration for multilayered metallizations. *26th IEEE Proc. of International Reliability Physics Symp.*: 179–184.

Onoda, H., Kageyama, M., Tatara, Y. and Fukuda, Y. (1993). Analysis of electromigmtion-induced failures in multilavered interconnects. *IEEE Trans. on Electron Devices* **40**(9): 1614–1620.

Pasco, R. W. and Schwarz, J. A. (1983). The application of a dynamic technique to the study of electromigration kinetics. *Proc. of IEEE IRPS*, 10–23.

Peirce, Charles Sanders (1986). "On the Theory of Errors of Observation", in Kloesel, Christian J. W., *et alia. Writings of Charles S. Peirce: A Chronological Edition* **3**: 1872–1878; Bloomington, Indiana: Indiana University Press, pp. 140–160.

Pierce, D. J. and Brusius, P. J. (1997). Electromigration: A review. *Microelectronics Reliability* **37**: 1053–1072.

Pinto, M. (1991). The effect of barrier layers on the distribution function of interconnect electromigration failures. *Quality and Reliability Engineering International* **7**: 287.

Poate, J., Tu, K., and J. W. Mayer (1978). *Thin Flims: Interdiffusion and Reaction.* New York, Wiley.

Quintard, V., Dilhaire, S., Phan, T. and Claeys, W. (1999). Temperature measurements of metal lines under current stress by high-resolution laser probing. *IEEE Transactions on Instrumentation and Measurement* **48**(1): 69–74.

Vook, R. W., Cheng, C. Y. and Park, C. W. (1993). *Proc. of SPIE Conference 1805*: 232.

Riege, S. P., Prybyla, J. A. and Hunt, A. W. (1996). Influence of microstructure on electromigration dynamics in submicron Al interconnects: real time imaging. *Applied Physics Lett.* **69**: 2367–2369.

Root, B. J. and Turner, T. (1985). Wafer level electromigration test for production monitoring. *Proc. of 23rd IEEE International Reliability Physics Symposium*, 100–107.

Rosenbaum, R. and Berenbaum, L. (1968). *Applied Physics Lett.* **12**: 201.

Rosenberg, R. (1971). *Journal of Vacuum Science and Technology* **9**: 263.

Rosenberg, R. and Berenbaum, L. (1968). Resistance monitoring and effects of non-adhesion during electromigration in aluminum films. *Applied Physics Lett.* **12**: 201–204.

Rosenberg, R., Mayadas, A. F. and Gupta, D. (1972). *Surface Science* **31**: 566.

Rosenberg, R. and Ohring, M. (1971). Void formation and growth during electromigration in thin films. *Journal of Applied Physics* **42**: 5671.

Ross, C. A., Drewery, J. S., Somekh, R. E. and Evetts, J. E. (1989). The effect of anodization on the electromigration drift velocity in aluminum films. *Journal of Applied Physics* **66**(6): 2349.

Rous, P. J., Yongsunthon, R., Stanishevsky, A. and Williams, E. D. (2004). Real-space imaging of current distributions at submicron scale using magnetic force microscopy: inversion methodology. *Journal of Applied Physics* **95**(5): 2477–2486.

Rousseeuw, P. and Leroy, A. (1996). *Robust Regression and Outlier Detection*. John Wiley & Sons., 3rd edition.

Schafft, H. A. (1987). Thermal Analysis of electromigration test structures. *IEEE Trans. Electron Devices* **34**(3): 664.

Schafft, H. A. (1992). The measurement, use and interpretation of the temperature coefficient of resistance of metallization. *Solid State Electronics* **35**(3): 403–410.

Schrag, B. D. and Xiao, G. (2003). Submicron electrical current density imaging of embedded microstructures. *Applied Physics Lett.* **82**(19): 3272–3274.

Schreiber, H. U. (1985). *Solid State Electronics* **28**: 617.

Schwarz, J. A. and Felton, L. E. (1985). Compensating effects in electromigration kinetics. *Solid State Electronics* **28**: 669–675.

Scorzoni, A., Cardinali, G. C., Baldini, G. L. and Soncini, G. (1990). A resistometric method to characterize electromigration at the wafer level. *Microelectronics Reliability* **30**: 123–132.

Scorzoni, A., Neri, B., Caprile, C. and Fantini, F. (1991). Electromigration in thin film interconnection lines: Models, methods and results. *Materials Science Reports* **7**: 143–200.

Shatzkes, M. and Lloyd, J. R. (1986). A model for conductor failure considering diffusion concurrently with electromigration resulting in a current exponent of 2. *Journal of Applied Physics* **59**: 3890.

Shaw, T. M., Hu, C. K., Lee, K. Y. and Rosenberg, R. (1995). The microstructural stability of Al(Cu) lines during electromigration. *Applied Physics Lett.* **67**(16): 2296.

Shen, Y. L. (1999). Analysis of Joule heating in multilevel interconnects. *Journal of Vacuum Science and Technology B* **17**(5): 2115–2121.

Shingubara, S., Nakasaki, Y., and Kaneko, H. (1991). Electromigration in a single crystalline submicron width aluminum interconnection. *Applied Physics Lett.* **58**(1): 42–44.

Skala, S. and Bothra, S. (1998). Effects of W-plug via arrangement on electromigration lifetime of wide line interconnects. *IEEE International Technol. Conf.*, 116–118.

Snyder, E. (1994). SWORD: Sandia wafer level reliability software for HP4062UX parametric test system. *Version 1.10, Hewlet-Packard.*

Strausser, Y. E., Euzent, B. L., Smith, R. C., Tracy, B. M. and Wu, K. (1987). The effect of metal film topography and lithography on grain size distributions and on electromigration performance. *25th Annual Proceedings: Reliability Physics*, 140–144.

T. Usui, T., Watanabe, T., Ito, S., Hasunuma, M., Kawai, M. and Kaneko, H. (1999). Significant improvement in electromigration of reflow-sputtered Al-0.5wt%Cu/Nb-liner dual damascene interconnects with low-k organic SOG dielectric. *37th IEEE International Reliability Physics Symposium Proceedings*. San Diego, CA, USA, 221–226.

Tan, C. M., Anand, V. A., Zhang, G., Krishnamoorthy, A. and Mhaisalkar, S. (2005). New analysis technique for time to failure data in copper electromigration. *Proc. of the JEDEX Conference*. San Jose.

Tan, C. M., Hou, Y., and Li, W. (2007). Revisit to the finite element modeling of electromigration for narrow interconnects. *Journal of Applied Physics* **102**: 033705.

Tan, C. M., Li, W., Tan, K. T. and Low, F. (2006). Development of highly accelerated electromigration test. *Microelectronics Reliability* **46**: 1638–1642.

Tan, C. M. and Lim, S. Y. (2002). Application of Wigner–Ville distribution in electromigration noise analysis. *IEEE Trans. Devices and Materials Reliability* **2**(2): 30–35.

Tan, C. M. and Roy, A. (2007). Electromigration in ULSI interconnects. *Material Science and Engineering: Review* **342**.

Tan, C. M., Roy, A., Tan, K. T., Ye, D. S. K. and Low, F. (2005). Effect of vacuum break after the barrier layer deposition on the electromigration performance of aluminum based line interconnects. *Microelectronics Reliability* **45**: 1449–1454.

Tao, J., Cheung, N. W. and Hu, C. M. (1995). Electromigration characteristics of TiN barrier layer material. *IEEE Electron Devices Lett.* **16**(6): 230.

Teal, V., Vaidya, S. and Fraser, D. B. (1986). *Thin Solid Films* **136**: 21.

Ting, L. M. and Graas, C. D. (1995). Impact of test structure design on electromigration lifetime measurement. *Proc. of International Reliability Physics Symposium*, 326–332.

Tobias, P. A. and Trindade, D. C. (1986). *Applied Reliability*. New York.

Towner, J. M., Dirks, A. G. and Tien, T. (1986). Electromigration in titanium doped aluminum alloys. *24th IEEE International Reliability Physics Symp.*, 7–11.

Vairagar, A. V., Mhaisalkar, S. G., and Krishnamoorthy, A. (2004). Electromigration behavior of dual-damascence Cu interconnect structures — structure, width and length dependences. *Microelectronics Reliability* **44**(5): 747–754.

Venkatrkrishnan, S. N. and Ficalora, P. J. (1992). *Stress Induced Phenomena in Metallization.* C. Y. Li, P. Totta and P. Ho, 236.

Vook, R. W. (1994). Transmission and reflection electron microscopy of electromigration phenomena. *Materials Chemistry and Physics* **36**: 199–216.

Wada, T., *et al.* (1973). *IEEE Trans. on Reliability* **34**: 2.

Wada, T., Higuchi, H. and Ajiki, T. (1985). The effect of oxygen plasma treatment against electromigration. *Proc. of IEEE Electronic Component Conference*, 32–36.

Wada, T., Sugimoto, M. and Ajiki, T. (1987). The influence of passivation and packaging on Electromigration. *Solid State Electronics* **30**(5): 493–496.

Walton, D. T., Frost, H. J. and Thompson, C. V. (1992). Development of near-bamboo and bamboo microstructures in thin film strips. *Applied Physics Lett.* **61**: 40.

Weiling, G., Zhiguo, L., Tianyi, Z., Yaohai, C., Changhua, C. and Guangdi, S. (1998). The electromigration and reliability of VLSI metallization under temperature gradient conditions. *Proc. of the 5th International Conference on Solid-State and Integrated Circuit Technology*, 226–229.

Zhang, G., Tan, C. M., Tan, K. T., Sim, K. Y. and Zhang, W. Y. (2004). Reliability improvement in Al metalization: A combination of statistical prediction and failure analytical methodology. *Microelectronics Reliability* **44**: 1843–1848.

CHAPTER 4
Experimental Studies of Cu Interconnections

With the aggressive device scaling due to the requirement of higher packaging density for high functionality and performances, Al based metallization was found to be unable to meet the circuit speed requirement around late 90s due to its higher electrical resistivity. Cu was proposed as it has the second lowest resistivity besides Ag, and Ag suffers from electrochemical migration easily (Christou 1994). Efforts to implement Cu based interconnects was accelerated after IBM's and Motorola's announcements of their efforts in 1997 to integrate Cu in CMOS technology.

Interestingly, in the history of EM, Cu was studied before Al (Grone 1961). The early work on Cu was performed on bulk samples which has little relevance to the problems of metallization reliability in ICs. The drift velocity measurement in these samples is attributed to the lattice diffusion dominated EM as the grain size is rather large in these samples. The activation energy for lattice EM in Cu is $\sim$2.3 eV (Grone 1961). These results encouraged the use of Cu to replace Al in IC interconnection. Also, Cu interconnects were expected to be more robust against electromigration failure due to the stronger Cu–Cu bonds as indicated by its higher melting temperature relative to that of Al.

However, unlike Al interconnects which are fabricated through subtractive etching, Cu interconnects and vias are fabricated through a Damascene processing method, which affects its resulting electromigration behavior significantly, and will be discussed in detail in this chapter.

Furthermore, the need to lower capacitance in the interconnect system to further decrease RC delay has motivated the implementation of low-k interlevel dielectric (ILD) materials in place of the traditional SiO_2-based material. Low-k materials are usually more compliant than SiO_2-based materials, but also require new processing methodologies. Such a change in processing methodologies and interconnect structure and materials lead to a change in the electromigration behavior of the Cu interconnect system as compared with its Al counterpart.

In short, the change in interconnect processing, structure and materials have led to the different in the EM behavior between Al and Cu. Although the basic physics of EM remain unchanged, the dominant physical processes that govern the observed EM failures in Cu are different from that of Al. In this Chapter, we will explore the dominant physical processes in Cu EM, with the help from our understanding on Al EM as obtained from Chapter 3, and reference to that in Al will be highlighted.

4.1 Different in Interconnect Processing and its Impact on EM Physics

The fabrication processes are drastically different for Cu and Al interconnects due to their differences in chemical properties. Al metallization is processed by subtractive etching in which the patterned lines are formed by etching the deposited blanket Al film. The Al architecture has thick refractory metal layers (made of TiN, Al_3Ti, or both), at the top and bottom of the lines serving as anti-reflection coating and seed layer for the via-fill process, respectively. Tungsten (W) filled vias are used to connect different levels of Al metallization as shown in Fig. 4.1(a) (Alam, Wei, Gan, Thompson and Troxel 2005).

In contrast, as suitable etchants for Cu thin films are unavailable, Cu interconnects are fabricated by the damascene method, in which a trench is etched into the inter-level dielectric before filling it with Cu by electroplating. Thin refractory metal layers consisting of Ta or TaN are placed at the sides and bottom of the lines to prevent Cu from diffusing into the device layer. After electroplating, the Cu and barrier over-burden are removed through chemical–mechanical-polishing (CMP). Once the

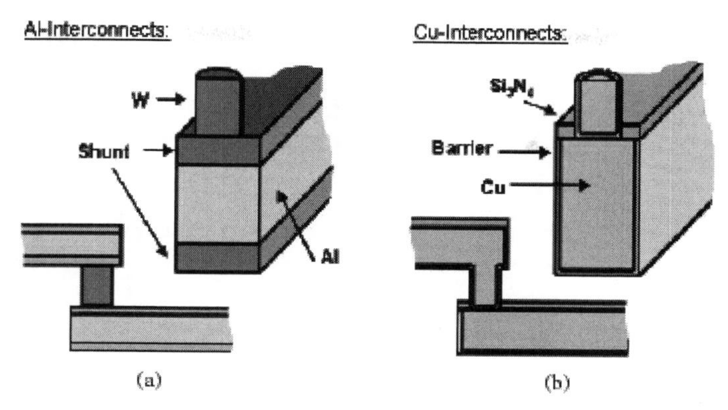

Fig. 4.1. (a) Al interconnects, with W-filled vias and conducting shunt layers at the top and bottom of a line. (b) Dual-damascene Cu interconnects, with Cu-filled vias, thin refractory liners at the side and bottom of a line, and a dielectric capping layer at the top. (Alam, Wei, Gan, Thompson and Troxel, 'Electromigration Reliability Comparison of Cu and Al Interconnects', Proc. of the 6th International Symposium on Quality Electronic Design), Copyright © 2005 IEEE.

overburden is cleared from the field, the exposed Cu surface is capped. The typical Cu-cap is a SiN-based dielectric, which serves as an inter-level diffusion barrier. This process is referred to as dual Damascene when the trench and via are processed together. Dual damascene (DD) Cu-filled vias are used to connect different levels of metallization in Cu interconnects as shown in Fig. 4.1(b) (Alam, Wei, Gan, Thompson and Troxel 2005).

Unlike the case of Al interconnections, Cu/cap layer is identified as the most dominant diffusing path during EM, which was not anticipated initially (Lloyd, Clemens and Snede 1999). This is believed to be due to the defects generated on the Cu top surface during CMP process (Hu and Harper 1998; Wang, Bruynserae and Maex 2004; Gan, Ho, Huang, Leu, Maiz and Scherban 2005). These defects are trapped near Cu/cap interface and provide a faster diffusion path for EM mass transport. Hence, the main focus of EM study on Cu based metallization were on the novel process technology development and on the Cu/cap interface engineering to reduce diffusion during EM.

Although the Cu/cap layer is the most dominant diffusion path, diffusion through the grain boundary is also present. Arnaud *et al.* (Arnaud, Berger and Reimbold 2003) performed electromigration experiments on copper/SiO_2

damascene interconnect lines and they found that electromigration activation energies values in the range of 0.65 to 0.8 eV were measured in wide lines (4 μm) with polycrystalline structures and failure analysis provided SEM pictures showing grain-boundary diffusion. In narrower lines (0.6 μm) where a quasibamboo microstructure was obtained, activation energies increased up to 1.06 eV and SEM pictures were consistent with interface diffusion at the copper top interface. In addition, in both linewidths, longer lifetimes were obtained for electroplated copper versus CVD copper.

This longer lifetime is attributed to two factors, namely the dual damascene process and the absent of the different material in the via. Filippi *et al.* (Filippi, Gribelyuk, Joseph, Kane, Suilivan, Clevenger, Costrini, Gambino, Iggulden, Kiewra, Ning, Ravikumar, Schnabel, Stojakovic, Weber, Gignac, Hu, Rath and Rodbell 2001) studied the effect of dual damascene on EM using the AlCu lines, and they compared the EM with the RIE AlCu (the conventional AlCu process). From their studies, they found several different EM behaviors for dual damascene and RIE AlCu samples as follows.

Firstly, they found that for 0.18 μm wide damascene samples, failures for resistance increase > 0.1% were never found for current density below 2 MA/cm^2, independent of the line thickness. This is in contrast to the RIE case. This is because the average grain size for damascene lines is approximately three times larger than that in RIE lines due to the higher deposition temperature in the Damascene process (<450°C) than in the RIE process (<150°C). Larger grains leads to a reduced number of grain boundaries and hence a fewer vacancy sources and diffusion paths (Filippi, Gribelyuk, Joseph, Kane, Suilivan, Clevenger, Costrini, Gambino, Iggulden, Kiewra, Ning, Ravikumar, Schnabel, Stojakovic, Weber, Gignac, Hu, Rath and Rodbell 2001). The fact that resistance increases are not observed in the 0.18 μm-wide Damascene samples for current densities <2.0 MA/cm^2 leads one to conclude that a back stress effect has occurred. The accumulation of Al atoms at the anode end of the line creates a stress gradient that opposes the electromigration driving force. The critical product of current density and line length $(jL)_c$ is nearly 40000 A/cm, more than 10 times higher than reported previously in RIE AlCu which is only 1020 A/cm (Proost, Maex and Delaey 1998) and 1260 A/cm (Blech 1976) for RIE lines. The huge difference in the above values can be explained in terms of the ability to pack atoms into the line. Dual Damascene samples in their study exhibit little tensile stress (partially due to minimal $TiAl_3$ formation), few grain boundaries and good bonding at the CVD TiN/CVD Al and SiO_2/PVD Al interfaces. There are not many places to move atoms and consequently a great deal of energy is required to add layers of atoms to grain boundaries or other interfaces. As long as this holds, the critical product of 40000 A/cm describes the strength of the Dual Damascene metallization system.

Secondly, the median resistance shift versus time for the two different samples are distinctly different as can be seen in Fig. 4.2. The first

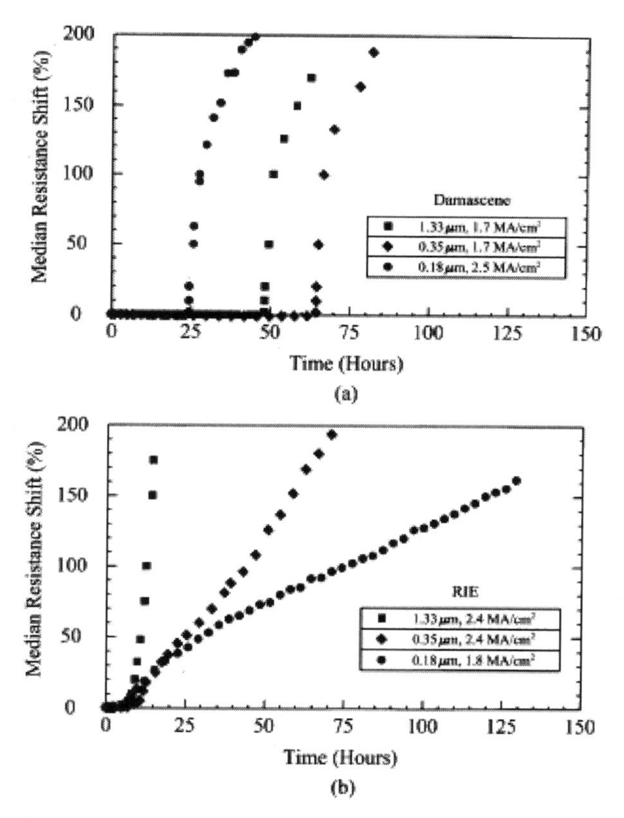

Fig. 4.2. Median resistance shift %. versus time for the (a). Dual Damascene samples and (b). RIE samples. Reprinted from Thin Solid Films, 388, Filippi, Gribelyuk, Joseph, Kane, Suilivan, Clevenger, Costrini, Gambino, Iggulden, Kiewra, Ning, Ravikumar, Schnabel, Stojakovic, Weber, Gignac, Hu, Rath and Rodbell, 'Electromigration in AlCu lines: comparison of dual damascene and metal reactive ion etching', pp. 303–314, Copyright © 2001, with permission from Elsevier.

observation to make is that the resistance incubation period, defined as the time before which the resistance starts to increase, is much longer for Damascene than for RIE samples. The second is that the Damascene samples exhibit an abrupt increase in resistance while the RIE samples (0.18 and 0.35 μm wide lines) gradually increase in resistance (Filippi, Gribelyuk, Joseph, Kane, Suilivan, Clevenger, Costrini, Gambino, Iggulden, Kiewra, Ning, Ravikumar, Schnabel, Stojakovic, Weber, Gignac, Hu, Rath and Rodbell 2001). The longer incubation time is believed to be the same as given previously due to the larger grain size and hence reduced vacancies sources and diffusion path.

The resistance versus time behavior illustrated in Fig. 4.2 can be explained in terms of the Ti/TiN layers. For Dual Damascene samples, a void in the M1 line forces the current to flow through the redundant Ti and TiN layers. These layers are quite thin, and have high resistivities ($\sim$ 55 $\mu\Omega$-cm for PVD Ti and $\sim$ 300–400 $\mu\Omega$-cm for CVD TiN) as compared with that of AlCu (3$\mu\Omega$-cm). In addition, since Ti is deposited by physical vapor deposition, the Ti thickness is not likely to be very uniform along the sidewalls and the bottom of the trench. Therefore, the Ti layer may be thinner than expected in certain regions. Once a void is formed, Joule heating of Ti and TiN reaches a critical level that causes adjacent Al regions to melt and rapidly solidify in certain areas. The combination of electromigration during the time the Al melts and subsequent solidification of melted Al may explain why islands of Al material lie between voided regions, as shown in Fig. 4.3(a). This process occurs very quickly, resulting in a rapid resistance increase for the Damascene samples.

For metal RIE, on the other hand, the Ti/TiN layers are thicker than in the Damascene case, and the resistivity of the PVD TiN layer ($\sim$ 200 $\mu\Omega$-cm) is significantly lower than that of CVD TiN. Also, the Ti thickness is expected to be more uniform along the length of the RIE line because all metal layers are put down in a blanket deposition. Once a void is formed, current flows predominately through the thick Ti/TiN layer and Joule heating is less likely to reach a critical level. Therefore, Al does not melt and the resistance increase is more gradual for 0.18- and 0.35-μm-wide RIE samples. The fact that the 1.33-μm-wide RIE structures show a rapid increase in resistance suggests that the electromigration test currents used for these wide lines generate sufficient Joule heating so as to cause Al to melt.

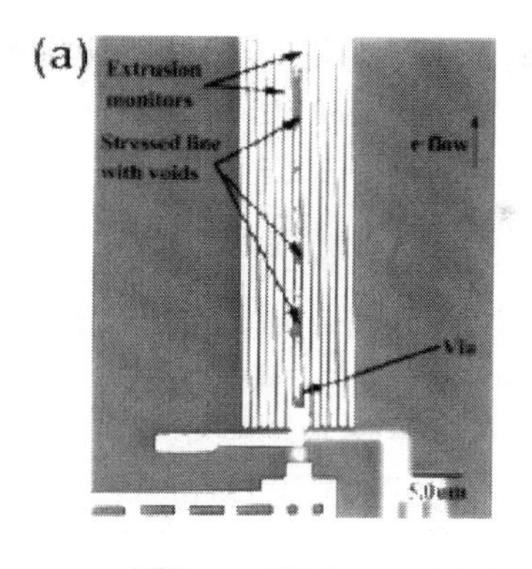

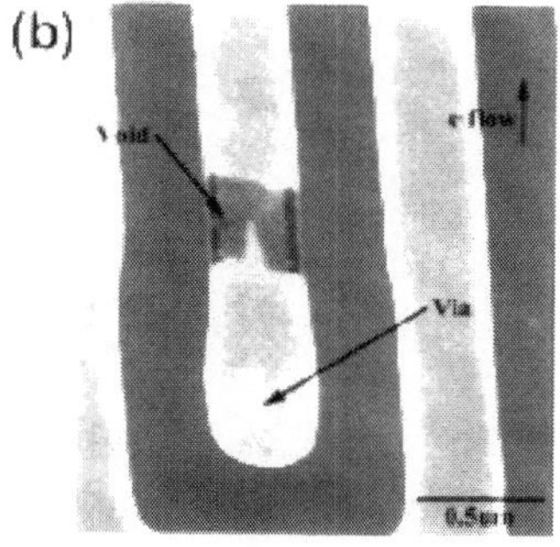

Fig. 4.3. Top-down micrographs showing electromigration-induced voids in 0.35-μm-wide lines: (a). Dual Damascene and (b) metal RIE. The Damascene sample is tested at 1.7 MA/cm^2, while the RIE sample is tested at 2.4 MA/cm^2. Reprinted from Thin Solid Films, 388, Filippi Gribelyuk, Joseph, Kane, Suilivan, Clevenger, Costrini, Gambino, Iggulden, Kiewra, Ning, Ravikumar, Schnabel, Stojakovic, Weber, Gignac, Hu, Rath and Rodbell, 'Electromigration in AlCu lines: comparison of dual damascene and metal reactive ion etching', pp. 303–314, Copyright © 2001, with permission from Elsevier.

Thirdly, resistance saturation effects are observed for Dual damascene samples, but this is not observed for the RIE samples. This is because as Aluminum melts due to excessive Joule heating and rapidly solidifies in certain areas, resulting in islands of solidified Al between voided regions as mentioned earlier, the melting of Al continues until a region is encountered in which the Joule heating is significantly reduced, such as a location along the line where the Ti liner is thicker. Additional voiding is prevented downstream from the last voided region because the jL product becomes comparable to the critical product of 40000 A/cm. Here, L is the distance from the last voided region to the anode end of the structure. For example, if $L = 160\,\mu$m (for a voided region $40\,\mu$m from the cathode end) and $j = 2.5\,\mathrm{MA/cm^2}$, the jL product is 40000 A/cm.

Fourthly, the current density dependence for the dual damascene is 1.95 and that for RIE samples is 1.03. This suggests that different mechanisms are responsible for void formation in the two cases. In the past, a current density exponent of $n = 2$ has been attributed to electromigration lifetimes limited by void nucleation (Shatzkes and Lloyd 1986) while a value of $n = 1$ has been associated with lifetimes controlled by void growth (Kirchheim and Kaeber 1991). More recently, however, n has been shown to vary from 2 to 1 as the lifetime progresses from an incubation period to a steady-state period in the case of two-level, multiple-grained AlCu structures (Hu 1995). The migration of Cu controls the incubation period and results in a value of $n = 2$ for failure times close to the incubation time. Following this incubation period, a steady-state regime is obtained in which the drift velocity becomes constant and the electromigration-induced void grows at a steady rate. A value of $n = 1$ is obtained for failure times much greater than the incubation time (Hu 1995). For the case of two-level, bamboo or near bamboo grained AlCu structures, a value of $n = 1$ that is obtained for the 0.35-μm-wide RIE lines is consistent with a void growth failure mechanism. The fact that the RIE lines exhibit a short resistance incubation period followed by a gradual increase in resistance is also in agreement with lifetimes controlled by void growth, and further suggests that Cu migration is less significant in these samples. On the other hand, a value of $n = 2$ that is obtained for the 0.35-μm-wide Damascene lines would seem to indicate that the lifetime of these samples is limited by Cu migration.

Furthermore, the absence of a fast diffusion path along extended segments of the Damascene lines produces a long resistance incubation period by reducing the Cu migration rate. Once Cu has migrated away, the time until the resistance starts to increase is relatively short since melting of Al begins as soon as a void becomes large enough to cause significant current to flow through the redundant Ti/TiN layer.

A larger value of n for Dual Damascene samples means that the relative lifetimes of these lines compared with metal RIE will improve as the current density decreases. Therefore, the ratio of the Damascene t_{50} to RIE t_{50} may be much higher at chip operating conditions at low current density.

Fifthly, TEM cross-section analysis shows that the formation of $TiAl_3$ intermetallic is less frequent in the dual damascene sample while RIE sample shows significant $TiAl_3$ formation near the bottom of the line. Recent TEM analyses indicates that CVD TiN can completely suppress the formation of $TiAl_3$ in Dual Damascene structures if the PVD AlCu deposition temperature is less than 430°C (Clevenger, Costrini, Dubuzinsky, Filippi, Gambino, Hoinkis, Gignac, Hurd, Iggulden, Lin, Longo, Lu, Ning, Nuetzel, Ploessl, Rodbell, Ronay, Schnabel, Tobben, Weber, Chen, Chiang, Guo, Mosley, Voss and Yang 1998; Iggulden, Clevenger, Costrini, Dobuzinsky, Filippi, Gambino, Gignac, Lin, Longo, Lu, Ning, Nuetzel, Rodbell, Ronay, Schnabel, Stephens, Tobben and Weber 1998). It should be noted that the formation of $TiAl_3$ causes a volume decrease in the metal line, which leads to a tensile stress and possible enhancement of void formation during electromigration testing. The presence of $TiAl_3$ has been recently shown to promote electromigration in metal RIE lines by providing a fast diffusion path for Al migration along the interface between Al and $TiAl_3$ (Hosaka, Kouno and Hayakawa 1998; Kouno, Hosaka, Niwa and Yamada 1998). This diffusion path, however, depends on a continuous $TiAl_3$ zone along the entire length or at least along extended segments of the conductor line. Therefore, although $TiAl_3$ is more prevalent in metal RIE lines than in Dual Damascene lines, the fact that it is not continuous in either case probably means that the electromigration lifetimes are not significantly affected by the frequency of intermetallic formation.

Sixthly, it is interesting to note that Dual Damascene samples with a random Al grain orientation exhibit a longer electromigration lifetime than that of RIE samples with <111> dominated texture. In RIE lines, the metal is deposited at a low temperature, patterned and then annealed. The Al grains can orient themselves during this subsequent anneal and the surface normal texture tends to improve. In contrast for Damascene lines, the metal is deposited at a high temperature into trenches where there is substantial growth on both the sidewalls and the bottom of the trench. The surface normal texture is poor, but the texture along the sidewalls is strong. Therefore, the film is textured in a manner with <111> grains present along all of the internal interfaces. This, on the one hand, gives rise to a weak surface normal texture, but, on the other hand, leads to a more perfect line by yielding fewer diffusion paths in the form of large angle grain boundaries. Combined with large grain size and smooth interfaces, the Damascene samples yield long lifetimes.

Seventhly, the texture of the dual damascene lines decreases as the line width is reduced, while the texture of the RIE lines is always stronger than that of the damascene lines and is essentially independent of line width. The fact that the RIE grain size is relatively independent of line width is to be expected since the RIE lines are derived from a blanket deposition. The grains in the 1.33-μm-wide Damascene lines are larger than those in the 0.35-μm-wide Damascene lines because there is more area for grain growth during the 'hot' AlCu deposition.

From the above discussion, one can therefore see that the change in the interconnect process from RIE to dual damascene alone can change the EM behavior. With a further change in the interconnect materials, the dominant diffusion path has changed as have been seen. With these background information, we will examine the manifestation of the various EM governing factors that were discussed for Al interconnect in Chapter 3 in this current chapter.

Also, although short circuit failure mode was identified as due to the hillocks formation in EM test on Cu based metallization, this failure mode is found to be not prominent. The reasons are as follows. Firstly, because of the presence of hard material as cap/passivation layer on the top and metallic barrier layer covers the bottom and side walls of the metallization, the hillocks formation is suppressed (Hu and Harper 1998). Secondly, probably due to the presence of defects, accumulation of vacancies to form voids is easier. As the line width is in the micron regime or smaller, the size of the void that is capable to increase the resistance at least up to the failure criterion in EM tests is small. As a result, failure due to void formation was the most dominant failure mode in Cu based metallization, and only this failure mode will be discussed in this book.

4.2 Process-induced Failure Physics

4.2.1 *Interface between Cu and surrounding materials*

This is unique to Cu as the faster diffusion path for Cu EM mass transport is at the interface. One of the reasons for which this interface becomes vulnerable to EM is due to the defect generation in the mandatory CMP process step which is required to remove the over burden Cu from the top of the wafer during pattering. These defects remain trapped at the interface along the length of the interconnect (Besser, Marathe, Zhao, Herrick, Capasoo and Kawasaki 2000; Gan, Ho, Huang, Leu, Maiz and Scherban 2005). Also, the adhesion between Cu and non metallic cap layer is not good, and Cu oxide is not a self-protective layer as Al oxide is. Thus the EM reliability for copper in comparison with aluminum is not as much as anticipated initially.

Since the conventional damascene technology does not protect copper top surface, tremendous effort is given to improve the Cu/cap layer by interface engineering against EM resistance. The interface engineering in copper based metallization can be divided into two categories, namely the interface modification or surface engineering and the use of different materials as cap layer.

4.2.1.1 *Surface engineering*

The plasma treatment is normally performed to remove the native oxide from the top surface of Cu metallization. The native oxide formation occurs during CMP and/or by reacting with the environment. This step is performed just before the cap layer deposition step and the wafers are shifted for cap layer deposition without breaking the vacuum.

Vairagar *et al.* (Vairagar, Gan, Shao, Mhaisalkar, Li, Tu, Chen, Zschech, Engelmann and Zhang 2006) studied surface treatment before SiN cap layer deposition by remote H_2 plasma treatment, NH_3 in-situ flush and Silane in-situ flush for Cu DD and NIST test structures. Their results are shown in Table 4.1.

It is clearly observed from Table 4.1 that the electromigration failure times are improved due to surface treatments. NH_3 treated samples showed similar failure times as control samples with no surface treatment. Significant improvement was observed for hydrogen plasma and silane treatment structures.

Contrary to the concerns over electromigration behavior of lower layer structures (Kaanta, Bombardier, Cote, Hill, Kerszykowski, Landis, Poindexter, Pollard, Ross, Ryan, Wolff and Cronin 1991), it was observed that the surface treatments improved electromigration performance of lower-layer structures similar to upper-layer structures. Hydrogen plasma and silane treatment showed significantly reduced early failures.

The slight improvement observed from the NH_3 treatment can be attributed to the reduction of copper oxide as well as the formation of Cu-N

Table 4.1. Surface treatments performed after M1 and M2 CMP, and the corresponding MTF and sigma (σ) data (Vairagar, Gan, Shao, Mhaisalkar, Li, Tu, Chen, Zschech, Engelmann and Zhang 2006).

		M-1		M-2	
Surface treatment	Description	MTF	σ	MTF	σ
Control samples	No surface treatment	80–90 hours	0.53	~125 hours	0.86
NH_3 treatment	In situ NH_3 flush before SIN cap deposition	80–90 hours	0.39	~150 hours	1.04
H_2 treatment	Remote H_2 plasma treatment	~180 hours	0.34	~190 hours	0.82
Silane treatment	In situ SiH_4 flush before SIN cap deposition	~200 hours	0.17	~210 hours	0.44

leading to a better Cu/SiN interface (Leong 2001; Parikh, Educato, Wang, Zheng, Wijekoon, Chen, Rana, Cheung and Dixit 2001), and this was further supported by the XPS analysis performed by Vairagar *et al.* (Vairagar, Gan, Shao, Mhaisalkar, Li, Tu, Chen, Zschech, Engelmann and Zhang 2006).

The improvement due to hydrogen plasma treatment is known to form hydride on the Cu surface with no change in microstructure (Chan, Chuang and Chuang 1998; Beyer, Baklanov, Conard and Maex 2000) and to clean the Cu surface (Parikh, Educato, Wang, Zheng, Wijekoon, Chen, Rana, Cheung and Dixit 2001) providing better adhesion between Cu and subsequently deposited silicon nitride dielectric-cap layer. Improvement in electromigration performance due to hydrogen has also been attributed to interfacial segregation of hydrogen and the reduction of defects by athermal annealing (Rodbell and Ficalora 1985; Rodbell, Ficalora and Koch 1987). Vairagar *et al.* (Vairagar, Gan, Shao, Mhaisalkar, Li, Tu, Chen, Zschech, Engelmann and Zhang 2006) further confirmed that hydrogen plasma treatment facilitates the formation of both Cu–N and Cu–Si bonds at the interface which directly improve the electromigration performance observed in Table 4.1 as compared with control sample with no treatment and NH_3 treated samples. Cross-section TEM before EM tests also indicated the copper silicide formation at the Cu/cap dielectric interface for both H_2 and silane treated specimens.

The difference in the amount of improvement between hydrogen and NH_3 plasma treatment was proposed by Vairagar *et al.* (Vairagar, Gan, Shao, Mhaisalkar, Li, Tu, Chen, Zschech, Engelmann and Zhang 2006) as the passivation effect on the kink sites on the Cu surface. For surface diffusion, a continuous flux of adatoms on the surface is necessary, and the kink sites are believed to be the source of the adatom flux because at the kink sites the transition to an adatom involves the breaking of the smallest number of bonds. Therefore, the key to stop or retard the surface interface electromigration in Cu interconnects is to reduce the number of kink sites or to increase the energy needed to dissociate Cu atoms from the kink sites. The formation of Cu–N and Cu–Si bonds can both passivate the kink sites. However, it is expected that the Cu–Si bond is much stronger than the Cu–N bond. From the thermodynamic point of view, Cu–Si has a bond formation enthalpy around −220 kJ/mol (2004) whereas that for the Cu–N bond is around −30 kJ/mol (Moreno-Armenta, Martinez-Ruiz and Takeuchi 2004). Therefore, it is the formation of copper silicide that will enhance the EM lifetime more, which is the case for H_2 and SiH_4 treated specimens. However, the present common industry practice on plasma surface treatment remain on the use of NH_3 instead of silane or hydrogen, probably due to the fact that high hydrogen content in plasma can degrade the transistors reliability as will be discussed later.

Similarly He-based plasma is found to be better (more than 40% improvement) than no plasma treatment (Wang, Su, Yang, Chen, Doong, Shih, Lee, Chiu, Peng and Yue 2002).

While surface plasma treatment is effective to improve EM lifetime, it is found to be detrimental to the flash memory devices due to the hydrogen effects on the core cell as measured by charge loss, charge gain and data retention bake package reliability tests (Brennan, Pangrle, Evans, You, Ngo, Qi, Baker, Baek, Romero, Stockwell and Tracy 2007). On the other hand, as have been seen, plasma treatment is necessary to improve Cu EM lifetime, and Brennan *et al.* (Brennan, Pangrle, Evans, You, Ngo, Qi, Baker, Baek, Romero, Stockwell and Tracy 2007) also showed that Cu film delamination and peeling can occur due to interface contamination on the Cu surface without a plasma treatment. Therefore, Brennan *et al.* (Brennan, Pangrle, Evans, You, Ngo, Qi, Baker, Baek, Romero, Stockwell and Tracy 2007) suggested that a low hydrogen content in plasma must be used so as to ensure good interconnect integrity and flash memory reliability as they have also shown experimentally.

Besides the hydrogen content, the intensity of the plasma is also important. Glasow *et al.* (Glasow, Fischer, Bunel, Friese, Hausmann, Heitzsch, Hommel, Kriz, Penka, Raffin, Robin, Sperlich, Ungar and Zitzelsberger 2003) found that the plasma can induce crystal defects in the copper, enabling vacancy generation and enhances stress voiding process, another interconnect reliability failure. They found that if the plasma energy is low, only the copper surface is affected and the bulk remains intact. Since only the modification of surface is needed, low plasma intensity is sufficient for the enhancement of EM lifetime without sacrifice the stress voiding reliability. The effect of plasma induced damage due to this plasma treatment to improve Cu EM lifetime is also studied by Wang *et al.* (Wang, Su, Yang, Chen, Doong, Shih, Lee, Chiu, Peng and Yue, 2002).

Similarly, an aggressive NH_3 plasma treatment is not advisable as this type of treatment is found to be detrimental in terms of plasma-induced-damage, gate oxide integrity, BEOL defectivity etc. (Ang, Lu, Yap, Goh, Goh, Lim, Chua, Ko, Tan, Toh and Hsia 2004), though it is beneficial for EM life-time improvement. This is especially true for low-k dielectric integration where plasma treatment can be the cause of significant damage to the surrounding dielectric and hence special care is needed to resolve this issue (Ernur, Iacopi, Carbonell, Struyf and Maex 2003).

4.2.1.2 *Alternative cap-layer materials*

An alternative to improve the Cu/cap layer diffusion is by the use of different cap layer materials. From the above discussion, it is emerging that a cap layer material which can form copper compound at the interface should offer a better interface against EM. As a result, apart from the plasma treatment technique, various cap layer materials have been studied in order to improve the adhesion between copper and cap layer. In this category, two types of materials are considered, namely metallic and dielectric.

Dielectric cap-layer

H_2 and NH_3 plasma treatment are studied for SiC_x as cap-layer (Usuni, Oki, Miyajima, Tabuchi, Watanabe, Hasegawa and Shibata 2004; Usui, Miyajima, Masuda, Tabuchi, Watanbe, Hasegawa and Shibata 2006). The

EM activation energy for H_2 and NH_3 plasma treated surfaces is found to be 0.67 eV and 0.72 eV respectively with one order of magnitude difference in the EM drift velocity at identical test condition, concluding that NH_3 based plasma is better, which is opposite to the case of SiN cap-layer as discussed earlier. On the other hand, Lin *et al.* (Lin, Lin, Chen, Tsai, Yeh, Liu, Hsu, Wang, Sheng, Chang, Su and Chang 2004) reported that hydrogen-rich pre-treatment can provide a much significant improvement for SiCN cap-layer, and that SiCN cap-layer performed better than SiN. Also, the increase in the Cu line resistance will be smaller for SiCN cap-layer with hydrogen pre-treatment. Hence, it seems that the effectiveness of hydrogen and NH_3 plasma treatment to improve EM lifetime depends on the material of the cap-layer. The exact dependence remains unknown.

Cheng *et al.* (Cheng and Wang 2008) further studied three different types of dielectric materials for passivation, namely the SiCO, SiC and SiCN. Again, they found that SiCN capping has the highest activation energy of EM, and it is attributed to the best adhesion as obtained from the 4-point bending test. The cap dielectric layers were deposited after a N-containing pre-treatment on Cu surface to remove the Cu oxide.

Cu silicidation was shown to be a main contributing factor for the enhancement of EM lifetime as discussed earlier. Chhun *et al.* (Chhun, Gosset, Michelon, Girault, Vitiello, Hopstaken, Courtas, Debauche, Bancken, Gaillard, Bryce, Juhel, Pinzelli, Guillan, Gras, Schravendijk, Dupuy and Torres 2006) showed that Cu silicidation is dependent both on Cu crystallographic orientation and on the availability of Si atoms at Cu surface. Dense (111) and (100) are less sensitive to silicidation, thus requiring higher amount of Si atoms at Cu surface to be fully silicided, leading to Cu resistivity degradation. As an alternative, an adequate surface preparation was shown to increase the rate of silicided Cu grains by introducing low silane content in the hydrogen plasma for the SiCN cap-layer. With this process, Cu resistivity is preserved while the Cu/cap interface is enhanced which improve the EM lifetime up to 10 times.

However, while SiCN cap-layer performed better than SiN_x, the EM lifetime distribution for SiCN passivation was found to be broader than that from SiN, and that the early failure time can be smaller for the case of SiCN passivation (Lin, Lin, Chen, Tsai, Yeh, Liu, Hsu, Wang, Sheng, Chang, Su and Chang 2004).

Metallic cap-layer

Another type of cap-layer is metallic. Lane *et al.* (Lane, Linger and Lloyd 2003) reported a comparison study using different cap layers such as SiN, BloK^{TM}, SiC and an electrolessly deposited film of CoWP, and found that CoWP is the best cap layer in comparison to others as can be seen in Fig. 4.4. The EM activation energy in CoWP capped samples is reported to be 2.4 eV, which is almost close to the bulk activation energy for copper.

Hu *et al.* (Hu, Gignac, Liniger, Herbst, Rath, Chen, Kaldor, Simon and Tseng 2003; Hu, Gignac, Rosenberg, Herbst, Smith, Rubino, Canaperi, Chen, Seo and Restanio 2004) also investigated the EM performance of copper damascene lines with bamboo-like grain structure, and capped with either Ta/TaN, SiN_x, $SiC_xN_yH_z$ layers, or without any cap. They found that a thin Ta/TaN cap on top of the Cu line surface significantly improves EM lifetime over lines without a cap and with lines capped with SiN_x, or $SiC_xN_yH_z$. The activation energy for EM is shown to increase from 0.87 eV for lines without cap to 1.0–1.1 eV for samples with SiN_x or $SiC_xN_yH_z$ caps, and to 1.4 eV for Ta/TaN capped samples. The same with CoWP capped samples is reported to be 2.4 eV.

Unlike the case of dielectric cap, the use of metallic cap-layer has also an added advantage that the line resistance is not expected to increase significantly. Also, the capping process is relatively easy to be introduced in a damascene flow by just adding one electroless plating step.

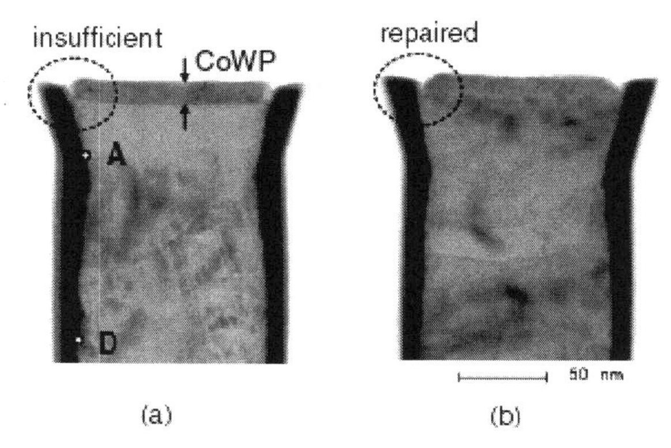

Fig. 4.4. TEM image of Cu line cross-section and Co detection (a) before EM test and (b) after EM test at 380°C. Reproduction from APEX/JJAP, vol (47), 6, pp. 4475–4479, 2008, Kakuhara, Kawahara, Ueno and Oda.

The mechanism for the EM lifetime enhancement with CoWP cap-layer is related to Co diffusion along the Cu/barrier metal interface, instead of diffusion in the bulk Cu under the CoWP. Kakuhara *et al.* (Kakuhara, Kawahara, Ueno and Oda 2008) showed a TEM observation in Fig. 4.4 that CoWP coverage is found to be insufficient at the Cu/barrier metal interface before EM test. However, CoWP coverage is repaired at the Cu/barrier metal interface after the EM test as shown in Fig. 4.4(b). In addition, the contrast of the TEM image at the CoWP/Cu interface becomes obscure after the EM test, which indicates Co alloys with Cu. The Co alloys with Cu at the top of the Cu interconnects should cause the resistance increase during the EM test.

With the above-mentioned observations, Kakuhara *et al.* (Kakuhara, Kawahara, Ueno and Oda 2008) proposed the possible mechanisms for the EM enhancement. The mechanisms are the Co diffusion into the Cu/barrier metal interface, Co allying with Cu at the CoWP capped Cu surface and CoWP coverage repair as shown in Fig. 4.5.

Among the three mechanisms, they (Kakuhara, Kawahara, Ueno and Oda 2008) found that Co diffusion into the Cu/barrier metal interface should not be effective in the EM lifetime enhancement. Because the dominant Cu diffusion path is considered to be the interface between Cu and the capping layer that is indicated by the EM void generation at the poor CoWP coverage site, therefore, Co alloying with Cu and coverage repair should be considered as the possible mechanisms for the observed EM lifetime enhancement. Hence, thermal treatment after the CoWP deposition at the high temperature is an effective way to improve the CoWP capped Cu interconnects by the Co alloying and the CoWP coverage repair effects. But it is not considered as a practical method because the interconnect

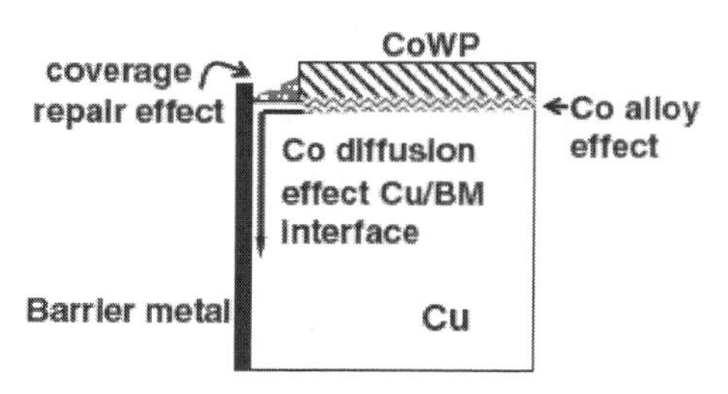

Fig. 4.5. Potential Mechanisms for EM enhancement. Reproduction from APEX/JJAP, vol (47), 6, pp. 4475–4479, 2008, Kakuhara, Kawahara, Ueno and Oda.

resistance increases and the dispersion of EM lifetime becomes larger. Since the poor coverage was observed at the EM failure site, the coverage repair should eliminate the void nucleation site and it leads to the EM lifetime enhancement. Therefore, initial uniform CoWP coverage on the Cu surface is important for the EM lifetime enhancement of the CoWP capped Cu interconnects in practical applications.

Further experiments by Kakuhara *et al.* (Kakuhara, Kawahara, Ueno and Oda 2008) also found that the coverage is repaired better without the NH_3 plasma treatment than with the NH_3 plasma treatment at the poor coverage site near the Cu/barrier metal interface. It is speculated that the repair of the CoWP coverage during the EM test at the high temperature is suppressed by the NH_3 plasma treatment, such as the Cu nitride at the insufficient coverage site.

The deposition of CoWP is made using electroless method. In order for the metal ions to be reduced at the metal surface, the reducing agent must have a lower (more negative) redox potential than the metal surface that is being plated. The early studies on CoWP deposition used hypophosphite as the reducing agent (O'Sullivan, Schrott, Paunovic, Sambucetti, Marino, Bailey, Kaja and Semkow (1998); Kohn, Eizenberg, Shacham-Diamond and Sverdlov 2001). However, Cu is a poor catalyst for the oxidation of hypophosphite, so it is not possible to reduce Co^{+2} on the Cu surface (O'Sullivan, Schrott, Paunovic, Sambucetti, Marino, Bailey, Kaja and Semkow 1998; Nakano, Itabashi and Akahoshi 2005). Cobalt deposition on Cu can occur by using Pd seeding, because Pd has a more positive potential than Cu ($Pd/Pd^{+2} = 0.95\,V$ versus $Cu/Cu^{+2} = 0.34\,V$) (Paunovic, Schlesinger and Weil 2000). However, in so doing, new contaminants are introduced into the VLSI fabrication lines.

Since the Cu surfaces after CMP are oxidized and both the Cu and dielectric surfaces are contaminated, leakage current through the Inter-level Dielectric (ILD) will increase after the electroless plating step due to the Pd ions adsorbed onto the ILD surface (Ishigami, Kurokawa, Kakuhara, Withers, Jacobs, Kolics, Ivanov, Sekine and Ueno 2004). Also, during the activation process, Cu atoms are partially displaced with Pd, thereby creating additional damage to the copper line (Petrov, Valverde, Chen, Xu, Paneccasio, Stritch and Witt 2005). Therefore, the Pd activation process may provoke the following defects: increase in additional cladding layer roughness (Gandikota, Padhi, Ramanathan, McGuirk, Naik, Parikh, Musaka, Yahalom and Dirish Dixit 2002); intermixing Cu with Pd; unwanted CoWP deposition onto ILD, which could reduce overall selectivity of the process (Gandikota, Padhi, Ramanathan, McGuirk, Naik, Parikh, Musaka, Yahalom and Dirish Dixit 2002; Fang, Weidman, Shanmugasundram, Naik, and Kapoor 2004). The defects specified above may result in line resistance increase and large leakage current distribution compared with the CMP baseline, especially for the interconnect features with critical sizes below 0.1 μm (Fang, Weidman, Shanmugasundram, Naik and Kapoor 2004; Wang, Owatari, Takagi, Fukunaga and Tsujimura 2004).

In order to eliminate challenges associated with the CoWP, Ishigami *et al.* (Ishigami, Kurokawa, Kakuhara, Withers, Jacobs, Kolics, Ivanov, Sekine and Ueno 2004) developed a CoWP process without using Pd activation and alkaline-metal free plating solution, and the results are promising. Other processes of selective Co alloy cap on Dual Damascene Cu interconnect were also proposed (Itabashi, Nakano and Akahoshi 2002; Fang, Weidman, Shanmugasundram, Naik and Kapoor 2004; Nakano, Itabashi and Akahoshi 2005). Also, the use of dimethylamine borane (DMAB) as the reducing agent in electroless Co–W alloy plating solution was suggested (Itabashi, Nakano and Akahoshi 2002; Nakano, Itabashi and Akahoshi 2005), and CoWB was obtained. Diffusion barriers of 160 nm CoWB films were compared with that of 360 nm CoWP films. Thin films were annealed at 400°C for 30 min. and Cu content in the films was measured. Approximately 6.2 at. % Cu was detected by Auger Electron Spectrometry (AES) on CoWP film surface while no Cu was detected on the surface of CoWB films, showing the superior barrier properties of CoWB versus CoWP. The process characteristics and material properties of electroless deposited CoWP and CoWB can be found in (Petrov, Valverde, Chen, Xu, Paneccasio, Stritch and Witt 2005).

Petrov *et al.* (Petrov, Valverde, Chen, Xu, Paneccasio, Stritch and Witt 2005) found that the amorphous structure of the formed boron-based Co alloys possesses superior thermal stability and better barrier properties than that of phosphorous-based Co alloys, which are nanocrystalline. They demonstrated that electroless CoWB films can be selectively deposited onto Cu interconnects reproducibly with high selectivity. Self-initiated process capping layers did not change Cu line resistance as opposed to the CoWP caps. Also, they demonstrated the capability of tight process control of the CoWB process. Thus, the use of CoWB instead of CoWP as cap-layer might be one of the future trends in Cu line.

4.2.2 *Microstructure*

While interface diffusion is considered as the most dominant diffusion path for Cu EM, grain boundary (GB) diffusion is also present in Cu EM. It is now accepted that interfaces are principal EM paths at the ~ 100°C and grain boundaries become preferential EM paths at high temperature ~ 300°C (Chen, Wu, Liao, Chen and Tu 2008). Also, as the interface between cap-layer and Cu is improved through surface engineering and alternative cap layer materials, GB diffusion can be the more active EM paths in the future.

Furthermore, it is found that the bamboo-like Cu grain structure observed above 65 nm node technology interconnects does not always exist in interconnects below 65 nm node and a polycrystalline grain structure or mixture of bamboo-polycrystalline grain structure have often been reported (Hinode, Hanaoka, Tahed and Konda 2001; Steinhogl, Schindler, Steinlesberger, Traving and Engelhardt 2005; Hu, Gignac, Neal, Liniger, Yu, Flaitz and Stamper 2007; Zhang, Brongersma, Li, Li, Richard and Maex 2007). Also, sections of multiple small grains are observed through the thickness, especially in the lower half of the line (Hu, Gignac, Neal, Liniger, Yu, Flaitz and Stamper 2007). Arnaud *et al.* (Arnaud, Tartavel, Berger, Mariolle, Gobil and Touet 1999) also found that grain boundaries are mainly aligned perpendicular to the bottom of the damascene trench for wide damascene lines with width above 65 nm, and grain boundaries rotated perpendicular to the sidewalls of the damascene trench when line width reduces below 65 nm. Therefore, the effect of microstructures of Cu line on EM is increasingly important as we advanced in the technology node, and interface strengthening will not be sufficient to improve Cu EM.

The effect of microstructures in Cu line on its EM has been reported. Gladkikh *et al.* (Gladkikh, Karpovski, Palevski and Kaganovskii 1998) studied the effect by having the Cu line surrounded by Ta with thickness of 10 nm, and they performed EM experiments with different microstructures in Cu lines. They found that the EM damages observed are very different for different microstructures. Figure 4.6 shows the EM damages for line with average grain sizes of 500, 150 and 120 nm respectively. One can

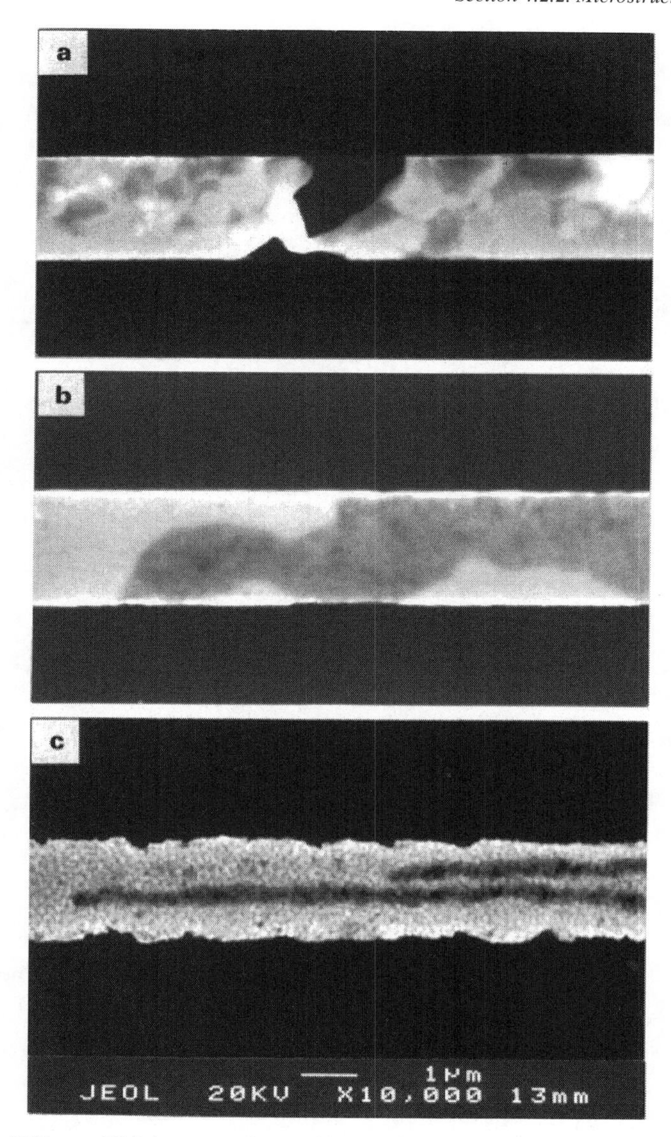

Fig. 4.6. Different EM damages observed in thin film Cu with different microstructures. The thickness of the line is 160 nm and width of 2 μm and 1 mm length. The Cu lines are covered with Ta layer of 10 nm thick. The average grain size of the Cu lines for the three micrographs are (a) 500 nm; (b) 150 nm and (c) 120 nm. Reprinted with permission from Gladkikh, Karpovski, Palevski and Kaganovskii. Copyright © 1998, Institute of Physics and IOP Publishing.

see that for large grain size as in Fig 4.6(a), void evolved across the line similar to the common observed surface diffusion dominated EM damage. However, when the grain size is down to 150 nm, global thinning of large areas is observed as in Fig. 4.6(b). When the grain size is further reduced, stripe-like voids propagating along the conductor line is observed as in Fig. 4.6(c). However, the last case is of no practical importance as its corresponding EM lifetime is very short. From these figures, it is evident that microstructures in the Cu line have dominant effects on Cu EM.

Gladkikh *et al.* (Gladkikh, Karpovski, Palevski and Kaganovskii 1998) explained the above experimental results using a theoretical model of EM developed by (Glickman and Klinger 1995; Klinger, Glickman, Fradkov, Mullins and Bauer 1995; Klinger, Glickman, Fradkov, Mullins and Bauer 1995) which considers coupled GB mass transport and surface diffusion, and indeed predicted the experimental results. According to the theory, two different modes of surface evolution can exist depending on the value of a dimensionless parameter $\alpha = Id^2/8B$, where $I = D_{gb}\delta j\rho Ze/kT$ is the GB mass flux induced by EM, and $B = D_s\gamma_s\omega^2 N_s/kT$ is known as the Mallins constant. D_{gb} and D_s are the GB and the surface self-diffusion coefficients. δ is the GB width, j is the current density, ρ is the specific resistivity, Ze is the effective charge of diffusing atoms, k is the Boltzmann constant, T is the absolute temperature, γ_s is the surface tension, ω is the atomic volume, N_s is the surface atomic density and d is the grain diameter. For large α, surface diffusion can not maintain the homogeneous surface motion, and voids or hillocks must form at GBs. As a result, the usual morphology of damage, as shown in Fig. 4.6(a), occurs.

However, for small α, surface diffusion becomes important and a global steady state regime of thermal grooving takes place when the surface moves homogeneously. In this case, sink or source of atoms at the GB can only slightly change the shape of GB grooves. This means that surface flux occurring due to a curvature gradient along the surface, may redistribute material between the GBs. Such kinetics will result in global thinning as shown in Fig. 4.6(b).

The interconnect microstructure depends on many parameters such as (a) core material deposition technique (current examples: electroplating, CVD, PVD, sputtering), (b) barrier material types (current examples: Ta, TaN, Ti, TiN and the combinations), (c) barrier material deposition technique (current examples: CVD, PVD, ALD), (d) Cu seed layer deposition technique and thickness, (e) line width, (f) line aspect ratio, and (g) deposition conditions. There is also cross interaction between these influential parameters on the resultant microstructure. As an example, for a given line width, the grain size may differ significantly when the Cu deposition technique changes from CVD to electroplating.

Ryu *et al.* (Ryu, Kwon, Loke, Lee, Nogami, Dubin, Kavari, Ray and Wong 1999) reported the effect of texture and grain structure on the EM life-time of Cu interconnects. Using different seed layers, (111) and (200) textured CVD Cu films with similar grain size distributions are obtained, and the EM life-time of (111) CVD Cu is about four times longer than that of (200) CVD Cu.

In contrast, electroplated Cu has relatively large grains in the damascene structure, resulting in longer EM life-time than CVD Cu in the deep submicron range. Typical grain size distributions for CVD Cu and electroplated Cu are shown in Fig. 4.7, and about two times EM life-time improvement is observed for electroplated Cu in comparison to that for CVD Cu. An annealing (350°C, 1 h) step after electroplating and before

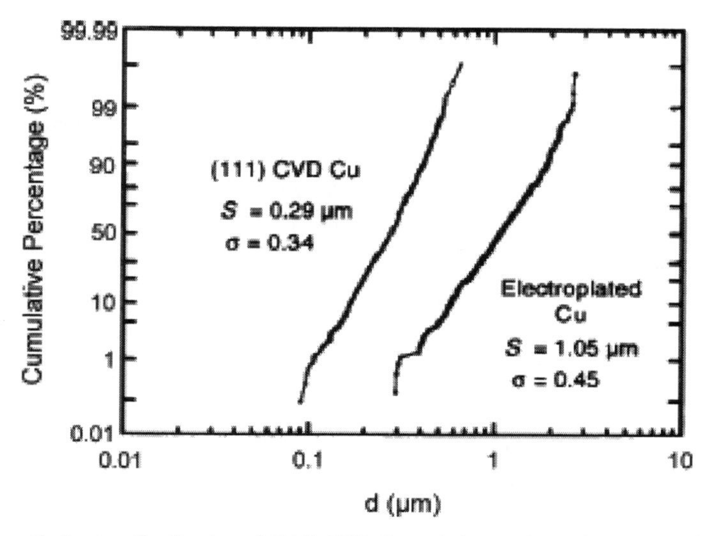

Fig. 4.7. Grain size distribution of (111) CVD Cu and electroplated Cu after 400°C anneal in forming gas for 1 h. (Ryu, Kwon, Loke, Lee, Nogami, Dubin, Kavari, Ray and Wong, 'Microstructures and Reliability of copper interconects', IEEE Trans. on Electron Devices), Copyright © 1999 IEEE.

CMP step forces grain recrytallization and growth, resulting in an increase in the grain size which is turn increase the EM life-time (Field, Dornisch and Tong 2001).

A significant effect of barrier material is found on the microstructure and EM performance of electroplated Cu interconnects (Lee and Lopatin 2005). A stronger (111) texture of Cu seed layer develops a more adhesive and better wetting interface of Cu/barrier, and in turn enhances the texture of electroplated Cu.

It is found that different interface interaction with different bond strength produces different degree of wetting of the Cu films on the barrier surface after thermal annealing. The effect of barrier on EM, including its effect on line microstructure will be discussed in the next section.

It is found that the voiding started from the sidewall of the interconnect trenches mostly at places where (111) grains with high angle grain boundaries were located, indicating the significant impact of microstructure on the localized EM induced damage in Cu interconnections (Wendrock, Mirpuri, Menzel, Schindler and Wetzig 2005).

Meyer (Meyer 2007) provided experimental evidence that voids seem to be formed at points where small grains are located next to large grains. In sub-100 nm structures with a certain fraction of small grains, some of these structures will have a component along the current flow direction and thus contribute to the mass transport and affect the EM lifetime (Zschech, Ho, Schmeisser, Meyer, Vairagar, Schneider, Hauschildt, Kraatz and Sukharev 2009). As a result, two different microstructure-driven failure mechanisms are possible, namely the formation of slit-like voids at nearly vertical grain boundaries away from the via, and the agglomeration of voids at the cathode end of the line. Figure 4.8 illustrates the two failure modes for a "downstream" test configuration. If no nearly vertical grain boundaries are encountered by the moving voids, the slit formation will not occur, and the voids will agglomerate at the line end (Meyer and Zschech 2007).

Also, in-situ scanning electron microscopy experiments on Cu based interconnects showed that voids often form away from the cathode and then drift toward the cathode to cause open circuit failure by diving into via (Meyer, Herrmann, Langer and Zschech 2002; Zschech, Meyer and Langer 2004; Choi, Monig, Thompson and Burns 2006). Recently, Choi *et al.* (Choi, Monig, Thompson and Burns 2008) performed an extensive study on the

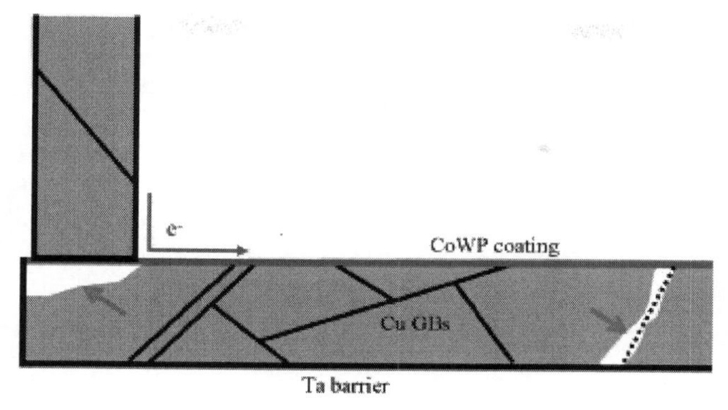

Fig. 4.8. Possible failure modes in CoWP-coated interconnects. Reprinted with permission from Meyer and Zschech, 'New microstructure-related EM degradation and failure mechanisms in Cu interconnects with CoWP coating'. Proc. 9th Int. Workshop Stress Induced Phenom, Metallization. Copyright@2007, American Institute of Physics.

effect of the microstructure on the formation, shape and motion of voids during EM, and their results are summarized in Table 4.2.

From Table 4.2, the following observations were made by Choi *et al.* (Choi, Monig and Thompson 2008):

(1) In all the 1.0 μm and 2.25 μm-wide samples, open-circuit failures were detected after the voids reached the end of the cathode (except one sample in 1 μm sample).
(2) In the case of the 0.3 μm-wide interconnects, voids that formed away from the cathode did not drift all the way to the end of the cathode. Instead, after drifting for only 1 or 2 μm, the voids grew to span the width and thickness of the line and caused failure.
(3) In all cases, the drift velocity of a void varied substantially as it drifted toward the cathode. Voids in 2.25 μm-wide samples drifted significantly faster than voids in 1.0 and 0.3 μm-wide samples. Drift velocities for 2.25 μm-wide samples also have a very wide range, from 8.1 to 100 nm/min.
(4) In general, the length of voids increased as the voids drifted closer to the cathode end, and the increase was especially pronounced near the cathode.

They also observed that voids initially grow in the locations at which they first appeared. The locations of void appearance and initial growth were

Table 4.2. Summary of the results by Choi *et al.* (Choi, Monig and Thompson 2008).

Structure width	Grains per width	Likely to form void away from cathode	Number of voids away from cathode	Location of first void away from cathode	Distance void drifted	Void drift velocity (nm/min)	TTF of samples with void away from cathode
0.3 μm	Mostly 1	Low (2 of 9)	1 or 2	12–13 μm away	1–2 μm	3.31	−1000 min
1.0 μm	1–3	Intermediate (3 of 7)	1 or 2	More than 13 μm away	To or close to cathode end	7.62	1000–2750 min
2.25 μm	2–5	High (4 of 4)	3–7	Within 3 μm away	To cathode end	46.22	<300 min

found to correlate with the locations of grain boundaries. Subsequently, voids will be depinned and drift toward the cathode, and the shape of a void changed as it passed through or along grains with different orientations. When a void moves within a single grain, the shape of the void is relatively stable.

To understand the experimental results, one can look at the activation energies for Cu diffusion on Cu surfaces which are known to vary on surfaces with different crystallographic orientations of the surface (Karimi, Tomkowski, Vidali and Biham 1995). It is reasonable to expect that this will also be true for Cu-passivation layer interfaces. Therefore, assuming a bamboo structure, if a void is pinned at a grain boundary and the grain on the anode side of the void has a higher diffusivity than that on the cathode side of the void, the void will grow in size due to the flux divergence. After growing to a certain size, the void may depin from the grain boundary and drift toward the cathode (Borgesen, Korhonen, Brown and Li 1992; Zaporozhets, Gusak, Tu and Mhaisalkar 2005). This depinning will occur after a void reaches a critical size (Borgesen, Korhonen, Brown and Li 1992).

As a void drifts, it can become pinned at another grain boundary. The grain size for the Cu used in their study was about 0.6 μm, so that a void can become pinned or change velocity frequently as it drifts toward the cathode. Once a void is pinned at a new grain boundary, it can continue to grow and eventually depin again. However, the flux divergence at the new boundary might also cause the void to shrink and eventually disappear.

Voids drift because of the difference in diffusivities on the surface of the void and in the Cu-passivation layer interface next to the void (D_{int}). This is proven by Choi *et al.* (Choi, Monig and Thompson 2007) where no drift was observed in in-situ SEM observations of unpassivated samples because the ($z^* D$) within the void and along the surfaces next to the voids are similar. Here $z^* D$ is the product of the effective charge and diffusivity, and it represents the driving force for atom diffusion (Hau-Riege 2004).

The Cu surface diffusivity (Ds) is a few orders of magnitude larger than the diffusivity in the interface (Hu and Reynolds 1997; Arnaud, Tartavel, Berger, Mariolle, Gobil and Touet 1999). Thus, at the cathode edge of the void there is a local flux divergence between the Cu surface of the void and the Cu-passivation layer interface next to the void, where the outgoing flux on the surface is larger than the incoming flux in the interface, causing the void to move toward the cathode end.

At the anode edge of the void, a flux divergence of opposite sign exists, and the outgoing flux through the interface is smaller than the incoming flux on the surface. These flux divergencies lead to gain (loss) of Cu on the anode (cathode) side of the void and therefore to a motion toward the cathode end.

They show that the void drift velocity is directly proportional to the surface diffusivity, Ds. Because Ds depends on the crystallographic orientation of the surrounding grain, the velocity of a void will vary with the crystallographic orientation of the grains it passes through. Therefore, the substantial variation of the drift velocity observed for drifting voids is likely due to changes in the diffusivities on different exposed grain surfaces as the voids moved from grain to grain.

The fact that the overall velocity of voids is higher in wider lines could be due to the preferential motion of voids along grain boundaries that are oriented parallel to the direction of the electron wind, as predicted by Borgesen *et al.* (Borgesen, Korhonen, Brown and Li 1992). This indicates that this particular region had a different orientation after the first void had passed through.

Choi *et al.* (Choi, Monig and Thompson 2008) found that more voids form away from the cathode as the width of the line increases. The observation that the density of the grain boundaries was higher in the wider lines supports the idea that voids preferentially form at grain boundaries or grain-boundary triple junctions (Korhonen, Borgesen, Brown and Li 1993).

The critical size for void nucleation is on the order of 1 nm (Gleixner, Clemens and Nix 1997) so that voids as small as 1 nm could grow due to an EM flux divergence, without the requirement that new voids be nucleated. Such small pre-existing voids, or void "embryos," would generally be hard to detect. Li *et al.* (Li, Tan and Raghavan 2009) modeled the void nucleation process due to the movement of this void "embryos" and their coalescence, and they derived a analytical time to failure equation that agrees well with the experimental results at different test conditions. Detail of the modeling will be described in Chapter 5.

In order to improve the microstructure quality for better EM performance, rapid thermal annealing (RTA) treatment is proposed for Cu films (Kwon, Park and Lee 2005). The texture quality as a function of RTA temperature for different ambient is shown in Fig. 4.9. For different line

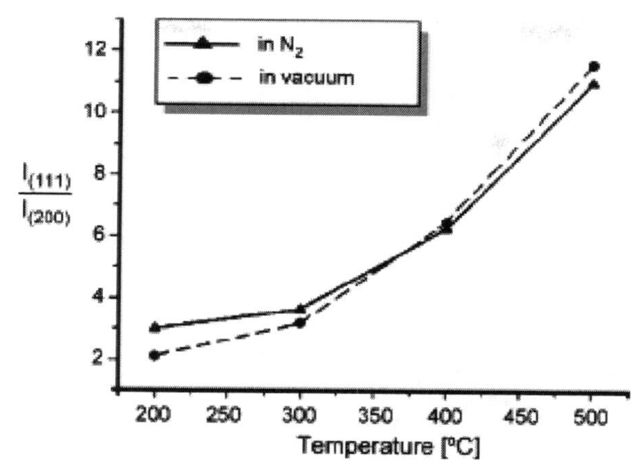

Fig. 4.9. The texture quality as a function of the RTA temperature for electroplated Cu interconnection. Reprinted from Thin Solid Films, 475, Kwon, Park and Lee, 'Electromigration resistance-related microstructural change with rapid thermal annealing of electroplated copper films', pp. 58–62, Copyright © 2005, with permission from Elsevier.

width structures, the variation of the ratio of line width (W) over Grain size (S) as a function of RTA temperature is shown in Fig. 4.10. In the RTA process, nitrogen should be used as the ambient instead of vacuum because nitrogen ambient offers lower resistivity and smoother film surface.

Resulting from the RTA treatment, the EM performance is found to improve significantly (Kwon, Park and Lee 2005). It can be seen from Fig. 4.10 that high temperature RTA is more effective for polycrystalline structure. As the thermal budget is limited in BEOL processing due to the various constrains from the FEOL, the improvement in microstructure quality by the RTA is limited to a temperature below 350°C. Fortunately, it is found that structure of line width less than 0.1 μm becomes bamboo by the RTA treatment at any temperature above 200°C as can be seen in Fig. 4.10, hence this RTA treatment is expected to be useful for 65 nm technology node or below.

Another way to improve the microstructure-driven EM performance is the use of twin-modified grain boundaries. Lu *et al.* (Lu, Shen, Chen, Qian and Lu 2004) have synthesized a high density of nanotwins in pure Cu foils by pulsed electrodeposition. The Cu foil shows a 10-fold improvement of the mechanical strength relative to a large-grained Cu, and the foil remains

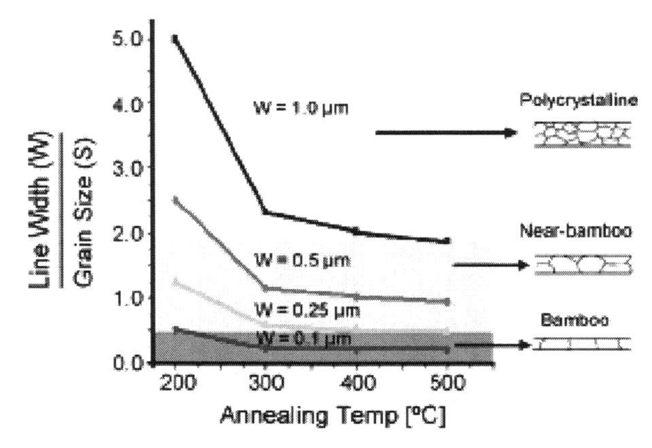

Fig. 4.10. Dependence of the line width/grain size ratio (W/S) of the electroplated Cu interconnect on the RTA temperature for different line widths. Reprinted from Thin Solid Films, 475, Kwon, Park and Lee, 'Electromigration resistance-related microstructural change with rapid thermal annealing of electroplated copper films', pp. 58–62, Copyright © 2005, with permission from Elsevier.

ductile but its electrical resistance did not change significantly. As high mechanical strength and low electrical resistivity are desired properties for interconnecting wires in integrated circuits from the consideration of the resistive-capacitive delay, electromigration (EM), and stress migration (Borgesen, Lee, Gleixner and Li 1992; Frankovic and Bernstein 1996; Hu and Harper 1998), Chen *et al.* (Chen, Wu, Liao, Chen and Tu 2008) studied the atomic diffusion behavior at twin-modified grain boundaries. They found that an atomic step appeared at a triple point of the twin boundary and it can move rapidly on the (111) plane toward the other triple point of the twin boundary. Once the atomic step reached the second triple point, they observed that the atomic step was trapped for a while before moving out onto the (422) plane. This slow down of the EM-induced atomic migration is expected to reduce the void growth rate by an order of magnitude, thus improve the Cu EM due to microstructure.

The slow down of the atomic step is believed to be due to the requirement of a new step and a kink site on the (422) surface. The nucleation of the steps from the triple points would be the rate limiting step, resulting in a time lag for the atomic diffusion across the triple points (Chen, Wu, Liao, Chen and Tu 2008).

4.2.3 *Presence of impurity*

Impurity levels in Cu interconnects have a strong influence on the microstructural properties of the copper. The rate of recrystallization with anneal was found to be significantly slower with higher impurity concentration (Alers, Lu, Sukamto, Kailasam, Reid and Harm 2004). Slower recrystallization can impact the average copper grain size for a given thermal history of copper lines. A strong correlation was also found between copper purity and the formation of small "microvoids" at the grain boundaries in copper (Sekiguchi, Koike, Kamiya, Saka and Maruyama 2001; Sekiguchi, Koike and Maruyama 2003).

The impurity level of the bulk copper can be introduced during the plating process. The plating conditions such as the applied current density and the additive content in the electrolyte determine the quality and the purity grade of the Cu metallization (Stangl, Acker, Oswald, Uhlemann, Gemming, Baunack and Wetzig 2007; Stangl, Lipták, Fletcher, Acker, Thomas, Wendrock, Oswald and Wetzig 2008). Alers *et al.* (Alers, Lu, Sukamto, Kailasam, Reid and Harm 2004) studied the effect of impurity level in Cu by intentionally introduce impurities in the plating process as shown in Table 4.3. They found that in order for the sheet resistance of Cu film to reduce by 10%, the time taken for annealing at room temperature is $>$ 2000, 600, and 200 hours for the low, medium, and high purity films respectively. With the slower anneal rate for low purity film, the grain size will be smaller for the same thermal history, thus degrade its EM

Table 4.3. Impurity concentration levels for copper deposited with different chemistries. High purity would be close to the impurity level of a typical copper interconnect (Alers, Lu, Sukamto, Kailasam, Reid and Harm 2004).

Impurity (ppm)	High purity	Medium purity	Low purity
S	0.3	7.4	16.5
Cl	2.1	28.4	61.4
C	4.1	42.5	135.6
O	2	1.7	24
N	11.7	25.3	15.6
Ag	0.4	0.6	0.5
Others	<1	<1	<1

Table 4.4. Purity level and number of microvoids observed in a 25 μm FIB/SEM cut (Alers, Lu, Sukamto, Kailasam, Reid and Harm 2004).

	Microvoids
High purity	0
Medium purity	18
Low purity	72

performance. Using FIB/SEM analysis, they found that the average number of grain/μm of line length was 3.8 for the high purity copper and 4.4 for the low purity copper.

Also, they found that low purity copper film has a higher tendency to form microvoids as shown in Table 4.4. The mechanism through which the impurity level modulates microvoid formation could be related to precipitates at the grain boundaries (Sekiguchi, Koike and Maruyama 2003). Furthermore, they also found that microvoids in the bulk can migrate and coalesce at the Cu/SiN interface Hence, smaller grain size with higher microvoids whch will coalesce at the interface will greatly reduce the EM lifetime.

Recently, Stangl *et al.* (Stangl, Liptak, Acker, Hoffmann, Baunack and Wetzig 2009) also studied the effect of non-metallic impurity in the electrolyte on Cu EM and again confirmed that lower activation energy of EM for Cu with higher impurity concentration. They further analyzed the mechanisms of the poorer EM performance, and they found that S has a strong affinity to crystal defects in terms of vacancies and forms a S-vacancy bound complex. As S is known to be very mobile within the Cu matrix, it may segregate to an interface, to a grain boundary or to a dislocation and brings at least one vacancy with it. The accumulation of S-vacancy pairs results in a more extended defect arrangement and adds up to a void. As a consequence, a higher amount of S impurities enhances the void formation and decreases the lifetime of Cu interconnects.

On the other hand, if the dominant diffusion path for Cu EM is grain boundary instead of interface, which could become possible through interface engineering as described earlier, it is found that the incorporation of S impurity can decrease Cu grain boundary diffusion due to the strong

S–Cu bonds in the grain boundaries, thus enhances EM lifetime (Surholt and Herzig 1997).

Besides S, a number of metallic impurities are found to enhance Cu EM lifetime significantly, even for the case where interface diffusion path is dominant. Following the work of adding Cu into Al film which enhances Al EM lifetime as mentioned in Chapter 3, one can apply the same principle of having precipitation of solute metals and/or compounds of solutes at the grain interiors, the grain boundaries, the surface of an alloy film and the interface between an alloy film and surrounding materials to act as obtstacles to the diffusion of Cu. To achieve this, the solutes used must satisfy the following requirements (Igarashi and Ito 1998):

(a) First, precipitation must occur at low temperatures, and considering the metallization process for device application, the temperature for precipitation annealing must be less than 500°C.
(b) Second, solubilities of solutes in Cu must be low enough for the formation of adequate amounts of precipitates in the grain boundaries, the surface and the interfaces after precipitation annealing. Low solubilities of solutes are also important for realizing highly reliable interconnects, because a high concentration of solutes in the grain interiors enhances lattice diffusion in EM (Stoebe, Gulliver, Ogurtani and Huggins 1965; Kim and Morris 1992)
(c) Third, resistivities of Cu alloys after annealing to induce precipitates must be lower than those of Al alloys ($<3\ \mu\Omega$ cm).

Solutes satisfying the above requirements are Sn (Hu, Luther *et al.* 1995), Cr (Jr., Harper, Holloway, Smith and Schad 1992), Mg (Ding, Lanford, Hymes and Murarka 1994; Hu, Luther *et al.* 1995) and Zr (Taubenblat, Marino and Batra 1979; Hu, Luther *et al.* 1995).

Igarashi *et al.* (Igarashi and Ito 1998) studied the Cu-Zr alloy film and they found that the film resistivity becomes lower than $3\mu\Omega$ cm after annealing at 500°C for 5 mins in vacuum (1×10^{-7} Torr) with Zr concentration up to 1.4 at. %. Depth profiles obtained using SIMS showed that Zr is concentrated at the surface and at the interface between Cu-Zr film and SiO_2 after annealing as shown in Fig. 4.11, explaining the low resistivity observed after annealing. EDS analysis also shows that Zr content inside Cu grains is below the detection limit and Zr can also be found in the grain boundaries between Cu grains.

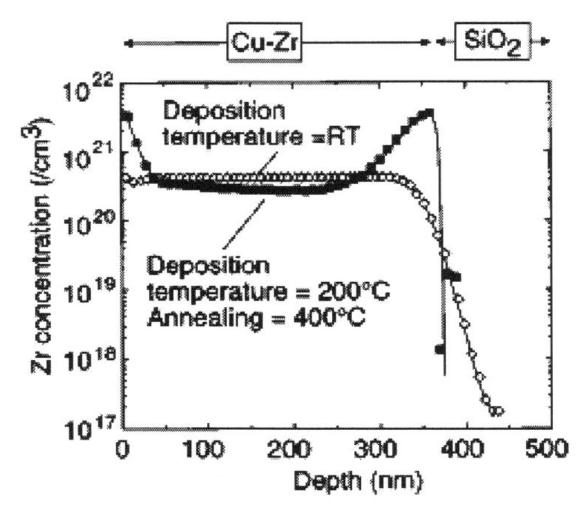

Fig. 4.11. Depth profile of Zr concentration in Cu-Zr films before and after annealing. Reprinted with permission from Igarashi and Ito, 'Electromigration properties of copper-zirconium alloy interconnects', Journal of Vacuum Science and Technology B, vol. 16(5), pp. 2745–2750. Copyright © 1998, American Institute of Physics.

For 1.4 at. % of Zr, the size of the precipitates is about 0.1 μm in diameter, comparable to the linewidth of the sub-micron line width. Hence, 0.14 at. % of Zr was used by Igarashi *et al.* where the size precipitates were so small to be observed in the TEM analysis. With this Cu-Zr film, they found that EM lifetime is three time longer than the pristine Cu film, and the activation energies are in the range of 1.37–1.46 for Cu-Zr film.

For 0.2-μm-wide Cu interconnects near-bamboo-type structure, the dominant diffusion path is an interface between Cu and the materials surrounding it. However, Igarashi *et al.* (Igarashi and Ito 1998) found that voids and hillocks were observed over the entire pattern in the case of the Cu-Zr interconnects which was different from the Cu interconnect. This leads them to suggest that the precipitation of Cu-Zr compounds prevents diffusion of Cu atoms through the interfaces, rendering voids and hillocks formed mainly by grain boundary diffusion in the Cu-Zr interconnects.

Braeckelmann *et al.* (Braeckelmann, Venkatraman, Capasso and Herrick 2000) studied the Cu-Mg alloy film. As Magnesium is an alkaline earth metal, which readily reacts with oxygen, fluorine, nitrogen, and carbon, typical constituents of dielectric films, Mg will react at the interface or surface to form compounds such as MgO or MgF (potentially altering adhesion properties). Cu on the other hand, has inherently poor adhesion to dielectrics. It reacts only slowly with oxygen and does not form a self-limiting oxide. Alloying Cu with Mg can thus provide a means of improving the adhesion of Cu. Braeckelmann *et al.* indeed showed the enhanced adhesion of Cu to the cap layer in their work (Braeckelmann, Venkatraman, Capasso and Herrick 2000).

The mechanism of enhancing Cu EM through Mg solute was found to be similar to Zr, and Mg also segregates to the surface just like the case of Zr. However, no report on the EM enhancement factor due to Mg inclusion was reported.

Park and Vook (Park and Vook 1993) studied the Cu-Pd alloy film and activation energy can go as high as 1.25 at 1 wt % of Pd in the film with resistivity below $3\mu\Omega$ cm. The mechanism of EM enhancement is expected to be similar to that of Zr in that it suppresses the interface diffusion.

Hu *et al.* (Lee, Hu and Tu 1995) found that improvement in EM resistance for Cu(Sn) alloy was better than Cu(Mg) and Cu(Zr). They used the drift velocity test samples to study the mechanism of Sn alloy in EM improvement. They found that the edge displacement for the Cu(Sn) alloy stripe is measured to be ~ 3.5 times smaller than that of the Cu strip after 99 hours of testing, showing that Cu(Sn) alloy stripe has a higher electromigration resistance than the Cu stripe under identical testing conditions.

Also, in contrast to Cu stripe, no obvious voids and hillocks are observed in Cu(Sn) stripe along the top surface of the test stripe, and this is the case even for those stripes as long as 300 μm and after hundreds of hours of electromigration testing. Most of the hillock formation in the Cu(Sn) alloy test stripe occurs along the side and at the anode end of the stripe. The better integrity of the top surface layer and the absence of voids and hillock formation along the top surface of the Cu(Sn) alloy test stripe are believed to be due to the formation of a tin oxide layer beneath the Ta layer. The tin oxide is a protective layer which may have prevented the voids and hillocks formation by eliminating the sources and sinks for vacancies (Lee, Hu and Tu 1995).

They also found that Sn atoms move at a constant rate in the same direction as Cu mass transport. The Sn atoms have to move beyond a finite distance away from the cathode end before the mass transport rate of Cu in Cu(Sn) alloys reaches a value comparable to that for pure copper stripe. They observed that the average drift velocity of Cu motion in Cu(Sn) alloys (i.e. equal to the slope of the edge displacement) is small at the beginning of the EM test, and it increases slowly with time and eventually reaches a value comparable to that of Cu motion in pure copper stripe. Using this

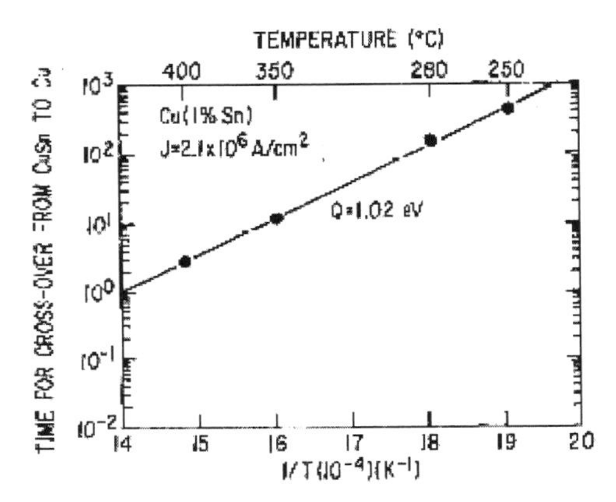

Fig. 4.12. Arrhenius plot of the time at which the Cu(Sn) stripe depletes almost like copper versus 1/T. T is the absolute temperature. Reprinted with permission from Lee, Hu and Tu, 'In-situ scanning electron microscope comparison studies on electromigration of Cu and Cu(Sn) alloys for advanced chip interconnects', Journal of Applied Physics, vol. 78(7), pp. 4428–4437. Copyright © 1995, American Institute of Physics.

measurement, they obtained a plot of the time to cross over from Cu(Sn) to Cu versus temperature as shown below, and the activation energy computed agreed well with that of the diffusion of Sn in Cu. The suppression of grain boundary migration of Cu is also confirmed by the experimental works by Michael *et al.* (Michael and Kim 2001).

Thus, although the addition of a small percentage of Sn (0.5–2 wt %) decreases the average grain size ($-0.1\ \mu$m) of the Cu stripe and hence increase the number of paths for electrotransports, this effect is evidently far outweighed by the reduction in effective Cu diffusivity with the presence of Sn in the copper stripe.

Cu(Al) alloy film is also studied by several workers. In fact, Cu(Al) is probably the first alloy Cu film studied by D'Heurle and Gangulee (D'Heurle and Gangulee 1975) in 1975. However, they found that the effect of Al in Cu was not significant. Recently, Yokogawa and Tsuchiya (Yokogawa and Tsuchiya 2007) found that the EM lifetime of the CuAl line is longer than that of the pure Cu line by a factor of 40, and Tada *et al.* (Tada, Abe, Furutake, Ito, Tonegawa, Sekine and Hayashi 2007) reported an improvement by a factor of 50. The improvement in EM is due to the increase in incubation time and reduction of the Cu drift velocity through the suppressing of both the void nucleation and growth for CuAl line as

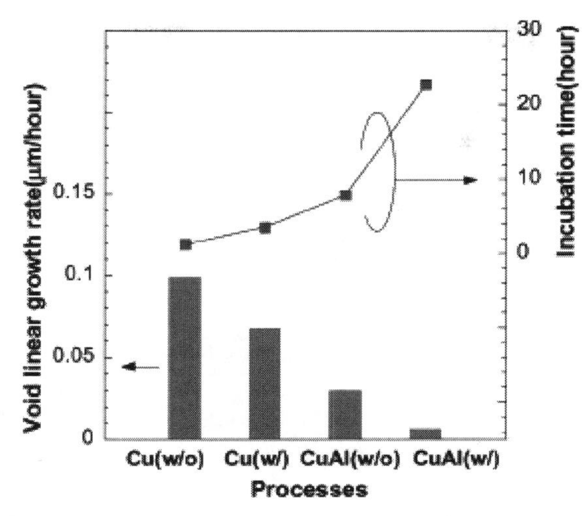

Fig. 4.13. (a) EM-induced void growth designed by via-to-via distance as a function of time to resistance step. (b) VLRG and incubation times of each interconnect estimated by using the "VLGR-TEG" under stress of 0.5 mA at 300°C. (Tada, Abe, Furutake, Ito, Tonegawa, Sekine and Hayashi, 'Improving reliability of copper dual-damascene interconnects by impurity doping and interface engineering', IEEE Trans. on Electron Devices), Copyright © 2007 IEEE.

determined by Tada *et al.* (Tada, Abe, Furutake, Ito, Tonegawa, Sekine and Hayashi 2007) as shown in Fig. 4.13. However, the resistivity of the CuAl line is higher, though it is still below $3\mu\Omega$ cm. A detail study by Yokogawa *et al.* (Yokogawa, Tsuchiya, Kakuhara and Kikuta 2008) demonstrated that the segregation of Al dopant atoms to the interface of the lines increases the resistivity through increases surface scattering.

The suppression of Cu drift velocity by Al along the Cu-cap layer interface is found to be different from that of Zr or Mg or Sn as discussed earlier. Tada *et al.* (Tada, Abe, Furutake, Ito, Tonegawa, Sekine and Hayashi 2007) analyzed the oxygen concentration at the interface between Cu-SiCN cap layer using SIMS and they found that there is a surface reductive treatment for CuAl line, rendering an increase in the adhesion strength.

Recently, with the interesting properties of carbon nanotube in term of its excellent electrical conductivity and high EM resistance, Chai and Chan (Chai and Chan 2008) use CNT as the alloying element to reinforce the EM resistance of Cu interconnect. With a weight ratio of 4.6% of CNT in Cu, they found that the EM lifetime is improved 5 times with a minor decrease in the conductivity of the Cu.

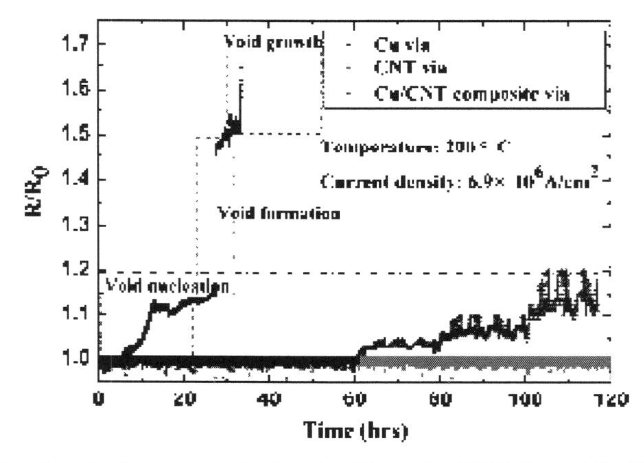

Fig. 4.14. The electrical resistance for the vias filled with CNT, Cu and Cu/CNT composite as a function of stressing time. The electrical resistances are normalized to the initiate resistance. (Chai, Chan, Fu, Chuang and Lu, 'Electromigration studies of Cu/CNT composite using Blech structure', IEEE Electron Devices Lett.), Copyright © 2008 IEEE.

Unlike the situation reported for Cu(Sn) and Cu(Al), the CNT itself is rooted on the seed catalyst, and did not move in the direction of the electron flow even under high current density and high temperature. The CNT acts as trapping center and causes a decrease in the diffusion of EM-induced migrating Cu atoms (Chai, Zhang, Zhang, Chan and Yuen 2007; Chai, Chan, Fu, Chuang, and Lu 2008). Also, for the Cu/CNT composite via, the resistance increases slowly after the incubation. After the voids form in the Cu/CNT composite, the current shunts into the neighboring CNTs. Therefore, the resistance shows gradual increase instead of the abrupt jump in the stressing process. A typical resistance change curve is as shown in Fig. 4.14.

4.2.4 *Mechanical stress*

Hydrostatic stress will be developed in interconnect line either due to processing because of differential thermal expansion of the metal film with its surrounding materials or evolution due to electromigration. If the tensile stress in the line exceeds a crtical value (σ_{crit}), a void will nucleate at that site which could lead to interconnect failure. A high value of σ_{crit} is desirable for longer electromigration lifetimes for a given interconnect system.

Hau-Riege *et al.* (Hau-Riege, Hau-Riege and Marathe 2004) derived an expression of σ_{crit} assuming the void is formed at the Cu-cap layer interface, and is given by

$$\sigma_{crit} = \sqrt{\frac{A_{void}\gamma}{V_{relaxation}\xi}}, \tag{4.1}$$

where A_{void} is the area of the void, assuming rectangular, γ is the interface energy per unit area, $V_{relaxation}$ is the interconnect volume where the stress is relieved over, and ξ is the strain-energy proportionality constant which depends on the direction of the strain and the elastic properties of the material, and can be determined through finite element modeling. They have computed the critical stress and found a good agreement with the experimental data. Equation (4.1) also shows that σ_{crit} increases with increasing adhesion energy between the Cu and cap layer.

The effect of mechanical stress can also be understood from the weakening of the chemical bonding between metal atoms as shown by Li *et al.* (Li, Tan and Hou 2007). They considered the pairing energy which is the amount of energy required to bring an atom from its lowest energy state and free it out of its surrounding atoms, in the absence of external excitations such as elevated temperature and stress. In the presence of stress, the energy state of the atoms will be raised by an amount equal to the strain energy, and thus the minimum energy required for the movement of the atoms is lowered, and the probability of movement is higher. Using Monte Carlo modeling, they model the movement of vacancies and void formation and thus provide a dynamic simulation of electromigration in polycrystalline interconnect thin film which agrees with experimental findings.

The material properties of Cu also render the mechanical stress effect more significant than the electron wind force for electromigration as shown by Tan *et al.* (Tan, Zhang and Gan 2004). By substituting the material properties into the flux divergences equations due to electron wind force, temperature gradient induced driving force and hydrostatic stress gradient induced driving force, one has the following equations for Al:

$$\begin{aligned} div(\vec{J}_A) &= 8.1 \times 10^{-5}\vec{j} \bullet \nabla T \\ div(\vec{J}_{th}) &= -3.196 \times 10^4 \times \nabla T^2 + 4.19 \times 10^{-17} j^2 \\ div(\vec{J}_{st}) &= 2.343 \times 10^4 \times \nabla\sigma \cdot \nabla T \\ &\quad -3.182 \times 10^3 \times \nabla T^2 + 3.667 \times 10^{-17} j^2 \end{aligned}, \tag{4.2}$$

where j is the current density with the range from 3.1×10^{10} to 8.23×10^{10} pA/μm^2, ∇T represents the temperature gradient varying from 0.04 to 3.03 K/μm, and $\nabla\sigma$ is the hydrostatic stress gradient ranging from 0.026 to 0.26 MPa/μm.

For Cu, the flux divergences equations are as follows:

$$\begin{aligned} div(\vec{J}_A) &= 1.511 \times 10^{-11}\vec{j} \bullet \nabla T \\ div(\vec{J}_{th}) &= -5.522 \times 10^{-2} \times \nabla T^2 + 1.84 \times 10^{-23} j^2 \\ div(\vec{J}_{st}) &= 1.297 \times 10^{-2} \times \nabla\sigma \cdot \nabla T \\ &\quad -1.236 \times 10^{-3} \times \nabla T^2 + 6.596 \times 10^{-24} j^2 \end{aligned}, \quad (4.3)$$

where the range of j is from 3.19×10^{10} to 2.43×10^{10} pA/μm^2,0.02 to 2.64 K/μm for ∇T, and -0.089 to 0.19 MPa/μm for $\nabla\sigma$.

Figure 4.15 shows the % of flux divergence due to different driving forces for Al and Cu. One can see that while the electron wind force dominates in Al throughout the entire process of EM, mechanical and thermal induced driving forces are actually dominating the Cu EM initially. Thus, mechanical stress is more crucial in Cu with respect to EM damage.

As the line width decreases due to the advances in IC technology, Tan *et al.* (Tan and Roy 2006) also showed the increasing contribution of the mechanical stress on the EM performance for Cu as shown in the Table 4.5. Figure 4.16 shows the ratio of atomic flux due to stress gradient induced force over that due to electron wind force, and one can see the significance of the stress gradient when the line width decreases.

Besides the increasing important of the mechanical stress in Cu EM due to the material properties, the presence of the mechanical stress can also cause damage to the barrier metal as it becomes very thin as we progress into smaller technology node. As shown in the schematic drawn by Hsu *et al.* (Hsu, Fang, Chiang, Chen, Lin, Chou and Chang 2006) which is reproduced in Fig. 4.17, one can see that during the EM testing, current crowding effect and current flow induced heating on via hole generate thermal stresses for various layers are indicated in Fig. 4.17. Under a constant testing temperature, the thermal stress is larger than the intrinsic stress and becomes dominant. Its magnitude can be calculated with (Park,

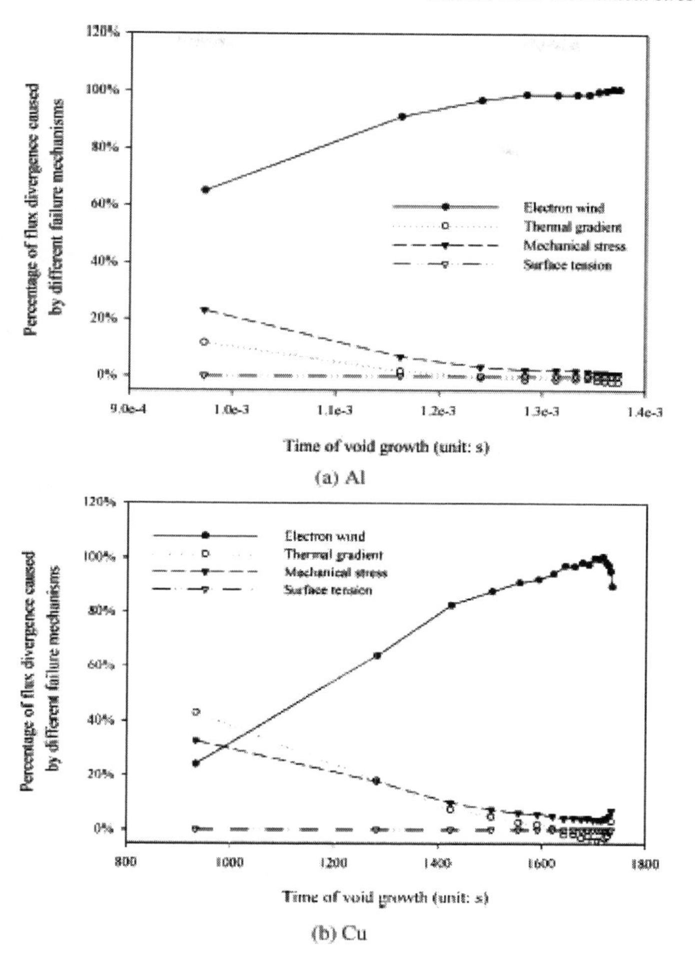

(a) Al

(b) Cu

Fig. 4.15. Change of flux divergences caused by various driving forces during EM with $j = 3\,\text{mA/cm}^2$ and $T = 400°\text{C}$ for (a) the intrinsic aluminum and (b) the intrinsic copper. Reprinted from Thin Solid Films, 462–463, Tan and Zhang, 'Overcoming intrinsic weakness of ULSI metallization electromigration performances', pp. 263–268, Copyright © 2004, with permission from Elsevier.

Lee and Hunter 2005)

$$\sigma_T = (T_0 - T)\Delta\alpha \times E, \tag{4.4}$$

where σ_T, T_0, T, $\Delta\alpha$ and E are thermal stress, stress free temperature, EM testing temperature, the difference of thermal expansion coefficient,

Table 4.5. AFD contributions from different sources for Cu DD via EM.

Stress condition $T = 300°C$, $j = 0.8\,mA/cm^2$	Test type: M1-text line width $0.4\,\mu m$	$0.7\,\mu m$	Stress condition $T = 300°C$, $j = 0.8\,MA/cm^2$	Test type: M2-test line width $0.4\,\mu m$	$0.7\,\mu m$
EWM	9.2%	21.8%	EWM	8.1%	16.5%
TM	10.9%	9.9%	TM	12.5%	18.0%
SM	79.9%	68.3%	SM	79.4%	65.5%
Total AFD	100%	100%	Total AFD	100%	100%

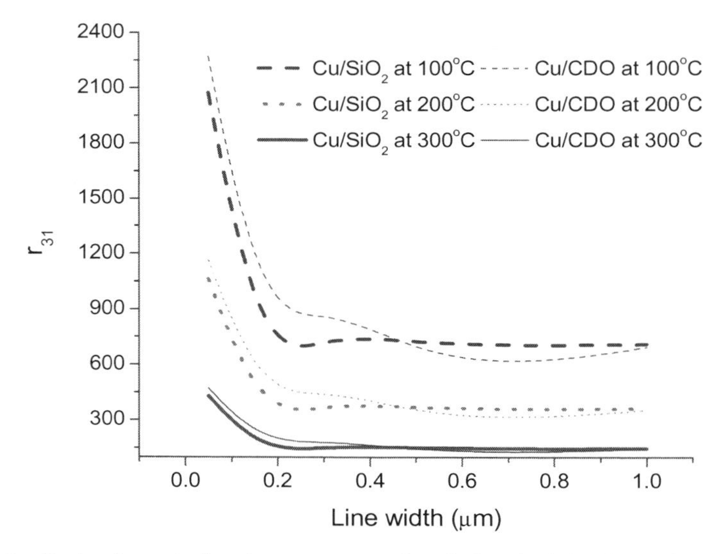

Fig. 4.16. Ratio of atomic flux due to stress gradient induced driving force (SGIDF) over that due to electron wind force.

and the Young's modulus, respectively. The thermal expansion coefficient (CTE)/the young's modulus of Cu, TaN, and FSG 3are 16.6×10^{-6}°C^{-1}/125, 6.5×10^{-6}°C^{-1}/457, and 0.5×10^{-6}°C^{-1}/80, respectively (Orain, Barbé, Federspiel, Legallo and Jaouen 2004). Using these data in Eq. 4.4, one finds that Cu possesses the largest thermal stress and the smallest for FSG, thus generating an unbalanced stress upon the TaN barrier layer which cause it to break at the weakest point. Simultaneously, the Cu atoms energized by heat diffuse into FSG through the broken barrier and create a void in via

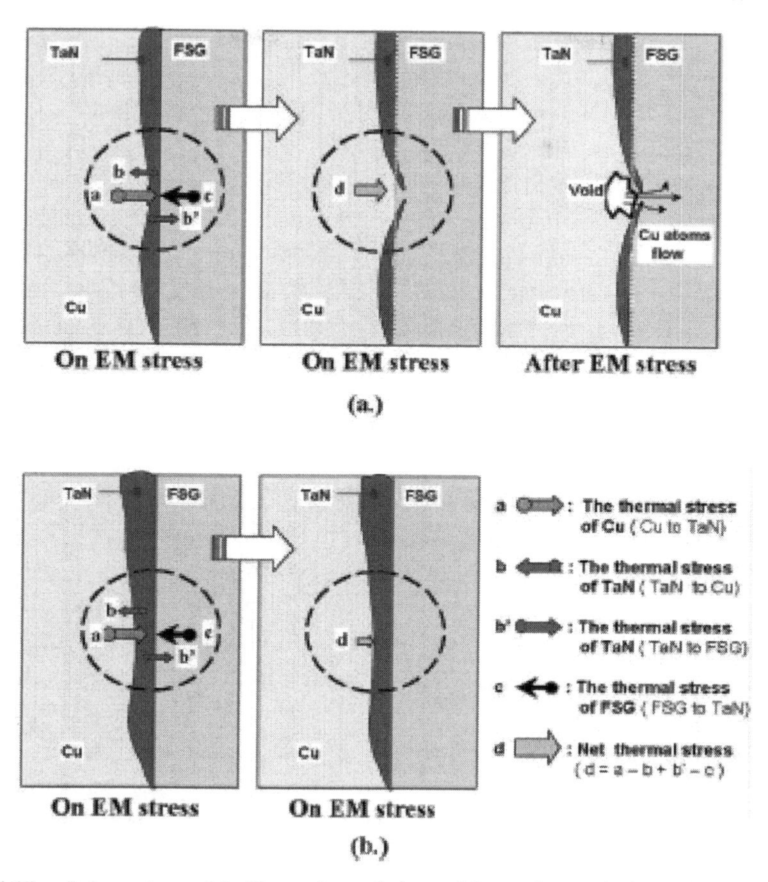

Fig. 4.17. Schematic models illustrating unbalanced thermal stress induced EM failure in the sidewall of copper via for (a) thin and (b) thick TaN barrier. Reprinted from Journal of Electrochem. Soc., 153, Hsu, Fang, Chiang, Chen, Lin, Chou and Chang, 'Failure Mechanism of Electromigration in Via Sidewall for Copper Dual Damascene Interconnection', pp. G782–G786, Copyright © 2006, with permission from The Electrochemical Society.

hole as shown in Fig. 4.17a. However, the same phenomenon was not found for a thick and conformal TaN barrier layer (Fig. 4.17b) because it is strong enough to resist the cracking.

The difference of thermal expansion coefficient between TaN layer and Cu layer also causes a tensile stress to the Cu layer and bends the layers in outbound direction under the high-temperature EM testing, especially for thinner TaN barrier layer. The heat-enhanced tensile stress then assists the diffusion of Cu atoms, creating small vacancies in the Cu/TaN interface during EM testing. At the same time, these vacancies transport along with the interface and coalesce into voids in via hole as reported (Hsu, Fang, Chiang, Chen, Lin, Chou and Chang 2006).

The major mechanical stress in the Cu film is due to process temperature of the film which is also dependent on the stress free temperature (SFT). Stress free temperature is an idealistic temperature at which all the thermo-mechanical stress component become zero. To stabilize the grains in the interconnection and for thermo-mechanical stabilization, a thermal annealing is performed on the fabricated wafers as a final process step. The temperature range and time of this annealing process are 300 to 400°C and 30 to 60 minutes respectively.

It is believed that the stresses in the entire structure become zero at the above mentioned annealing temperature after the annealing. Thus the SFT is the final annealing temperature of the fabrication process for Cu DD interconnects. There are several experimental evidences showing interconnect stress becomes zero around the SFT (Du, Wang, Merrill and Ho 2002; Reimbold, Sicardy, Arnaud, Fillot and Torres 2002; Onishi, Fujii, Yoshikawa, Munemasa, Inoue and Miyagaki 2003; Shen and Ramamurty 2003).

The impact of SFT on EM is anticipated to lead to a different failure mechanism in normal operating condition in comparison to that of accelerated test condition (Roy, Tan, Kumar and Chen 2005). For a given SFT of 360°C, Roy *et al.* (Roy, Tan, Kumar and Chen 2005) studied the total AFD distribution in the structure and the possible failure sites using finite element analysis. The results are summarized in Table 4.6. From the FEA results, it can be seen that as the line temperature decreases, the dominant driving force changes from electron wind force dominate to mechanical stress dominate. Moreover, the maximum AFD site changes from failure Site-D to Site-E with the decrease in line temperature where the different failure sites are shown in Fig. 4.18. One thus expects Site-D failure to occur in the accelerated test while Site-E failure will emerge at the normal operating condition. In the intermediate range of test temperature, a bimodal failure is expected.

From experimental observation, Roy *et al.* (Roy, Kumar, Tan, Wong and Tung 2006) also reported an increase in EM life-time when SFT is reduced

Table 4.6. Effect of EM test condition on M1 test (SFT = 360°C, $j = 0.8$ MA/cm^2).

Test temperature (°C)	Failure site	Dominant migration factor
325	Site-D	EWM
300	Sits-D and E	EWM and SM
250	Site-E	SM
200	Site-E	SM
150	Site-E	SM
100	Site-E	SM

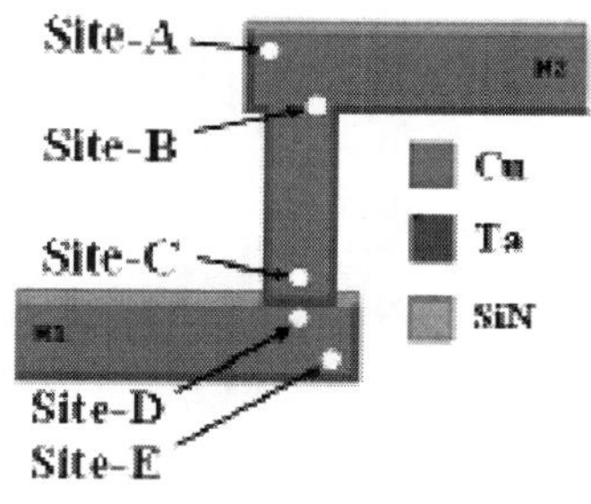

Fig. 4.18. Expected failure sites in line-via Cu structure. Reprinted from Microelectronics Reliability, 45, Roy, Tan, Kumar and Chen, 'Effect of test condition and stress free temperature on the electromigration failure of Cu dual damascene submicron interconnect line-via test structure', pp. 1443–1448, Copyright © 2005, with permission from Elsevier.

as anticipated above. Apart from the final annealing step, they performed additional annealing in the temperature range of 100–200°C for 48 hrs prior to EM test to reduce the SFT, and they found that the EM life-time can be improved by 26% in comparison with the case of no additional annealing. The important observations here is the delay in the step formation in the resistance change (see Fig. 4.19) due to the reduction in SFT.

Since SFT is important, an accurate determination of SFT is necessary. However, experimental determination of SFT is available only with the use of high brilliance X-ray source. Using this X-ray source, one can determine the stress in the interconnect film at different line temperatures from which SFT can be deduced.

Due to the fact that SFT is higher than the normal operating condition and even the test temperature in general, the interconnect films usually possess hydrostatic stress, and the stress will relax during operation.

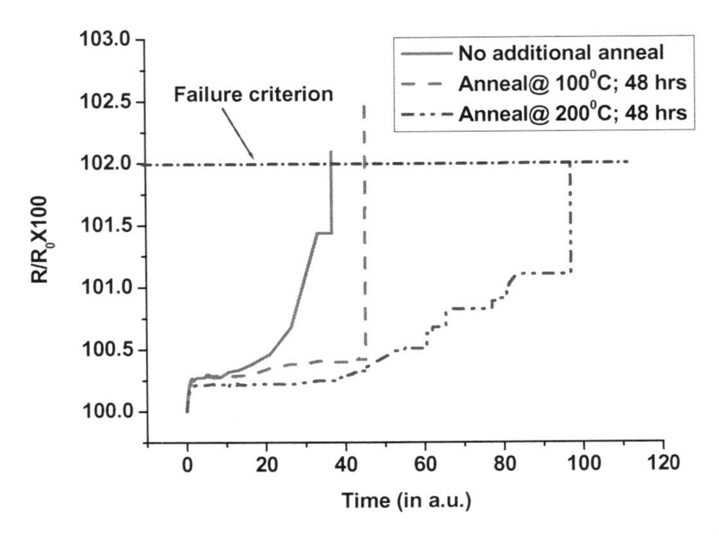

Fig. 4.19. Resistance change profiles. Reprinted from Microelectronics Reliability, 46, Roy and Tan, 'Experimental investigation on the impact of stress free temperature on the electromigration performance of copper dual damascene submicron interconnect', pp. 1652–1656, Copyright © 2006, with permission from Elsevier.

Glickman *et al.* (Glickman, Osipov, Ivanov and Nathan 1998) showed that instantaneous large scale dislocation glide is not possible for a supported thin film such as that in interconnections, and the time dependent EM stress relaxation process is more likely that involves diffusion and viscous flow, i.e. creep. This is also verified by Gan *et al.* (Gan, Huang, Ho, Leu, Maiz and Scherban, 2004).

Gan *et al.* (Gan, Huang, Ho, Leu, Maiz and Scherban, 2004) found that the initial stresses in the passivated Cu film depends on the SFT, and upon heating from room temperature, a higher initial stress will relax more as shown in Fig. 4.21. The stress relaxation consists of an initial stage of sharp decrease of stress with time and then a steady stress relaxation over a long period. The plastic strain rates deduced from Fig. 4.20 are shown in Fig. 4.21 as a function of the stress. The significance of the transient behavior at the initial stage is clearly observed from Fig. 4.21, and we can see that it consists of an initial transient regime and a steady state regime which is a typical behavior of creep (Josell, Weihs and Gao 2002). Within the transient regime, the plastic strain rate not only depends on the current

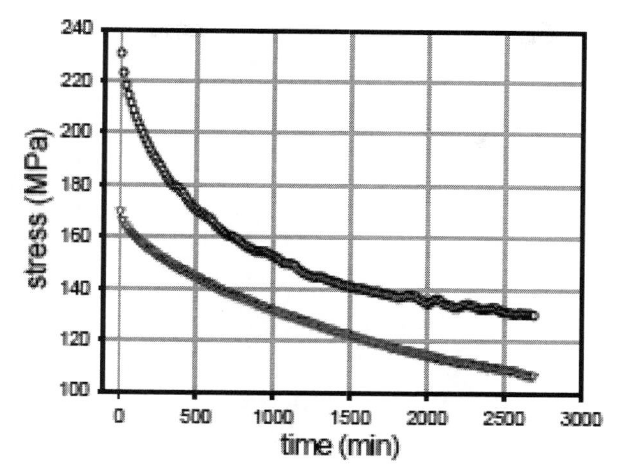

Fig. 4.20. Isothermal stress relaxation of Cu films from different initial stresses at 200°C:) passivated Cu film. Reprinted with permission from Gan, Huang, Ho, Leu, Maiz and Scherban, 'Effects of Passivation Layer on Stress Relaxation and Mass Transport in Electroplated Cu Films'. Proc. of International Workshop on Stress-induced Phenomena in Metallization. Copyright © 2004, American Institute of Physics.

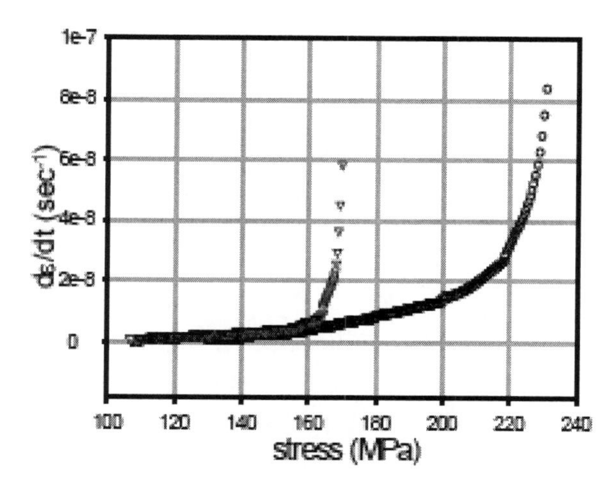

Fig. 4.21. Plastic strain rate as a function of stress in isothermal stress relaxation at 200°C for the passivated Cu film. Reprinted with permission from Gan, Huang, Ho, Leu, Maiz and Scherban, 'Effects of Passivation Layer on Stress Relaxation and Mass Transport in Electroplated Cu Films'. Proc. of International Workshop on Stress-induced Phenomena in Metallization. Copyright © 2004, American Institute of Physics.

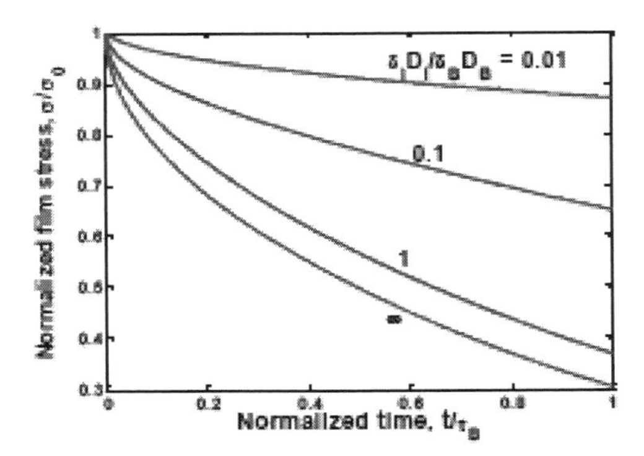

Fig. 4.22. Effects of interface diffusivity on stress relaxation of passivated films. Reprinted with permission from Gan, Huang, Ho, Leu, Maiz and Scherban, 'Effects of Passivation Layer on Stress Relaxation and Mass Transport in Electroplated Cu Films'. Proc. of International Workshop on Stress-induced Phenomena in Metallization. Copyright © 2004, American Institute of Physics.

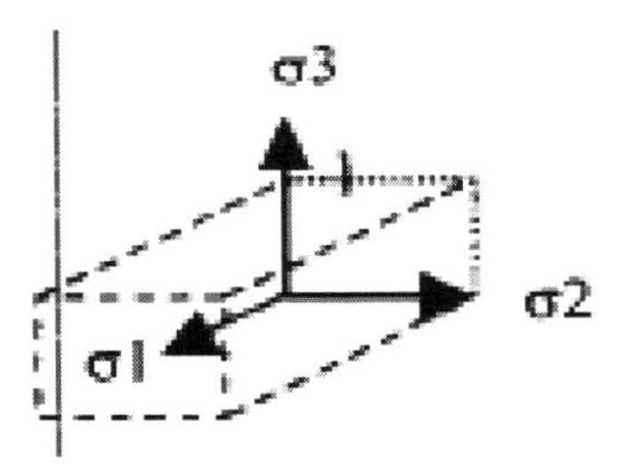

Fig. 4.23. The three different stresses in a metal film.

stress but also on the stress history or, in particular, the initial stress and the time since the relaxation started.

Using a coupled diffusion model, Gan *et al.* (Gan, Huang, Ho, Leu, Maiz and Scherban 2004) studied the effect of interface diffusivity on stress relaxation of passivated Cu film and the results are shown in Fig. 4.22. One can see that the stress relaxation is sensitive to the interface diffusivity for passivated films especially when the diffusivities follow the general trend, i.e. $\delta_B D_B > \delta_i D_i$ where the subscript *B* represent grain boundary diffusion and *i* represents interface diffusion.

The mechanical stress in the film can also be measured directly using X-rays diffraction method as described by Reimbold *et al.* (Reimbold, Sicardy, Arnaud, Fillot and Torres 2002). Denote σ_1, σ_2 and σ_3 as the stresses in the axial, transverse and normal directions as shown in Fig. 4.23, they found that the stress state is biaxial, i.e. $\sigma_1 = \sigma_2$ and σ_3 is near zero when the line width is large with respect to the thickness, and it becomes triaxial when the line width decreases with $\sigma_3 \neq 0$. They explained the results by assuming Cu is stress-free just after the 400°C annealing step, and the elastic accommodation of the differences in thermal expansion of Cu and surroundings is the main cause of stress. Also, the stresses increase when copper linewidth decreases.

When low k material is used instead of SiO_2, Young's modulus of dielectric decreases, and the triaxial stresses of Cu are substantially decreased. This is believed to be due to the absorption of deformation by material with low Young's modulus (Reimbold, Sicardy, Arnaud, Fillot and Torres 2002). Therefore, low k materials present lower density of nanodefects generated by plastic deformation as the stresses are smaller, and they should also relax the stresses build up in Cu during EM more easily. On the other hand, they also have lower Blech length, i.e. lower back stress. Also, they have lower mechanical strength and thus may induce possible crack. A good mechanical compromise should be found between the different dielectric materials.

4.2.5 *Barrier metal*

It is well known that Cu can diffuse easily into Si and SiO_2 and form Cu silicide compounds at very low temperatures (McBrayer, Swanson and Sigmon 1986; Hong, Comrie, Russell and Mayer 1991) and drift through oxide under electric field acceleration (McBrayer, Swanson and Sigmon 1986). So an appropriate barrier layer between Cu and SiO_2/Si is required in order to overcome these disadvantages. Many kinds of barrier layers had been studied, such as Ta, TaN, Ti, Ta-SiN, TiN, etc.

As a continuation from Al interconnects, Ti and TiN barrier metals were experimented, however it was found that the EM activation energy of TaN is larger than that of Ti based barrier metals, and the interfacial condition between Cu and Ti based barrier metals is also less stable than that between Cu and TaN (Hayashi, Nakano and Wada 2003). Thus Ta based barrier metal is used for the subsequent investigation.

Since the Cu EM is mainly through interface diffusion, the interface bonding strength of Cu/cap layer and Cu/barrier metal must be carefully studied. We have discussed the work on Cu/cap layer earlier; the Cu/barrier metal is an important topic to discuss for the EM performance of Cu.

Direct measurement of adhesion strength of Cu seed on barrier is difficult since the weakest interface, i.e. barrier/dielectric will fail earlier than the Cu/barrier interface during adhesion tests. Agglomeration study using AFM provides a possible way of gauging the interface bonding strength (Nogami, Joo, Lopatin, Romero, Bernard, Blum, Engelmann, Gray, Tracy, Chen, Lukanc, Brown, Besser, Morales and Cheung 1998). The agglomeration result is found to relate to the EM behavior with respect to this interface (Nogami, Romero, Dubin, Brown and Adem 1998).

Wang *et al.* (Wang, Lopatin, Marathe, Buynoski, Huang and Erb 2001) studied the adhesion strength of Cu on Ta and TaN, and they found that on the Ta barrier, no agglomeration was observed in either Cu or Cu alloy. However, on the TaN barrier, significant dewetting of the pure Cu seed was observed, while none of the three Cu-alloy seeds agglomerated. Hence, two implications to the results were made, namely a smoother and more agglomeration-resisiant Cu-seed is obtained using Cu alloy, thus extending the use of PVD seed to a thinner layer. Secondly, the alloy addition can intrinsically reduce Cu/barrier interface diffusion. With proper selection of the Cu alloy, a significant EM reduction is expected. If pure Cu is to be used, Ta barrier is superior than TaN.

The superiority of Ta over TaN is also demonstrated by Yang *et al.* (Yang, Zhang, Li, Tan and Prasad 2004) with the effect significant for line-via structure. The better performance due to Ta barrier is attributed to the stronger Cu [111] orientation and larger average Cu grain size. It is to be noted that the effect of barrier metal on Cu microstructure can not be observed for Cu blanket film as demonstrated by Muppidi *et al.* (Muppidia, Fielda and Woo 2005).

Chin *et al.* (Chin, Chiou and Wu1 2002) also studied the effect of Ta barrier layer on the EM resistance of PVD Cu interconnect, and the superiority of Ta on Cu EM can be seen from their results as summarized in Table 4.7.

Lee and Lopatin (Lee and Lopatin 2005) also studied the effect of Ta based barrier type on the microstructure of Cu, and they found that Cu(111) texture is strongest on Ta-SiN, medium in Ta and weakest on TaN. They also found that the texture of Cu film deposited on CVD barrier is much poorer than those on PVD barriers.

Furthermore, their AFM measurement on Cu surface after annealing at 400°C for 30 mins showed that the roughness value ranges from 2.8 to

Table 4.7. Peak ratio $I_{(111)}/I_{(200)}$ and grain size of Cu and Ta/Cu/Ta films (Chin, Chiou and W.-F. Wu1 2002).

		Texture				Grain size distribution	
		Relative intensity			Cu(111)	Median grain size	Lognormal
Specimen	Annealing condition	$I_{(111)}$	$I_{(200)}$	$I_{(111)}/I_{(200)}$	FWHM (deg)	(μm)	standard deviation
Cu	As-deposited	4810	720	6.68	7.84	—	—
	400°C, 1 h	5856	913	6.41	7.64	0.42	0.45
Ta/Cu/T_3	As-deposited	28516	546	52.23	6.18	—	—
	400°C, 1 h	35313	346	102.06	6.22	0.58	0.47

Table 4.8. AFM surface roughness of sputtered Cu films on various barrier materials after Cu agglomeration experiments (annealed at 400°C for 30 min) (Lee and Lopatin 2005).

Barrier	TaSiN	Ta	TaN	CVD-TaSiN	SiN
Surface roughness	2.8 nm	6.2 nm	11.8 nm	19.8 nm	40.1 nm

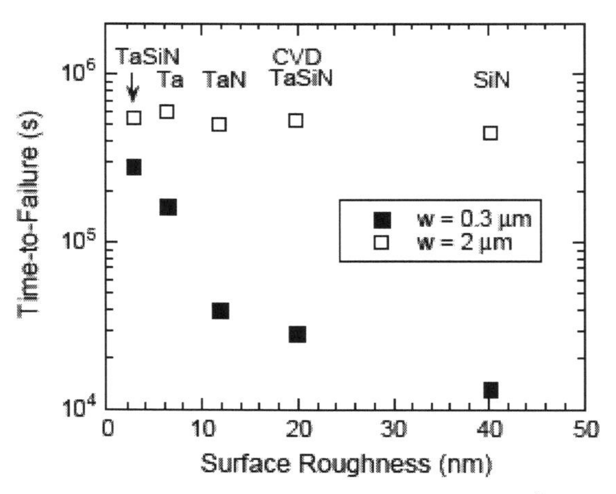

Fig. 4.24. Electromigration mean time-to-failure (j = 8 MA/cm^2 at 239°C) of electroplated Cu with various barrier types as a function of surface roughness after agglomeration experiment. Reprinted from Thin Solid Films, 492, Lee and Lopatin, 'The influence of barrier types on the microstructure and electromigration characteristics of electroplated copper', pp. 279–284. Copyright © 2005, with permission from Elsevier.

40.1 as shown in Table 4.8 (Lee and Lopatin 2005). Cu seed layer remains smooth and almost continuous on Ta-SiN, while the Cu agglomerates and forms islands on the other barrier systems. Agglomeration is driven by the minimization of surface free energy. Strong interface bonding at Cu/TaSiN will hold Cu atoms in position during annealing, resulting in a smooth surface without agglomeration. On the other hand, Cu atoms will be free to diffuse and form large islands through the interface with a weak bond strength. Since Cu EM is dominated by interface diffusion, strong interface implies better EM performance, and it is indeed observed by Lee *et al.* as shown in Fig. 4.24 (Lee and Lopatin 2005).

Also it is observed that EM, lifetimes in the sub-micron regime of linewidth are noticeably different depending on the barrier material, such that TaSiN>Ta>TaN>CVD–TaSiN>SiN despite the similar grain size. This again corresponds to the results of texture measurements and agglomeration experiments, and thus is likely attributed to the difference in the interface bond strength. The difference in TTF dependent on the barrier material becomes more pronounced as the linewidth further decreases, suggesting the interface diffusion becomes more dominant as the linewidth decreases. Note that the TTF of electroplated Cu on Ta is much longer than that of CVD Cu on Ta in wide lines probably due to much larger grain size.

While Ta can acts as a good barrier, Aubel *et al.* (Aubel, Bugiel, Krüger, Hasse and Hommel 2006) showed that by annealing the Cu/Ta at 500°C for 10 hours, Cu EM lifetime improved significantly. As they observed no influence of the annealing on the Cu grain size or the dislocation density, they postulated the enhancement in Cu EM lifetime is due to the out diffusion of Ta into the Cu layer and the diffusion of Cu into the Ta barrier. As Ta is nearly not soluble in Cu, Ta must be placed in the grain boundaries or the interfaces. Vacancies in the copper/liner and the copper/SiN-cap interface could be occupied by Ta atoms. Such reduction of vacancies leads to a repression of electromigration induced diffusion of copper atoms in these critical interfaces. For this reason, diffusion of Ta atoms out of the liners into the copper and the critical interfaces may be the cause for the drastically increased electromigration lifetime after high temperature storage observed in their experiments.

With further shrink of geometry as devices migrate from 130 nm to 90 nm and beyond, the contribution of barrier resistance to the overall interconnect resistance becomes larger thereby increasing the effective interconnect resistivity. It is therefore necessary to reduce the thickness of barrier layer along the sidewalls. However, deposition of very thin layers along sidewall using PVD technique may result in a discontinuity in the barrier film, which is unacceptable (Dixit, Padhi, Gandikota, Yahalom, Parikh, Yoshida, Shankaranarayanan, Chen, Maity and Yu 2003). Recently, researchers have shown that a thin conformal layer of TaN can be deposited by an ALCVD technique.

However, thin barrier metal will experience high mechanical stress that can render it fracture as mentioned in the early section. Schindler *et al.* (Schindler, Penka, Steinlesberger, Traving, Steinhogl and Engelhardt 2005) also found that the time to failure of Cu/Ta EM decreases significantly with the reduction in the barrier metal. Furthermore, Higashi *et al.* (Higashi, Yamaguchi, Omoto, Sakata, Katata, Matsunaga and Shibata 2004) compared ALD and PVD process for barrier metal deposition, and they found that Cu film on PVD-Ta was stable against the annealing while on ALD-TaN Cu agglomerated after annealing at 200°C for 60 mins.

In order to overcome the above-mentioned challenges, a new generation of underlayers with low resistivity and better adhesion property with Cu is needed. Ru is an air-stable transition metal with a high melting point of 2310°C and is nearly twice as thermally and electrically conductive (7.6 $\mu\Omega$ cm) as Ta (Kim, Koseki, Ohba, Ohta, Kojima, Sato and Shimogaki 2005). Additionally, Ru shows negligible solubility in Cu even at 900°C, and based on the binary phase diagram, there are no intermetallic compounds between Cu and Ru (Massalski 1990; 1995–1996). These properties of Ru show that Ru may be a good candidate for Cu glue layer between Cu and TaN and also for a Cu diffusion barrier layer.

Recently, there are several works to adapt Ru as an adhesion promoter between Cu and a diffusion barrier or seed layer for electroplating (Aaltonen, Alen, Ritala and Leskela 2003; Chyan, Arunagiri and Ponnuswamy 2003; Josell, Wheeler Witt and Moffat 2003; Chan, Arunagiri, Zhang, Chyan, Wallace, Kim and Hurd 2004; Cho, Park, Suh, Kim, Kang, Suh, Park, Ha and Joo 2004; Kwon, Kwon, Park and Kang 2004; Wang, Ekerdt, Gay, Sun and White 2004). Electroplating deposition of Cu on planar Ru substrate shows a smooth surface and accomplished the bottom-up filling deposition of Cu in trenches with Ru seed layers (Chyan, Arunagiri and Ponnuswamy 2003; Josell, Wheeler, Witt and Moffat 2003). In addition, Ru promotes 2D planar growth of metallorganic chemical vapor deposition of Cu without interfacial reactions between Cu precursor and Ru substrate (Cho, Park, Suh, Kim, Kang, Suh, Park, Ha and Joo 2004; Kwon, Kwon, Park and Kang 2004).

Kim *et al.* (Kim, Koseki, Ohba, Ohta, Kojima, Sato and Shimogaki 2005) performed experiments and they found that the average wetting angle of Cu on Ta substrate was 123.0 ± 8.5°, and that for Ru substrate was 43.0 ± 5.5°, indicating that Ru has a much better adhesion to Cu than to Ta. Also, the surface oxidation of Ta substrate has a more critical effect on Cu wettability than that of Ru. However, they showed that 20 nm Ru film was unsuitable as a Cu diffusion barrier due to Cu diffusion through the grain boundary of the Ru film as it showed columnar crystal structure. The same conclusion was also found by Yang *et al.* (Yang, Spooner, Ponoth,

Chanda, Simon, Lavoie, Lane, Hu, Liniger, Gignac, Shaw, Cohen, McFeely and Edelstein 2006). On the other hand, if Ru is used as a glue layer in between Cu and TaN, Cu EM performance was shown to be better than Cu/Ta/TaN (Yang, Spooner, Ponoth, Chanda, Simon, Lavoie, Lane, Hu, Liniger, Gignac, Shaw, Cohen, McFeely and Edelstein 2006).

Self-forming Mn barrier is another alternative barrier structure to the conventional Ta/TaN barrier layers. Koike *et al.* (Koike, Wada, Usui, Nasu, Takahashi, Shimizu, Yoshimaru and Shibata 2006) investigated the possibility of the self-forming barrier layer using Cu-Mn alloy thin films deposited directly on SiO_2. After annealing at 450°C for 30 min, an amorphous oxide layer of 3–4 nm in thickness was formed uniformly at the interface. The oxide formation was accompanied by complete expulsion of Mn atoms from the Cu-Mn alloy, leading to a drastic decrease in resistivity of the film. This is also verified by Koike *et al.* (Koike, Haneda Iijima and Wada 2006) as shown in Fig. 4.25. Also, no interdiffusion was observed between Cu and SiO_2 as shown in Fig. 4.26, indicating an excellent diffusion-barrier property of the interface oxide. Usui *et al.* (Usui, Nasu, Koike, Wada, Takahashi, Shimizu, Nishikawa, Yoshimaru and Shibata 2005) also found that the barrier formed is $MnSi_xO_y$ barrier layer, and no delamination was found in the chemical mechanical polishing process, probably because of better adhesion strength between the $MnSi_xO_y$ barrier and dielectric.

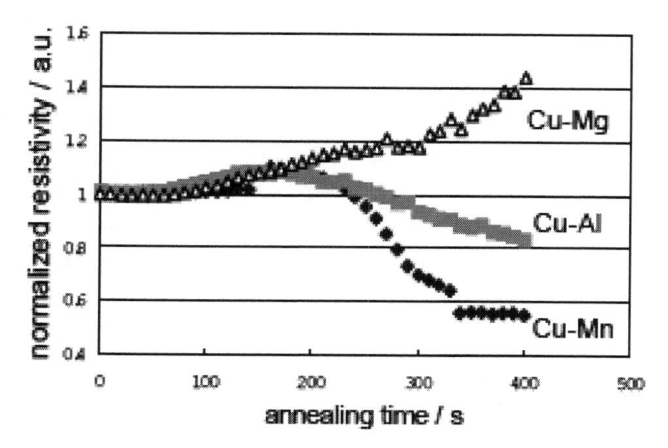

Fig. 4.25. Resistance change during heat treatment at 450°C. Resistance values are normalized by the initial value of each alloy sample. (Koike, Haneda, Iijima and Wada, 'Cu alloy metallization for self-forming barrier process', Proc. of International Interconnect Technology Conference), Copyright © 2006 IEEE.

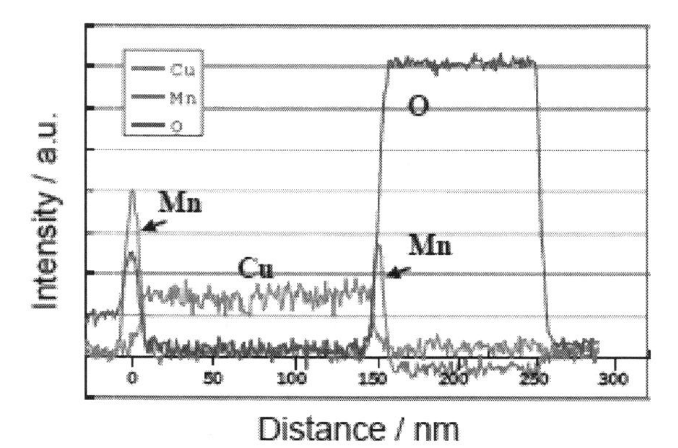

Fig. 4.26. Composition distribution in the heat-treated Cu-Mn/SiO_2. (Koike, Haneda, Iijima and Wada, 'Cu alloy metallization for self-forming barrier process', Proc. of International Interconnect Technology Conference), Copyright © 2006 IEEE.

Microstructure analysis by transmission electron microscopy showed that an approximately 2 nm thick and continuous $MnSi_xO_y$ layer is formed at the interface between Cu and dielectric of the via and trench and there is no barrier at the via bottom. This via structure without the bottom barrier provides several essential advantages: reduced via resistance; significant via-electromigration lifetime improvement due to there being no flux divergence site at the via; and excellent stress-induced voiding performance.

The reasons that the Cu-Mn alloy can self-form a stable barrier layer and reduce resistivity to the level of pure Cu are believed to be the following as proposed by Koike *et al.* (Koike, Haneda, Iijima and Wada 2006). In the first, the standard free energy of Mn oxide is slightly smaller than SiO_2. Therefore, Mn does not strongly reduce SiO_2 as Mg and Al do. This may be a reason for a slow growth rate of MnO_x in comparison to MgO_x. In the second, activity coefficient of Mn in Cu is close to unity, while that of Mg or Al is close to zero. The activity coefficient γ is a measure of deviation from a state of ideal solid solution and represents excess chemical potential. In an ideal solution without any chemical interaction between Cu and X, γ is equal to 1. In a non-ideal solution, γ is either larger or smaller than one. For $\gamma > 1$, chemical potential becomes larger than that of the ideal solution. The opposite happens for $\gamma < 1$. The element X of $\gamma > 1$ is less stable in a solid solution than the other element X of $\gamma < 1$. If a third element, such as oxygen, is present next to the Cu-X solid solution, the X atoms of $\gamma > 1$

have a greater tendency to migrate out from the solid solution to form the oxide.

For device that needs to be operate at above 600°C, most of the barrier metal will fail as Cu can diffuse through them. In such case, ZrN barrier metal will be a choice for such a device as is illustrated by Chen *et al.* (Chen, Liu, Yang and Tsao 2004).

4.2.6 *Presence of defects*

It is known that the presence of defects produce void embryos as well as atomic flux divergence as discussed earlier. The mechanism is of no difference than that in Al. On the other hand, the types and causes of defects in Cu are different.

The defects in Cu mainly come from the CMP process in the damascene process. Key defects observed during post CMP inspection are corrosion (spot or edge), missing Cu (pits), hillocks and scratches in Cu (Parikh, Educato, Wang, Zheng, Wijekoon, Chen, Rana, Cheung and Dixit 2001).

Corrosion can be eliminated by using dry in dry out process and by improving the post-polish clean. Scratches can be minimized by using a less abrasive slurry and by reducing agglomeration of the slurry particles. Chan *et al.* (Chan, Chen and Chang 2004) used IR thermal camera to characterize CMP slurry, and they found that non-Preston slurry is better in term of defect generation. For minimization of pits, in-situ annealing instead of furnace annealing is found to be effective (Parikh, Educato, Wang, Zheng, Wijekoon, Chen, Rana, Cheung and Dixit 2001). This was believed to be due to the mechanism for the void formation related to the difference in thermal expansion between Cu and oxide. Since Cu in the trenches is restricted from expanding, it may migrate upward. With slower cooling, as in the case of furnace anneals, there is considerable time for the voids to grow which may migrate to the sidewall of trenches.

Surface engineering is also a common way to remove or reduce the defects produced by CMP. We have discussed the various aspects of surface engineering in our earlier sections.

Besides the CMP process, the electroplating process will also introduce defects in copper, and the plating chemistry used has a large impact on the density of the defects (Alers, Dornisch, Siri, Kattige, Tam, Broadbent and Ray 2001). The impact of the plating chemistry is higher as the aspect ratio of the trench is increased and/or the post-plate anneal temperature is increased.

4.2.7 *Temperature gradient*

Temperature gradient induced driving force is more important for Cu than for Al as shown in Fig. 4.15. However, as line width decreases, the importance of the temperature gradient induced driving force become insignificant as shown in Fig. 4.27.

4.2.8 *Material differences*

Unlike Al, Cu interconnect system does not have W via. However, the dual damascene process may have etch stop layer at the bottom of the via for the via hole etching. The presence of the this etch stop layer, either the barrier metal or passivation layer creates a material difference and causes atomic flux divergence in the same way as in Al.

However, Kim *et al.* (Kim, Kim, Lee and Chang 2007) found that via with barrier metal at the bottom showed higher EM lifetime than those with direct contact, i.e. without any etch stop layer. They attributed the

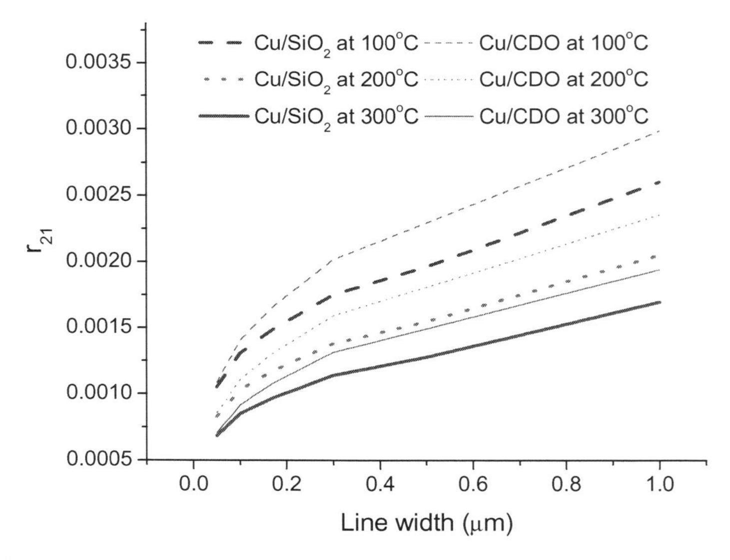

Fig. 4.27. Ratio of atomic flux due to temperature gradient induced driving force (TGIDF) over electron wind force as a function of Cu line width.

experimental results as the back stress in the via with barrier metal, thus enhanced its lifetime.

4.2.9 *Temperature*

As EM is basically atomic and vacancies diffusion, and diffusion process is very temperature dependent, the effect of temperature has an adverse effect on Cu EM just like the case of Al EM.

For Al based metallization, the difference between the activation energies for grain boundary and lattice diffusion is small. Therefore, even with the fine grain size present in the thin films, the "cross-over temperature", i.e. the temperature at which the major diffusion pathway changes from one to another, is about 275°C (Grone 1961; Chow 1988; Carr 1990; Christou 1994; Lloyd and Clement 1995; Lloyd, Clemens and Snede 1999). The procedure for estimating "cross-over temperature" can be found in ref. (Lloyd, Clemens and Snede 1999).

For Cu interconnection, however, there is no meaningful "cross-over temperature" due to the large difference between the interfacial and the lattice diffusion activation energies. The most conservative estimate calculates a "cross-over temperature" well above the melting temperature of Cu (Lloyd, Clemens and Snede 1999). It is to be noted here that the crossover from the interface to the grain boundary will not occur due to the near equivalence of their respective cross sectional areas. Thus unlike Al where we are limited to about 275°C, we can stress Cu based interconnects at extremely high temperature, and the only limitation will be from the phase diagrams for Cu alloys in principle. For pure Cu, this is not an issue as there are no phase changes before melting as the temperature is increased. However, Cu interconnect in real application is surrounded by various materials and inter-diffusion of these material may be significant during extremely high temperature EM tests. As such, copper interconnects are normally tested in the range of 250 to 350°C, although the higher limit of this temperature range is also accepted if the failure mechanism is conserved.

Since metallization temperature is important in determining the stress state of the metallization, and this stress state is strongly related to EM failure characteristics through stress gradient induced driving force, the test temperature is expected to have strong influence on the EM parameters (Hau-Riege and Thompson 2000; Tan and Zhang 2004). It is shown that for

Cu DD line-via M1 tests, 'log-normal sigma' of the EM failure distribution varies systematically with test temperature regardless of the test line-width (Roy, Kumar, Tan, Wong and Tung 2006), indicating the change in failure mechanism with the test temperature. Thus it is necessary to understand such influences of the test condition on the EM failure characteristics for accurate assessment of interconnect EM.

4.3 Design-Induced Failure Mechanism

In the design induced failure mechanisms, the key differences between Cu EM and Al EM are the line width dependence of the EM lifetime, the current direction dependence of EM lifetime, current crowding, line width transition, reservoir effect, Blech short length effect and the via structure design.

4.3.1 *Line width dependence of EM lifetime*

It is known that the EM lifetime of Al line decreases with line width and then increases as the line width decreases beyond 2 μm as shown in Fig. 4.28

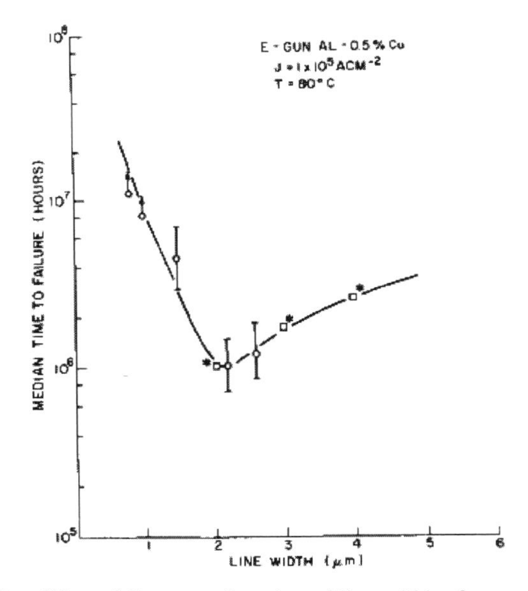

Fig. 4.28. Medium life to failure as a function of line width of e-gun Al-0.5% Cu lines at 80°C and 1 $\times$ 10^5 A/cm^2. Reprinted with permission from Vaidya, Sheng and Sinha, 'Linewidth dependence of electromigration in evaporated Al-0.5%Cu', Applied Physics Lett., vol. 30, pp. 464–466. Copyright © 1980, American Institute of Physics.

(Vaidya, Sheng and Sinha 1980). The reason for such a trend is that as line width decreases, the void size required to cause 10% increases in line resistance is smaller, and hence the EM lifetime decreases as line width decreases. However, when line width decreases beyond 2 μm, the grain size become comparable with the line width, and the micro-structure of the line become near-bamboo and bamboo if the line width is further reduced. Also, the grain texture becomes more (111) as line width reduces beyond 2 μm. As a result, the number of triple points for the grain boundary diffusion of Al atoms reduces, and the diffusivity of Al atoms also reduces with the increase in the (111) texture. Thus the EM lifetime increases with line width reduction for Al when it reduces beyond 2 μm (Vaidya, Sheng and Sinha 1980).

However, the line width dependence of Cu EM lifetime is found to be different as shown in Fig. 4.29 (Sato and Ogawa 2001). The reasons for the different line width dependency for Cu are the following.

(a) Cu EM is mainly due to Cu diffusion along interface as has been discussed, and hence its EM lifetime is less dependent on the grain structure of the line. For a given current density, the relative atomic flux flowing through the interface region is proportional to the interface

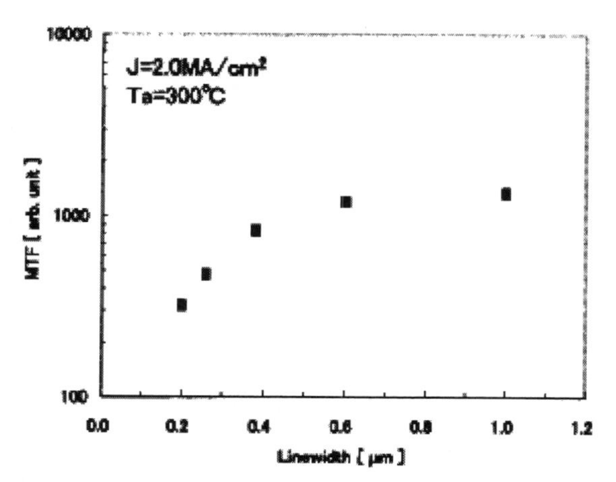

Fig. 4.29. Mean time to failure as a function of line width of Cu damascene lines with $J = 2\,\mathrm{MA/cm^2}$ at 300°C. (Sato and Ogawa, 'Mechanism of dependency of EM properties on linewidth in dual damascene copper interconnects', Proceedings of the IEEE 2001 International Interconnect Technology Conference), Copyright © 2001 IEEE.

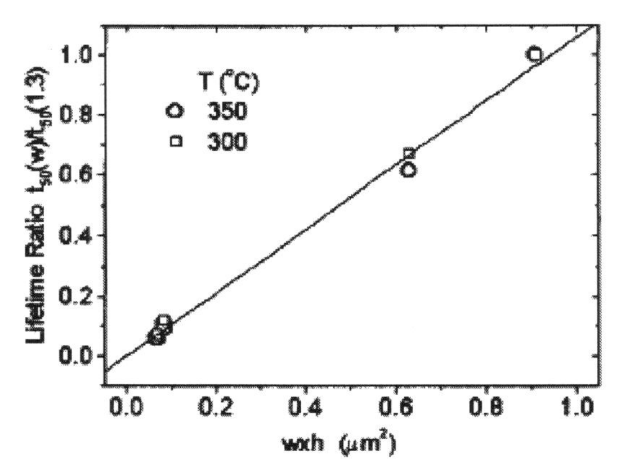

Fig. 4.30. $t_{50}(w)/t_{50}(1.3\ \mu m)$ as a function of line area. (Hu, Rosenberg, Rathore, Nguyen and Agarwala, 'Scaling effect on electromigration in on-chip Cu wiring', Proceeding of IEEE International Interconnect Technology Conference), Copyright © 1999 IEEE.

area to line area ratio, i.e. $\delta_s w/(wh) = \delta_s/h$ where δ_s is the effective thickness of the interface region, w is the line width and h is the thickness of the line. Thus, as line width and thickness are reducing, the atomic flux flowing through the interface region increases for a given current density, and thus its EM lifetime decreases (Hu, Rosenberg and Lee 1999), and experimental evidence is shown in Fig. 4.30.

Another way of looking at the same thing is the use of drift velocity concept. Hu *et al.* considered the case where grain boundary diffusion is negligible, and the drift velocity of atoms is given as (Hu, Rosenberg and Lee 1999)

$$v_d = \delta_s(2/w + 1/h)D_s Z_s^* e\rho j/(kT) + \delta_I(1/h)Z_I^* D_I e\rho j/(kT), \qquad (4.5)$$

where δ_I and δ_s denote the width of the interface and surface, D_I and D_s denote diffusivities along the interface and surface, Z_I^* and Z_s^* are the apparent effective charge number for electrons along the interface and surface, w and h are the width and thickness of the Cu line, e, ρ, k are elementary electronic charge, resistivity of the film and the Boltzmann's constant, T, j are the temperature and current density of the line.

From Eq. (4.5), one can see that the drift velocity in the bamboo-like line structure will be increased as the line width or thickness are decreased for a given current density and temperature. On the other hand, for the case of wider line where grain boundary diffusion cannot be neglected, the effective drift velocity is given as (Hu, Rosenberg and Lee 1999)

$$v_d = [(\delta_{GB}/d)(1 - d/w)D_{GB}Z^*_{GB} + \delta_s(2/w + 1/h)D_sZ^*_s]e\rho j/kT. \quad (4.6)$$

Here δ_{GB}, D_{GB}, Z^*_{GB} are the width of the grain boundary, diffusivity along the grain boundary and effective charge number along the grain boundary respectively. $d$is the grain size.

Combine Eqs. (4.5) and (4.6), Hu *et al.* obtained a plot of drift velocity vs. line width as shown in Fig. 4.31 which clearly explained the experimental observation of the line width dependence of the Cu EM lifetime.

(b) Muppidia *et al.* (Muppidia, Fielda and Woo 2005) found that the average grain size decreases with the line width for Cu due to the constraint imposed by the side walls of the trench. This is shown in Fig. 4.32. With smaller grain size, the EM lifetime will also decreases with the line width.

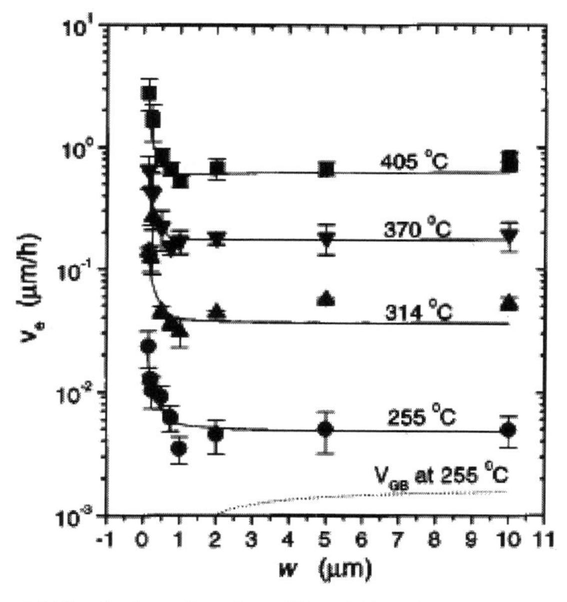

Fig. 4.31. Plot of drift velocity as function of linewidth and sample temperature. Reprinted with permission from Hu, Rosenberg and Lee, 'Electromigration path in Cu thin-film lines', Applied Physics Lett., vol. 74(20), pp. 2945–2947. Copyright © 1999, American Institute of Physics.

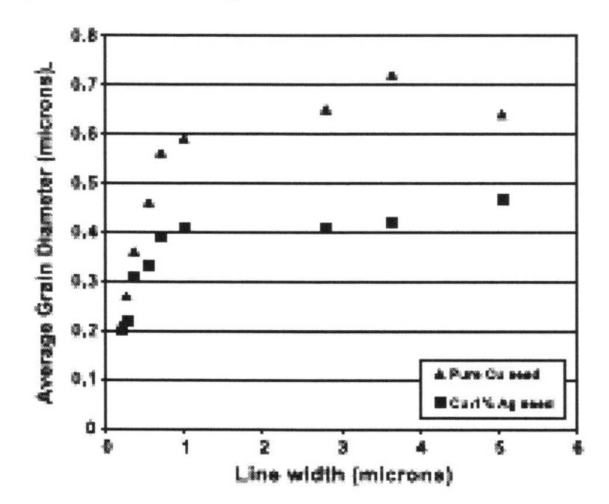

Fig. 4.32. The average grain diameter in pure Cu and Cu-Ag lines as a function of the line width. Cu-Ag lines have a lower average grain diameter than pure Cu. Reprinted from Thin Solid Films, 471, Muppidia, Fielda and Woo, 'Barrier layer, geometry and alloying effects on the microstructure and texture of electroplated copper thin films and damascene lines', pp. 63–70, Copyright © 2005, with permission from Elsevier.

While the EM lifetime decreases with the reducing line width for Cu line, this is not in contradiction to the work by Vairagar *et al.* (Vairagar, Mhaisalkar and Krishnamoorthy 2004) who reported smaller time to failure for wide line of a line-via structure. The seemingly contradiction is because they used the same current density in the lines for both the wide and narrow lines, but the via diameter was the same in both cases, hence wider line will have a much larger current in the line and larger current density in the via, this rendered higher peak current density at the line-via interface due to current crowding and resulted in early failure of the wide line. If multiple via is used to reduce the current crowding, the decreasing EM lifetime with reducing line width can also be observed with the line-via structure as shown by Cheng *et al.* (Cheng, Lin, Lee, Chiu and Wu 2007).

Besides the reduction of EM lifetime as line width reduces, we also discovered that the diffusion path for Cu changes as line width is reduced to 150 nm and below (Roy, Kumar, Tan, Wong and Tung 2006). Figure 4.33 shows the images of the failed 100 nm wide interconnect (AR = 5) where the failure is found to have occurred at the bottom Cu/Ta interface while the top Cu/cap layer interface remains unaffected. This is in contrast to the EM failure observed for the Cu line width above 0.28 μm. Many small voids

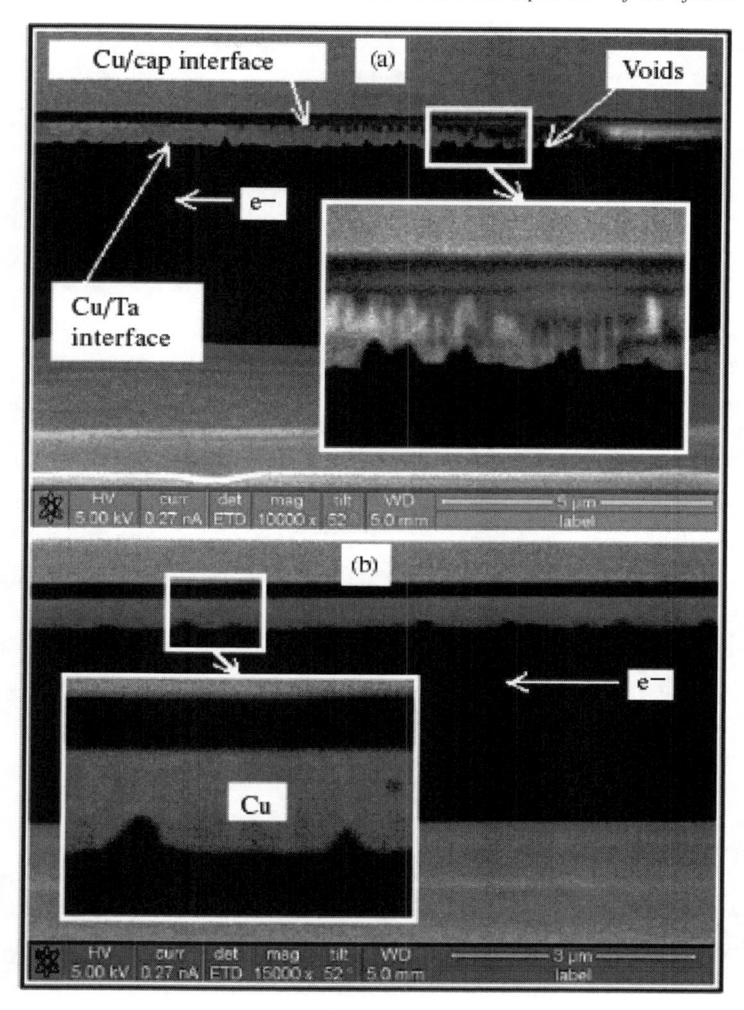

Fig. 4.33. Failed sample images of 100 nm wide lines: (*a*) near the cathode, $T = 325°C$, $j = 1.5\,MA\,cm^{-2}$ and (b) 20 μm away from the cathode, $T = 300°C$, $j = 1.5\,MA\,cm^{-2}$. Reprinted with permission from Roy, Kumar, Tan, Wong and Tung. Copyright © 2006, Institute of Physics and IOP Publishing.

are observed at the Cu/Ta interface rather than a catastrophic void in the failed line. It is also observed that the void density at the Cu/Ta interface decreases rapidly with the increase in distance from the cathode end. Thus, majority of the voids occur near the cathode end.

The Cu/Ta interface was found to be vulnerable to EM for both the 150 nm and 100 nm line width Cu interconnects, and our finite element analysis indeed shows that the Cu/Ta interface is the path of maximum atomic flux divergence, assuming the defects at Cu/cap layer is negligible. This assumption is justified because the number of defects due to CMP at the Cu/cap layer can indeed become small as the line width becomes smaller than 150 nm (Roy, Kumar, Tan, Wong and Tung 2006). Another possiblility is due to the difference in thermal expansion coefficient between Ta and Cu layer that causes Cu layer to buck out similar to the observation by Hsu *et al.* for Cu on TaN as described earlier (Hsu, Fang, Chiang, Chen, Lin, Chou and Chang 2006).

On the other hand, the failure mechanisms of the 150 and 100 nm line width Cu interconnects are quite different. For the 100 nm wide lines, the failure is due to the occurrence of very high void density at the bottom Cu/Ta interface near the cathode as seen in Fig. 4.34, indicating the possibility of grain boundary diffusion as discussed in the section on microstructure, and it is believed to be related to the fine grain size of Cu as Cu line width decreases as shown in Fig. 4.33. Cathode thinning and depletion from the bottom Cu/Ta interface is the main cause of failure for the 150 nm wide lines as shown in Fig. 4.35, indicating the interface diffusion along the Cu/Ta as the main diffusion path.

4.3.2 *Current crowding*

In multilevel Cu dual-damascene structure, current crowding is obvious at the inner corner of the via-bottom as in the case of Al. For a given line current density and via diameter, it is expected that current will be much crowded in a wide line-via structure in comparison with that in a narrow line-via because the total current is higher in the wide line-via structure. Hence, in the absence of any other line width dependent issues, a shorter EM life-time for wide line–via structure is expected with identical EM stressing in comparison with that of a narrow line–via which is indeed observed experimentally (Padhi and Dixit 2003; Vairagar, Mhaisalkar and Krishnamoorthy 2004; Tan, Roy, Vairagar, Krishnamoorthy and Mhaisalkar 2005).

The current density distribution at EM test condition for the structures under investigation was computed using coupled-field finite element method where the coupling of current and temperature was considered (Tan, Roy,

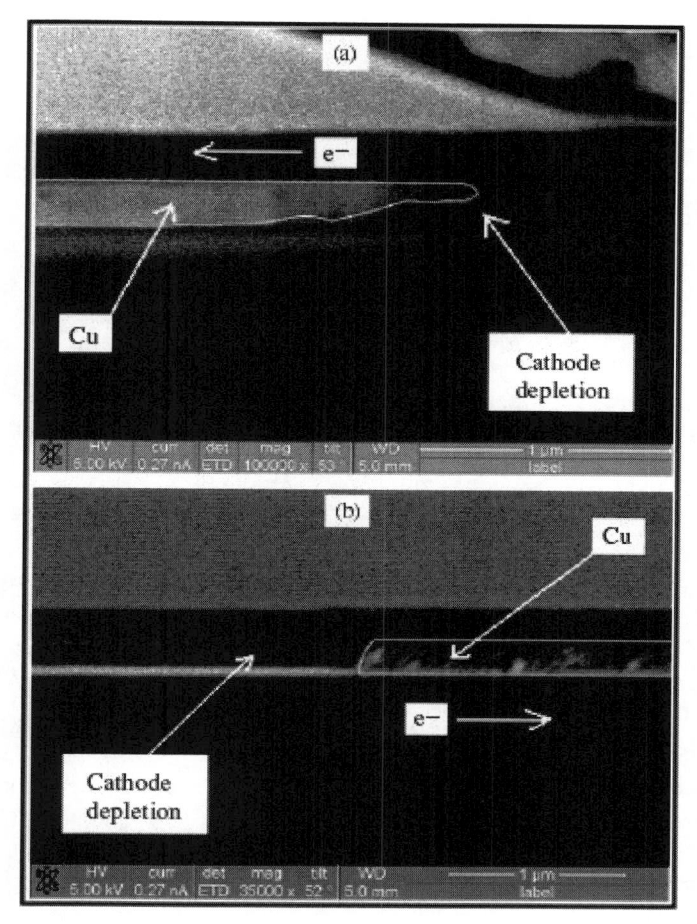

Fig. 4.34. FIB x-sections of failed samples of 150 nm wide lines (j = 3.7 MA cm^{-2}). (a) Test temperature: 325°C and (b) test temperature: 300°C. Reprinted with permission from Roy, Kumar, Tan, Wong and Tung. Copyright © 2006, Institute of Physics and IOP Publishing.

Vairagar, Krishnamoorthy and Mhaisalkar 2005) as shown in Figs. 4.34 and 4.35 for narrow and wide Cu lines with 0.28 and 0.7 μm width respectively.

Since the thickness of Ta barrier is 25 nm, Tan *et al.* (Tan, Roy, Vairagar, Krishnamoorthy and Mhaisalkar 2005) considered the Ta barrier as a perfect blocking boundary to the flow of Cu atoms, and thus interface diffusion can only occur at the interface between M1 line and the cap layer near the via or at the interface between M1 line and Ta barrier at the via bottom, and the

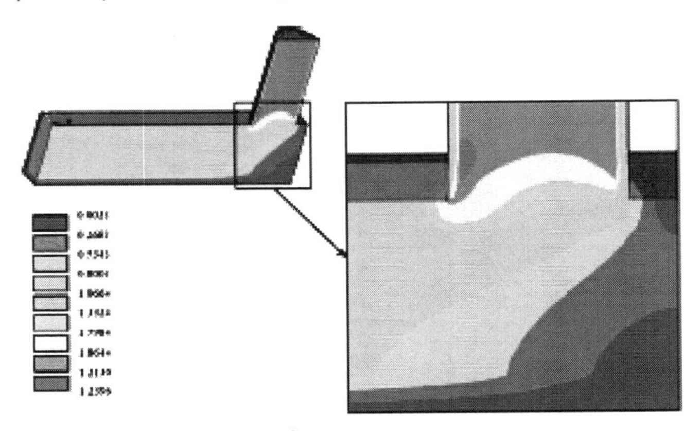

Fig. 4.35. Current density (MA/cm^2) distribution in narrow line-via structure. (Tan, Roy, Vairagar, Krishnamoorthy and Mhaisalkar, 'Current crowding effect on copper dual damascene via bottom failure for ULSI applications', IEEE Trans. Device Mater. Reliab.), Copyright © 2005 IEEE.

failure can therefore be considered as a line electromigration failure that follows the semi-classical Black's equation (Black 1969):

$$t_{50} = Bwj^{-n} \exp\left(\frac{E_a}{kT}\right), \tag{4.7}$$

where B is a constant, w the line width, j the applied current density, and n is the current density exponent. As current density distribution is not uniform as can be seen in Figs. 4.35 and 4.36, Tan *et al.* (Tan, Roy, Vairagar,

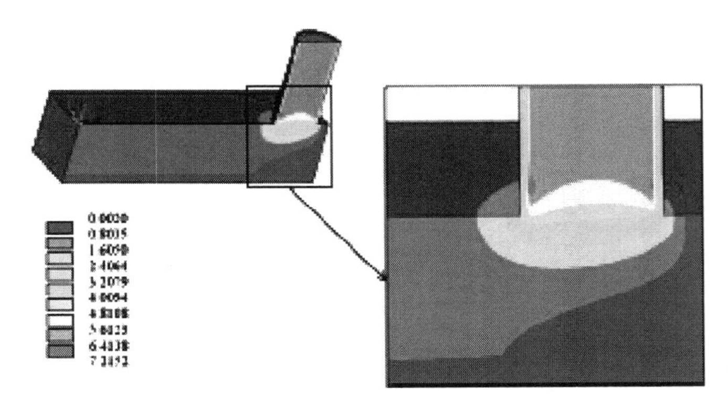

Fig. 4.36. Current density (MA/cm^2) distribution in wide line-via structure. (Tan, Roy, Vairagar, Krishnamoorthy and Mhaisalkar, 'Current crowding effect on copper dual damascene via bottom failure for ULSI applications', IEEE Trans. Device Mater. Reliab.), Copyright © 2005 IEEE.

Table 4.9. Comparison of experimental results of the ratio of $t_{50}$of narrow over wide lines with the computation using Eq. (4.7) and average current at the via-line interface (Tan, Roy, Vairagar, Krishnamoorthy and Mhaisalkar 2005).

Stressing temperature (°C)	Experimental	Computational
300	1.33	1.33
325	1.38	1.38
350	1.43	1.42

Krishnamoorthy and Mhaisalkar 2005) used the average current density at the via-line interface Eq. (4.7) and the results agreed excellently with their experiments as shown in Table 4.9.

4.3.3 *Line width transition*

As real interconnects often have junctions with narrow-to-wide transitions, the EM study of line with narrow-to-wide transition is necessary. Hau-Riege and Thompson (Hau-Riege and Thompson 2000) reported such a study for Al interconnects, and they found that the narrow-to-wide transition point is a site of atomic flux divergence as expected due to the discontinuity in diffusivities between the narrow and wide segments, which have bamboo and polygranular microstructures, respectively. They also found that the electromigration failure rate increases as the position of width transition is decreased below a critical distance (150 μm) from the electron-source via, and they attributed this to an interaction between evolution of the stress field at the electron-source via and the evolution of the stress field at the width transition. Narrow-to-wide transitions with positions greater than a critical distance from the electron-source via have much higher lifetimes which are not a function of the transition location, due to a lack of interaction between the electron–source via and the width transition. Also, changing the line width ratios did not affect interconnect reliability when the transition was 150 μm away from the electron–source via. Additional electromigration studies were also conducted on constant-width interconnects in which microstructural transitions were created using a scanned laser annealing, and they found that the experimental results are consistent with those on microstructural transitions caused by width transitions.

Similar observations are also found on Cu EM by Hau-Riege *et al.* (Hau-Riege, Marathe and Choi 2008) where two failure modes were observed, one at the transition point, and another at the cathode end with the former failed earlier. They found that voids that form in the wide region coalesce at the cathode vias at the line-end, while voids that form in the narrow region coalesce at the width transition site. Due to the $4x$ difference in width (and therefore current density), voids grow $4x$ slower in the wide region than the narrow region, thus failure mode at the transition point occurred much earlier than the other failure mode. Furthermore, they also observed the decreasing failure time of the early failure as the transition point is moving closer to the cathode end, and the independence of the position of the transition point for the later failure mode.

As Cu EM is mainly due to interface diffusion, the microstructure transition observed in Al should not be the cause for the width transition effect that was observed in Cu EM. It was likely to be due to stress evolution and interaction as proposed by Hau-Riege *et al.* (Hau-Riege, Marathe and Choi 2008). Roy *et al.* (Roy, Hou and Tan 2009) performed finite element analysis and determine the total atomic flux divergence distribution as shown in Fig. 4.37, and one can see that the flux divergence at the transition point is indeed very low if microstructure transition is not considered. As a result, void at the width transition point may not occur as indeed found in their experiments (Roy, Hou and Tan 2009). At present, there is no clear understanding on the different findings between Hau-Riege *et al.* (Hau-Riege, Marathe and Choi 2008) and Roy *et al.* (Roy, Hou and Tan 2009) because the detail of the structure used by Hau-Riege *et al.* was not given.

4.3.4 *Reservoir effect*

A reservoir can be defined as an extended region of metal line under or over a via in a multilevel interconnect system. The reservoir lengths for Cu dual damascene (DD) M1 and M2 test structures are schematically shown in Fig. 4.38. Here M1 test structure is also called downstream structure, and M2 test structure is also called upstream structure.

The reservoir hardly carriers any current because its area is not in the path of current flow. For Al-based metallization, it has been shown that the presence of a reservoir improves EM life-time and it is believed that this reservoir supplies atoms to the void areas and delays the failure process

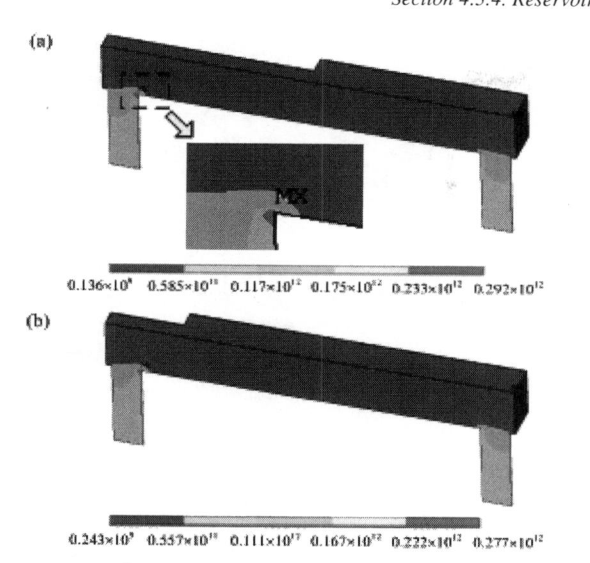

Fig. 4.37. AFD (atoms/μm^3-sec) distribution in width transition test structure. (a) $r = 0.5$ and (b) $r = 0.07$. Here r is the ratio of the length of the narrow segment over that of the wide segment. Reprinted from Microelectronics Reliability, Roy, Hou and Tan, 'Electromigration in width transition copper interconnect', Copyright © 2009, with permission from Elsevier.

(Domenicucci, Filippi, Choi, Hu and Rodbell 1996). A more systematic study was carried out by Dion (Dion 2001) and it was found that the increase in the EM life-time is related to the natural-log of the reservoir length for a given line-width. Le *et al.* reported that the EM life-time can be increased by 100% using a reservoir (Le, Ting, Tso and Kim 2002). Park *et al.* found an optimal length of the reservoir (Park and Jeon 2003). From the simulation model and experiments, Nguyen *et al.* concluded that reservoir-via layout such as square or row do not play a significant role in EM life-time but the reservoir area does (Nguyen, Salm, Wenzel, Mouthaan and Kuper 2002).

The reservoir effect is also observed in Cu metallization. Shao *et al.* investigated the reservoir effect for Cu DD M2 test structure (Shao, Vairagar, Tung, Xie, Krishnamoorthy and Mhaisalkar 2005). Similar to Al-based metallization, they also found an increase in EM life-time with the increase in the reservoir length. They found that the optimal reservoir length for the M2 test structure is around 0.08 μm.

To understand better the mechanism of the enhancement effect of the reservoir length, Vairagar *et al.* (Vairagar, Mhaisalkar and Krishnamoorthy

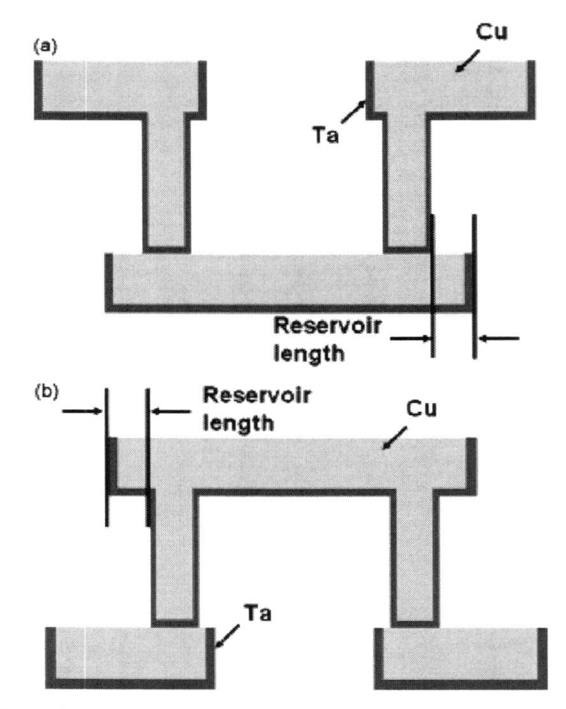

Fig. 4.38. Schematic representation of reservoir length in Cu DD structures for the case of (a) M1 test and (b) M2 test. Reprinted from Materials Science and Engineering Review, 58, Tan and Roy, 'Electromigration in ULSI interconnects', pp. 1–75, Copyright © 2007, with permission from Elsevier.

2005) performed an in-situ secondary electron microscopy observation of electromigration on Cu line-via structure with reservoir. They observed that voids were formed on the line away from the via, and these voids move toward the via at the cathode due to the electron wind force and pinned above the via as well as slightly into the reservoir. The void size was increased to span the via region only after agglomeration of additional voids moving along the Cu/SiN_x interface leading to failure.

With this observation, and since the movement of the voids along the Cu/SiN_x is due to electron wind force, and that the current in the reservoir is very small or even zero, Gan *et al.* (Gan, Shao, Mhaisalkar, Chen, Li, Tu and Gusak 2006) computed the current density distribution of the structure and construct a current density contour. They defined the contour of 0.4 MA/cm^2

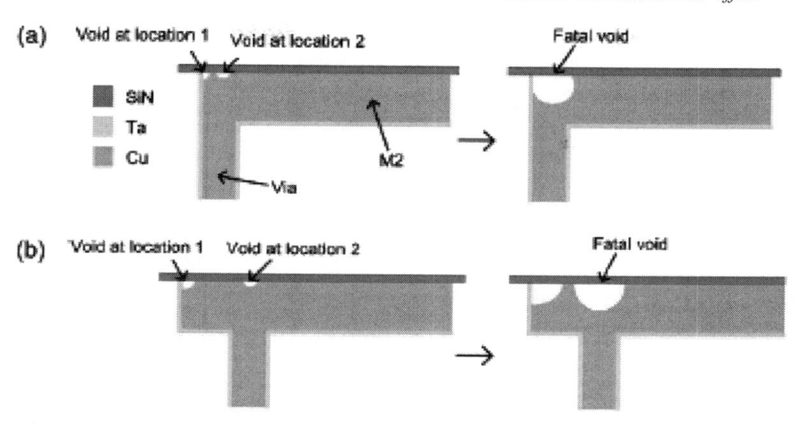

Fig. 4.39. Void growing (through void agglomeration) process in the reservoir region. (a) For Metal-2 structure with no reservoir extension, the initial two voids collapse into a bigger void and a critical void forms above via finally. (b) For Metal-2 structure with large reservoir extension, the critical void is solely developed from the initial void at Location 2. Reprinted with permission from Tan, Hou and Li, 'Revisit to the Finite Element Modeling of Electromigration for Narrow Interconnects', Journal of Applied Physics, vol. 102(3), p. 033705. Copyright © 2007, American Institute of Physics.

as critical current density since it corresponds to the saturation reservoir length of 80 nm as observed in their experiment, and they proposed that voids movement will slow down as they approach the cathode due to reduced current density and finally pinned at the critical current density contour where agglomeration of voids.

While the above explanation seems plausible, recent observation of the enhancement factor of the reservoir effect by Hou and Tan (Hou and Tan 2009) showed that the enhancement factor increases as current density decreases. This is in contradiction to the above postulation by Gan *et al.* (Gan, Shao, Mhaisalkar, Chen, Li, Tu and Gusak 2006) because for smaller current density, the contour of critical current density will be closer to the via and thus the effect of reservoir effect will be smaller. Tan *et al.* (Tan, Hou and Li 2007) performed finite element analysis and showed the presence of two maximum AFD sites above the via at the cathode as shown in Fig. 4.39. The main driving force for the maximum AFD is due to stress gradient. These maximum AFD sites are the locations where voids will be pinned and agglomerate as shown by the Monte Carlo modeling by Li *et al.* (Li, Tan and Hou 2007) due to energy minimization principle, causing voids to grow.

When the reservoir length increases, the distance between the two maximum AFD increases, and more voids can be accommodated without growing downward to the via. However, when the reservoir length grows beyond the optimal length, the distance becomes so large that void agglomeration will be so large to reach the via before they reach the other maximum AFD at the end of the reservoir length, rendering no further enhancement of EM lifetime as reservoir length continue to increase. With this postulation, they showed that the optimal reservoir length should be 80 nm which agrees with the experimental observation (Tan, Hou and Li 2007). They were also able to explain the increasing enhancement effect with reducing current density and/or temperature as observed experimentally, and an empirical equation is derived to describe the enhancement factor as a function of current density and temperature given below (Hou and Tan 2009), here the enhancement factor M is defined as the $MTTF_{reservoir}/MTTF_{noreservoir}$:

$$M = 0.443\left(T_0 - T_{oven} - \frac{1.785 \cdot T_{oven}}{573} j^2\right)^{0.9356}. \tag{4.8}$$

The dependences of the enhancement factor with current density and temperature are shown in Figs. 4.40 and 4.41.

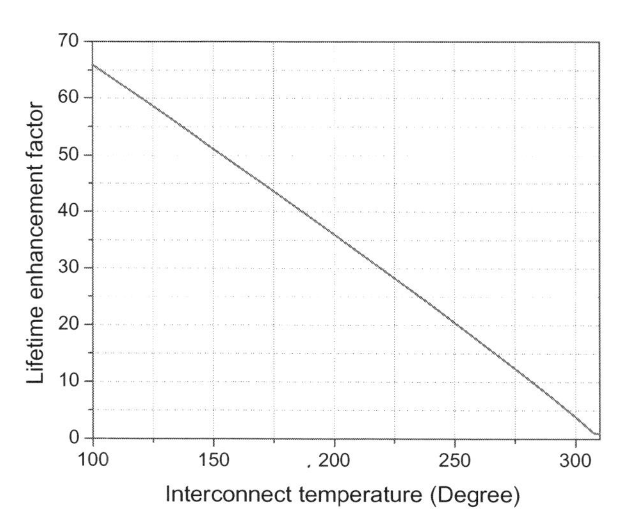

Fig. 4.40. Lifetime enhancement factor as a function of interconnect temperature. Current density is at 0.1 MA/cm^2. (Hou and Tan, 'Reservoir effect in Cu interconnects', submitted to IEEE Trans. on Devices and Materials Reliability), Copyright © 2009 IEEE.

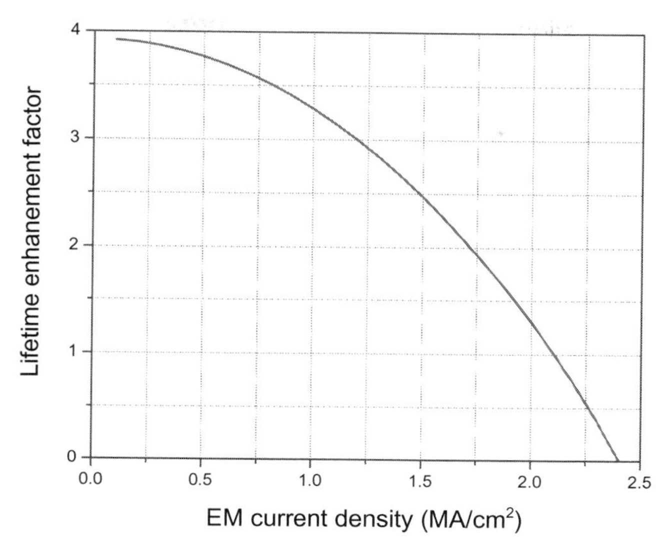

Fig. 4.41. Lifetime enhancement factor as a function of interconnect stress current. Oven temperature is at 300°C. (Hou and Tan, 'Reservoir effect in Cu interconnects', submitted to IEEE Trans. on Devices and Materials Reliability), Copyright © 2009 IEEE.

4.3.5 *Current direction dependence of EM lifetime*

For a line-via structure, it is found that the EM lifetime depends on the current direction for Cu, similar to Al (Kisselgof and Lloyd 1997). Cheng *et al.* (Cheng, Lin, Lee, Chiu and Wu 2007) found that Cu EM lifetime in downstream electron flow is smaller than that in upstream flow, similar to that of Al interconnect. Similar observations were also found for Cu interconnects by other workers such as Lin *et al.* (Lin, Chang, Su and Wang 2007), Gan *et al.* (Gan, Thompson, Pey, Choi, Tay, Yu and Radhakrishnan 2001) and Padhi *et al.* (Padhi and Dixit 2003).

The different EM performances for the upstream and downstream current flow is attributed to the different void locations for the two cases in Al (Kisselgof and Lloyd 1997) and in Cu interconnects (Gan, Thompson, Pey, Choi, Tay, Yu and Radhakrishnan 2001). While the void location for the downstream current flow is the same for both Al and Cu, the void locations are different in the case of upstream flow for Al and Cu. This is because for the case of upstream flow, the dominant diffusion path in Cu is the Cu/cap layer and the void nucleation location is at the upper cornet of the M2 line, and the void grows in the direction of the electron flow, resulting in a partially spanning void (Gan, Thompson, Pey, Choi, Tay, Yu and Radhakrishnan 2001). On the other hand, grain boundary diffusion is the dominant path in Al, and the void nucleate and grow on a line away from the via (Kisselgof and Lloyd 1997).

For the case of Cu interconnects, since the void grows along the Cu/cap layer interface as well as across the thickness of the line with the former proceeds at the faster rate, the void is spanning only partially across the line. As a result, the resistance of the test structure will increase only slightly while there is still a high-conductivity Cu path for conduction. An open-circuit failure will result only when the void grows to span the whole thickness of the metal line. When this happens, all the current is forced to flow through the thin Ta liner layer, inducing significant Joule heating in the liner and leading finally to an open-circuit failure (Gan, Thompson, Pey, Choi, Tay, Yu and Radhakrishnan 2001).

For the downstream current flow, voids will preferentially nucleate at the Cu/etch stop layer near the cathode end of the M1 line. As such voids grow, even a partially spanning void can block current through the via in the case where dielectric etch stop layer such as Si_3N_4 is used since it does not provide a conducting path to shunt current. Thus, a much smaller void volume is required for failure of M1 structures than M2 structures, resulting in the longer lifetimes for the M2 structures (i.e. upstream current flow) (Gan, Thompson, Pey, Choi, Tay, Yu and Radhakrishnan 2001).

With this understanding, one must observe the direction of current flow through a dual-damascene via in the assessment of the reliability of an integrated circuit with Cu metallization. Also, this effect should be accounted for in the design and layout of circuits with optimized reliability.

The mechanisms of the void locations observed in the upstream and downstream current flow can be understood from the total atomic flux divergence (AFD) distribution as computed using finite element method by Roy and Tan (Roy and Tan 2007) as well as Tan *et al.* (Tan, Hou and Li 2007). Figure 4.42 shows the AFD distribution for the downstream case, and Fig. 4.43 shows the AFD distribution for the upstream case.

With the different failure locations, Padhi *et al.* (Padhi and Dixit 2003) showed that the activation energies for both the upstream and downstream cases are the same, but their current exponents n are different, being 1.44 and 1.87 with upstream and downstream conditions respectively. This can

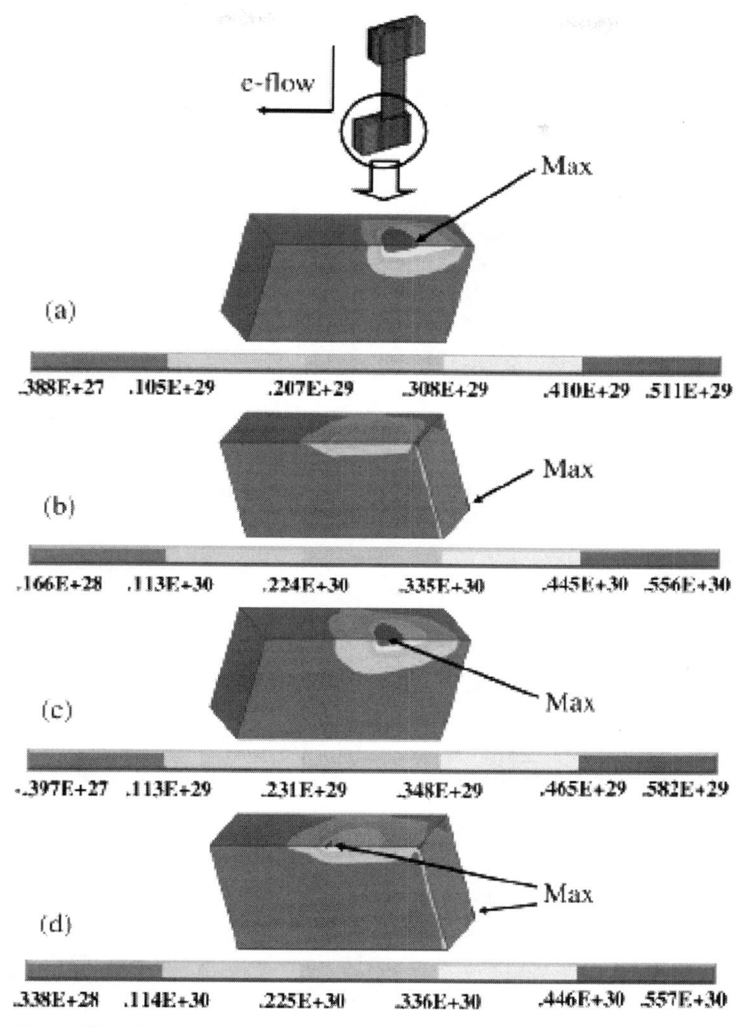

Fig. 4.42. AFD distributions due to (a) electron-wind force, (b) stress gradient, (c) temperature gradient and (d) total of them for Cu line with downstream current flow (i.e. M1 test). Reprinted from Thin Solid Films, 515, Roy and Tan, 'Probing into the asymmetric nature of electromigration performance of submicron interconnect via structure', pp. 3867–3874, Copyright © 2007, with permission from Elsevier.

be understood that for the upstream case, the void must grow to a large extent before significant resistance change can be observed as mentioned earlier, hence void growth is a major part for the upstream case, and n is closer to 1 as $n = 1$ represents the case of void growth (Lloyd 1991). On

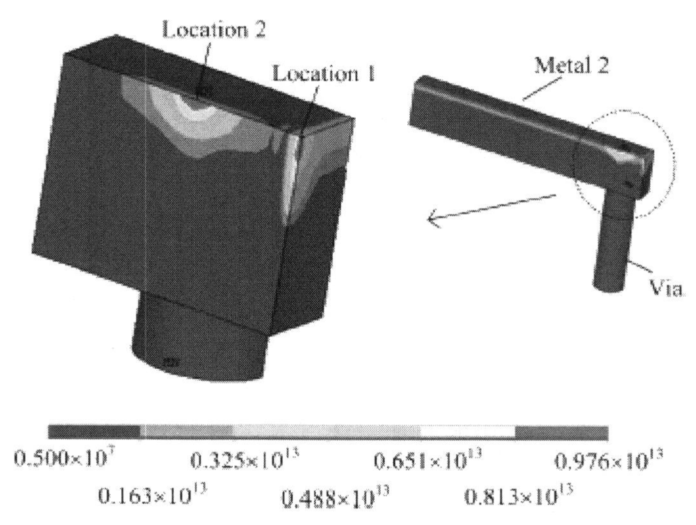

Fig. 4.43. AFD distributions and the maximum AFD sites are labeled as "Location 1" and "Location 2" at Cu/SiN interface with upstream current flow (i.e. M2 test). Reprinted with permission from Tan, Hou and Li, Copyright © 2007, American Institute of Physics.

the other hand, in the case of downstream, a small void can already severe the interconnect, and thus void nucleation is a major part of the voiding process in this case, and n is closer to 2 since $n = 2$ represents the case of void nucleation (Lloyd 1991).

Lee and Oates (Lee and Oates 2006) found that the time variation of resistance for upstream and downstream cases are different. For the case of downstream, the resistance change is either abrupt or progressively increasing whilst for upstream case, only progressive resistance increase is observed. The progressive resistance increases for the downstream case are associated with voids that are either under the via but displaced along the line or void in the line attached to the via. In these cases, the trench liners maintain electrical redundancy and hence result in progressive resistance change. The abrupt resistance changes are due to voids directly under vias with the void shape in the form of a narrow slit. Thus multi-model failure time distribution is commonly observed for the case of downstream EM test. For the case of upstream, the void can be either within the via due to non-optimal process or along the line connected to the via. If process is optimal, mono-model failure time distribution will be observed.

Another current direction dependence of EM lifetime is on the width transition structure. Roy *et al.* (Roy, Hou and Tan 2009) studied the effect of current direction on the EM of width transition structure, and they found that the EM lifetime is much shorter if current is flowing from the narrow segment to the wide segment. This is because of the following. Firstly, if narrow segment is on the cathode end, voids will be formed in the narrow segment. As a result, a smaller void will cause a significant increase in the line resistance, and hence a shorter lifetime. This is further accelerated as the current density in the narrow segment is higher which accelerate the formation and growth of the void. This finding will be useful for reliability assessment of metal line as well as circuit design and layout to ensure good reliability.

The effect of reservoir length in enhancing EM lifetime is also found to be dependent on the current direction. Hau-Riege *et al.* (Hau-Riege, Marathe and Choi 2008) found that while current is flowing away from the reservoir can enhance the EM lifetime as discussed in the earlier section, when the current is flowing toward the reservoir, there is a reduction of EM lifetime or loss of the lifetime enhancement benefit. This is because the presence of the reservoir length reduced the building up of the back stress to a slower rate as it does not serve as an effective blocking boundary. Hence care must be taken during IC layout so that the negative effect of the reservoir length may not be invoked.

4.3.6 *Blech short length effect*

If the line is bound by zero-flux boundaries, such as contacts and vias lined with an impermeable diffusion barrier, the stress along the length of the line will evolve until there is a uniform stress gradient, at which point the back force due to this gradient balances the electron wind force. This has been discussed in Chapter 3 for Al interconnects. In summary, Al-based interconnects with shunt layers exhibit three different modes of short-line effects, depending on the magnitude of jL and jL/B where B is the effective bulk modulus. For $jL < (jL)_{crit}$ where $(jL)_{crit} = \frac{\Omega\Delta\sigma_{crit}}{z^{*}e\rho}$ and $\Delta\sigma_{crit}$ is the minimum stress difference that leads to void nucleation, no void nucleation occurs, and the line is immortal. For $jL > (jL)_{crit}$, voids nucleate, and the current must flow through the higher resistivity refractory-metal-based shunt layer. The resistance of the line increases but eventually saturates due to back stress effects, and the maximum relative resistance increase at steady state is proportional to (jL/B). If the resistance increase is acceptable, the line is considered to be immortal; otherwise it is mortal (Hau-Riege 2002).

However, for the case of Cu interconnects, Hau-Riege *et al.* (Hau-Riege 2002) found that the above-mentioned criteria do not apply from their experimental observations. For the case of $jL < (jL)_{crit}$, the mode of immortality was not observed for Cu for jL values as small as 2100 A/cm, unless a much smaller value is used which is not useful for the IC design. This is because the critical stress of void nucleation in Cu is less than 41 MPa (Hau-Riege 2002) as compared with Al which is 600 MPa (Hau-Riege and Thompson 2000).

For $jL > (jL)_{crit}$, voids nucleate in both Al and Cu interconnects. In the case of Al, void growth eventually saturates, leading to a maximum resistance increase that is proportional to jL/B as mentioned earlier. However, this was not observed by Hau-Riege (Hau-Riege 2002). Instead, the resistance of the Cu lines either remained constant for the duration of the test, or the resistance increased rapidly, leading to abrupt open-circuit failure. Therefore, (jL/B) is not the correct figure of merit to describe the immortality of Cu lines. They found that the immortality of Cu is probabilistic and it depends on (jL^2/B) with the relationship as shown in Fig. 4.44.

Also, with the introduction of low-K dielectric, the effective bulk modulus for Cu interconnect is also much reduced especially at high test temperature. Hou and Tan (Hou and Tan 2007) computed the effective bulk modulus of Cu/low-K system using the bilinear model of the stress-strain relationship of Cu as shown in Fig. 4.45 (Shen and Ramamurty 2003), and

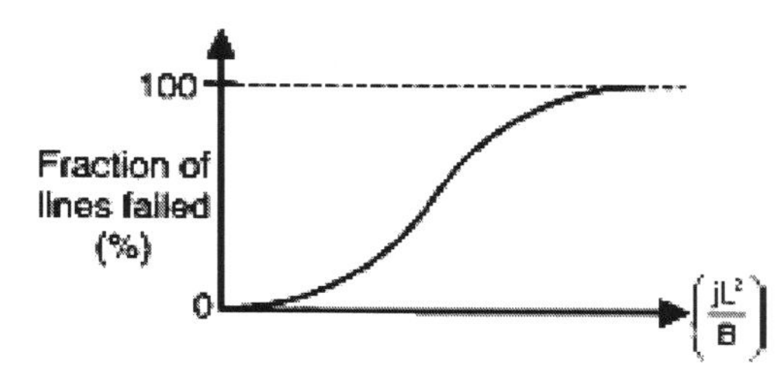

Fig. 4.44. Probabilistic immortality conditions for Cu-based interconnects. The fraction of lines failed increases monotonically with (jL^2/B). Reprinted with permission from Hau-Riege. Copyright © 2002, American Institute of Physics.

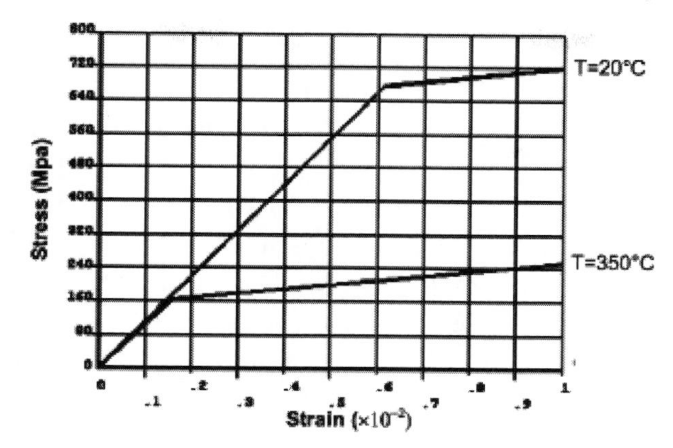

Fig. 4.45. Stress-strain behavior of Cu by bilinear model. (Hou and Tan, 'Blech effect in Cu interconencts with oxide and low-k dielectrics': 'Proceeding of 14th International Symposium on the Physics and Failure Analysis of Integrated Circuits, p. 65'), Copyright © 2007 IEEE.

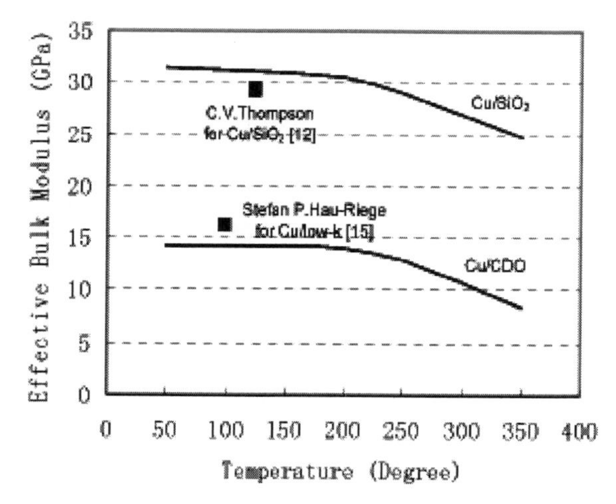

Fig. 4.46. Simulation results of effective bulk modulus for both Cu/oxide and Cu/CDO interconnections. (Hou and Tan, 'Blech effect in Cu interconencts with oxide and low-k dielectrics': 'Proceeding of 14th International Symposium on the Physics and Failure Analysis of Integrated Circuits, p. 65'), Copyright © 2007 IEEE.

the agree with other reported works (Korhonen, Borgesen, Brown and Li 1993; Hau-Riege and Thompson 2000) as shown in Fig. 4.46. With the reduced effective bulk modulus, the back stress in Cu/low-K metallization system will be weaker.

4.3.7 *Via structure design*

In order to improve the EM performance of interconnections, different via structures are sometimes employed. We have discussed this for Al based metallization. For the case of Cu, multiple vias were being proposed and studied. Marras *et al.* (Marras, Impronta, De Munari, Valentini and Scorzoni 2007) observed the increase of EM lifetime when higher number of vias are used as expected. However, they found that the mechanism of improved EM lifetime is not due to the "back-up" path provided for current flow, but rather the reduction of the current density through each via. From the plot of the median time to failure (MTF) versus the number of via as shown in Fig. 4.47, they found that the current does not split equally across the number of vias. Instead, the current was split in such a way that the effective current density follows the relationship given as below (Marras, Impronta, De Munari, Valentini and Scorzoni 2007)

$$J_{eff} = J_{test} \cdot \exp\left[\frac{\gamma}{n}\left(1 - \frac{1}{vias}\right)\right], \tag{4.9}$$

where γ is fitting parameter of the plot in Fig. 4.47, and $vias$ is the number of via, n is the current density exponent in the Black's equation. With Eq. (4.9), the Black's equation becomes (Marras, Impronta, De Munari, Valentini and

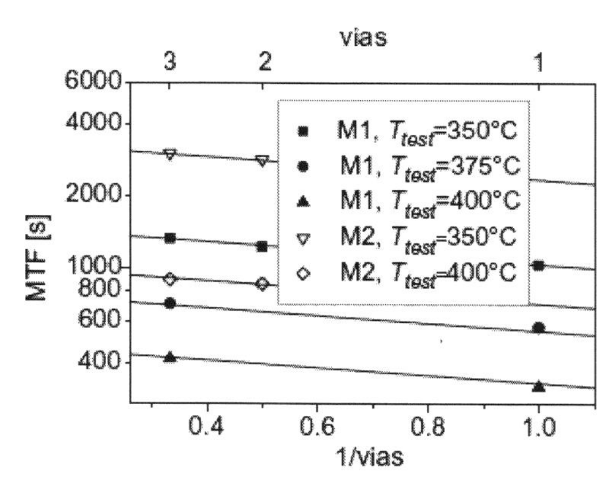

Fig. 4.47. MTF of downstream M1 and M2 structures featuring 1, 2, 3 vias, in different stress conditions. Reprinted from Microelectronics Reliability, 47, Marras, Impronta, De Munari, Valentini and Scorzoni, 'Reliability assessment of multi-via Cu-damascene structures by wafer-level isothermal electromigration tests', pp. 1492–1496, , Copyright © 2007, with permission from Elsevier.

Scorzoni 2007):

$$MTF = A \cdot J_{eff}^{-n} \cdot \exp \frac{E_a}{kT_{test}} \cdot \exp\left[-\gamma \cdot \left(1 - \frac{1}{vias}\right)\right]. \tag{4.10}$$

For double via structure, Orio *et al.* (Orio, Ceric, Carniello and Selberherr 2008) found from their modeling that either the outer via or inner via will fail depends on the dominant diffusion path during EM of either grain boundary or interface diffusion. For the case of interface diffusion dominated EM, inner via will fail, and for GB diffusion, outer via will fail. However, experimental results are yet to confirm the findings.

4.4 Electromigration Testing

Electromigration testing for Cu metallization is basically no different from that for Al metallization. However, in view of the absence of "cross-over temperature" as mentioned in Sec. 4.2.9, there is no upper limit on the temperature of EM test for Cu, in contrast to Al metallization. With this knowledge, Roy and Tan (Roy and Tan 2008) performed EM test at a very high current density of 8 MA/cm^2, and they found that the activation energy obtained has the same value as obtained at nominal test current density of 0.8 MA/cm^2. The rise in line temperature due to high current density therefore does not affect the diffusion path of EM as expected. However, the current density exponent n determined from very high current density is different from that at nominal current density, and a log-linear relationship is found between n and $\ln(j)$. The physics behind this log-linear relationship remain unknown at this point. Nevertheless, with the experimental report by Roy *et al.* (Roy and Tan 2008), one can shorten the test time by a factor of 60 to determine E_a and n.

Another way to determine E_a in a short time is to use the slope of the resistance change curve versus time as proposed by Federspiel *et al.* (Federspiel, Doyen and Courtas 2007). Similar method has been used on Al films by Rosenberg *et al.* (Rosenberg and Berenbaum 1968). However, this method was not used for Al films later on because line width scale down made the progressive part of the resistance change not observable. On the other hand, for Cu interconnects even down to 100 nm, such progressive part of resistance change remains observable, and hence the method remains applicable for Cu.

For a typical resistance change curve as shown in Fig. 4.48, Federspiel *et al.* (Federspiel, Doyen and Courtas 2007) found that $\ln(slope)$ has a linear relationship with $1/kT$ as shown in Fig. 4.49, and the gradient of the linear relationship gives the value of E_a. This is attributed to the fact that the slope

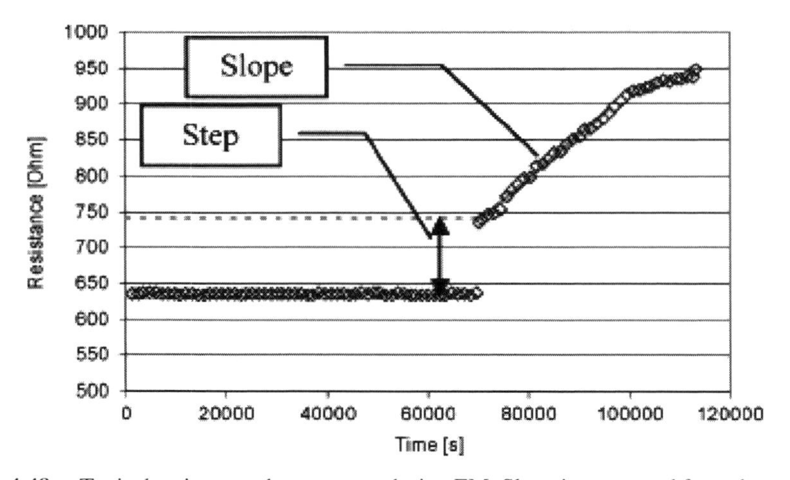

Fig. 4.48. Typical resistance change curve during EM. Slope is computed from the end of the step to 30% of resistance increase. (Federspiel, Doyen and Courtas, 'Use of Resistance-Evolution Dynamics During Electromigration to Determine Activation Energy on Single Samples', IEEE Trans. Device Mater. Reliab.), Copyright © 2007 IEEE.

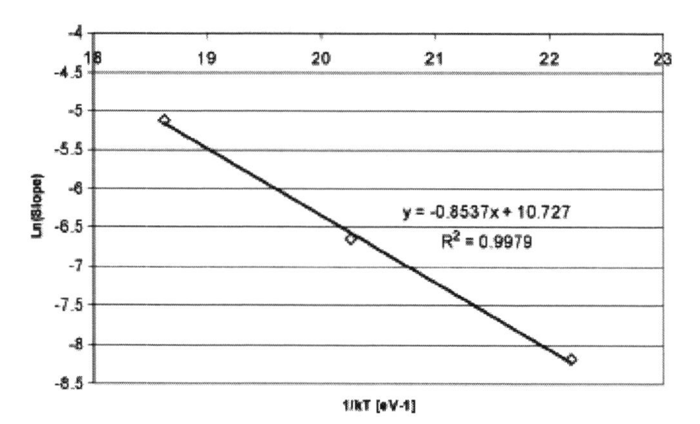

Fig. 4.49. Arrhenius plot of median slope as a function of temperature. (Federspiel, Doyen and Courtas, 'Use of Resistance-Evolution Dynamics During Electromigration to Determine Activation Energy on Single Samples', IEEE Trans. Device Mater. Reliab.), Copyright © 2007 IEEE.

is a function of the drift velocity of void growth, and drift velocity in term is related to the E_a of the atom diffusion (Doyen, Federspiel, Arnaud, Terrier, Wouters and Girault 2007).

Although the use of the slope of resistance profile to extract E_a is useful, one needs to take note that the slope depends on the barrier metal cross sectional area, barrier metal and Cu line resistivities, the Cu line length, and line thickness/width (Castro, Hoofman, Michelon, Gravesteijn and Bruynseraede 2007). Any variation in the above within sample set can render variation of the E_a values so determined. Furthermore, Lee and Oates (Lee and Oates 2006) showed that abrupt resistance change in Cu line is a more concerned via reliability issue than the failure mode with progressive resistance change. Hence, the practicality of the E_a extraction method proposed by Federspiel *et al.* (Federspiel, Doyen and Courtas 2007) may be questionable.

While Cu EM test provides way to reduce test time for obtaining E_a, one also needs to be extra careful in conducting EM test on Cu/low-K metallization system because of the effect of EM stress on inter-metal dielectric reliability. Li *et al.* (Li, Bruynseraede, Groeseneken, Maex and Tokei 2007) performed experiments and found that EM stress does degrades the lifetime distribution of the subsequent TDDB tests as shown in Fig. 4.50. Here TDDB test refers to Time Dependent Dielectric Breakdown test which measures the leakage current through the dielectric for a given applied voltage across it. They also found that the impact of EM stress on TDDB is neither the thermal annealing effect nor link to mechanical stress, but it is due to the pre-voltage stress of the inter-metal dielectric before the actual TDDB test. The potential difference between the EM stressed copper wire and its neighboring floating wire appears as a voltage stress over the inter-metal dielectric. A simple FEM simulation done by them showed that the induced voltage is on the order of 30 V, which is close to actual TDDB voltage.

Therefore, the reliability of the whole structure is a competition between copper line failure and dielectric breakdown, and BEOL failures during electromigration tests should be carefully analyzed to separate electromigration failures from dielectric breakdown especially for technology nodes beyond the 32 nm node using statistics that can differential different failure mechanisms as mentioned in Chapter 3.

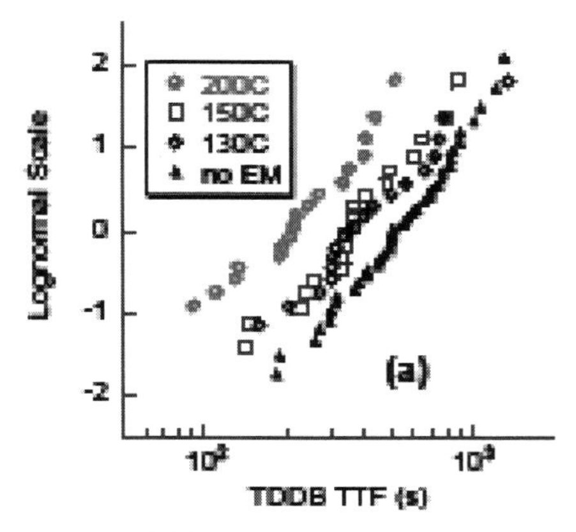

Fig. 4.50. TDDB test results of 90nm line width and spacing after 1.8 mA EM stress at 200°C for 900 seconds. (Li, Bruynseraede, Groeseneken, Maex and Tokei, 'On the Interaction Between Inter-Metal Dielectric Reliability and Electromigration Stress', Proceedings of 45th IEEE International Reliability Physics Symposium, pp. 642–643), Copyright © 2007 IEEE.

4.5 Statistics of Cu Electromigration

The statistical analysis of the EM test data of Cu is similar to that of Al, except that dual damascene Cu vias exhibit multiple electromigration failure modes (Gill, Sullivan, Yankee, Barth and Glasow 2002), and consequently, multi-modal is often observed in Cu electromigration failure distributions. For the current flowing downstream, two type of resistance time variation profiles are observed as mentioned earlier, and the one with abrupt resistance change have a larger σ which implies that it dominates at low failure percentiles, and thus determines via reliability.

As larger σ in the failure time distribution implies a shorter lifetime when extrapolating to normal operating condition, Hauschildt *et al.* (Hauschildt, Gall, Justison, Hernandez and Ho 2008) studied the factors that affect the value of σ in Cu interconnects. They found that interconnect thickness and cap layer material will not affect the σ value. However, the value of σ depends on the microstructure, line temperature variation, and the void shape and sizes at failure which is also microstructural dependence as expected. Also, dual damascene tends to have higher σ than single damascene line as voids can grow into the via, render a larger via size variation. Furthermore with multiple failure modes occur in Cu dual damascene, a boarder σ will be expected if the different failure modes are not identified and separated for analysis, and care must be taken to ensure these multiple failure modes are identified using the method as described in Chapter 3.

References

(1995–1996). *Handbook of Chemistry and Physics*, CRC Press.

(2004). *CRC Handbook of Chemistry and Physics,*, CRC Press.

Aaltonen, T., Alen, P., Ritala, M. and Leskela, M. (2003). *Chem. Vap. Deposition* **9**: 45.

Alam, S. M., Wei, F. L., Gan, C. L., Thompson, C. V. and Troxel, D. E. (2005). Electromigration reliability comparison of Cu and Al interconnects. *Proc. of the 6th International Symposium on Quality Electronic Design*.

Alers, G. B., Dornisch, D., Siri, J., Kattige, K., Tam, L., Broadbent, E. and Ray, G. W. (2001). Trade-off between reliability and post-CMP defects during recrystallization anneal for copper damascene interconnects. *Proc. of IEEE International Reliability Physics Symposium*, 350–354.

Alers, G. B., Lu, X., Sukamto, J. H., Kailasam, S. K., Reid, J. and Harm, G. (2004). Influence of Copper Purity on Microstructure and Electromigration. *Proceedings of the IEEE 2004 International Interconnect Technology Conference*, 45–47.

Ang, C. H., Lu, W. H., Yap, A. K. L., Goh, L. C., Goh, L. N. L., Lim, Y. K., Chua, C. S., Ko, L. H., Tan, T. H. S., Toh, S. L. and Hsia, L. C. (2004). A study of SiN cap NH3 plasma pre-treatment process on the PID, EM, GOI performance and BEOL defectivity in Cu dual damascene technology. *Proc. of IEEE ICICDT*, 119–122.

Arnaud, L., Berger, T. and Reimbold, G. (2003). Evidence of grain-boundary versus interface diffusion in electromigration experiments in copper damascene interconnects. *Journal of Applied Physics* **93**(1): 192–204.

Arnaud, L., Tartavel, G., Berger, T., Mariolle, D., Gobil, Y. and Touet, I. (1999). Microstructure and Electromigration in Copper Damascene Lines. *Proc. of 37th IEEE IRPS*, 263–269.

Aubel, O., Bugiel, E., Krüger, D., Hasse, W. and Hommel, M. (2006). Investigation of the influence of thermal treatment on interconnect-barrier interfaces in copper metallization systems. *Microelectronics Reliability* **46**: 768–773.

Besser, P., Marathe, A., Zhao, L., Herrick, M., Capasoo, C. and Kawasaki, H. (2000). Optimizing the electromigration performance of copper interconnects. *Proc. of IEEE IEDM*, 119–122.

Beyer, G. P., Baklanov, M., Conard, T. and Maex, K. (2000). *Material Research Society Symposium* **612**: D9171.

Black, J. R. (1969). Electromigration — A brief survey and some recent results. *IEEE Trans. Electron Devices* **16**(4): 338–347.

Blech, I. A. (1976). Electromigration in thin aluminum films on titanium nitride. *Journal of Applied Physics* **47**: 1203–1208.

Borgesen, P., Korhonen, M. A., Brown, D. D. and Li, C. Y. (1992). Stress induced voiding and electromigration. *AIP Conf. Proc.* **263**: 219.

Borgesen, P., Lee, J. K., Gleixner, R. and Li, C. Y. (1992). *Applied Physics Lett.* **60**: 1706.

Braeckelmann, G., Venkatraman, R., Capasso, C. and Herrick, M. (2000). Integration and reliability of copper magnesium alloys for multilevel interconnects. *Proceedings of the IEEE 2000 International Interconnect Technology Conference*, 236–238.

Brennan, B., Pangrle, S., Evans, A., You, L., Ngo, M. V., Qi, W. J., Baker, R., Baek, W. C., Romero, J., Stockwell, W. and Tracy, B. (2007). Damascene Cu: Dielectric Nitride Capping Surface Plasma Treatment Optimization for Flash Memory Devices. *Proc. of IEEE ISSM*.

Carr, J. W. (1990). *US Patent 4*, **954**, 142.

Castro, D. T., Hoofman, R. J. O. M., Michelon, J., Gravesteijn, D. J. and Bruynseraede, C. (2007). Void growth modeling upon electromigration stressing in narrow copper lines. *Journal of Applied Physics* **102**: 123515.

Chai, Y. and Chan, P. C. H. (2008). High electromigration-resistant copper/carbon nanotube composite for interconnect application. *Proc. of IEEE IEDM*, **4**.

Chai, Y., Chan, P. C. H., Fu, Y. Y., Chuang, Y. C. and Lu, C. Y. (2008). Electromigration studies of Cu/CNT composite using Blech structure. *IEEE Electron Devices Lett.* **29**(9): 1001–1003.

Chai, Y., Zhang, K., Zhang, M., Chan, P. C. H. and Yuen, M. M. F. (2007). Carbon nanotube/copper composites for Via filling and thermal management. *Proc. of 57th IEEE ECTC*, 1224–1229.

Chan, C.-H., Chen, J. K. and F.-C. Chang (2004). In-situ characterization of Cu CMP slurry and defect reduction using IR thermal camera. *Microelectronic Engineering* **75**: 257–262.

Chan, R., Arunagiri, T. N., Zhang, Y., Chyan, O., Wallace, R. M., Kim, M. J. and Hurd, T. Q. (2004). *Electrochem. Solid-State Lett.* **7**: G154.

Chan, Y. L., Chuang, P. and Chuang, T. J. (1998). *Journal of Vacuum Science and Technology* **A16**: 1023.

Chen, C.-S., Liu, C.-P., Yang, H.-G. and Tsao, C. Y. A. (2004). Influence of the preferred orientation and thickness of zirconium nitride films on the diffusion property in copper. *Journal of Vacuum Science and Technology B* **22**(3): 1075–1083.

Chen, K.-C., Wu, W.-W., Liao, C.-N., Chen, L.-J. and Tu, K. N. (2008). Observation of atomic diffusion at twin-modified grain boundaries in copper. *Science* **321**: 1066–1069.

Cheng, Y. L., Lin, B. L., Lee, S. Y., Chiu, C. C. and Wu, K. (2007). Cu Interconnect width effect, mechanism and resolution on down-stream stress electromigration. *Proc. of IEEE IRPS*, 128–132.

Cheng, Y. L. and Wang, Y. L. (2008). Effect of interfacial condition on electromigration for narrow and wide copper interconnects. *Journal of Nanosci. Nanotechnol.* **8**(5): 2494–2499.

Chhun, S., Gosset, L. G., Michelon, J., Girault, V., Vitiello, J., Hopstaken, M., Courtas, S., Debauche, C., Bancken, P. H. L., Gaillard, N., Bryce, G., Juhel, M., Pinzelli, L., Guillan, J., Gras, R., Schravendijk, B. V., Dupuy, J.-C. and Torres, J. (2006). Cu surface treatment influence on Si adsorption properties of CuSiN self-aligned barriers for sub-65 nm technology node. *Microelectronic Engineering* **83**: 2094–2100.

Chin, Y.-L., Chiou, B.-S. and Wu1, W.-F. (2002). Effect of the tantalum barrier layer on the electromigration and stress migration resistance of physical-vapor-deposited copper interconnect. *Japanese Journal of Applied Physics* **41**: 3057–3064.

Cho, S. M., Park, K. C., Suh, B. S., Kim, I. R., Kang, H. K., Suh, K. P., Park, H. S., Ha, J. S. and Joo, D. K. (2004). *IEEE VLSI Technical Digest*, **64**.

Choi, Z., Monig, R., Thompson, C. V. and Burns, M. (2006). Kinetics of void drift in copper interconnects. *Materials, Technology and Reliability of Low-K dielectrics and copper interconnects*. T. Y. Tsui, Y. C. Joo, L. Michaleson, M. Lane and A. A. Volinsky.

Choi, Z. S., Monig, R. and Thompson, C. V. (2007). Dependence of the electromigration flux on the crystallographic orientation of different grains in polycrystalline copper interconnects. *Applied Physics Lett.* **90**: 241913.

Choi, Z. S. and Monig, R. and Thompson, C. V. (2008). Effects of microstructure on the formation, shape and motion of voids during electromigration in passivated copper interconnects. *J. Mater. Res.* **23**(2): 383–391.

Chow, M. F. (1988). *US Patent 4*, **789**, 648.

Christou, A. (1994). *Electromigration and Electronic Degradation*. New York, John Wiley & Sons.

Christou, A. (1994). *Electromigration and Electronic Degradation*. New York, John Wiley & Sons.

Chyan, O., Arunagiri, T. N. and T. Ponnuswamy (2003). *Journal of Electrochem. Soc.* **150**: C347.

Clevenger, L. A., Costrini, G., Dubuzinsky, D. M., Filippi, R., Gambino, J., Hoinkis, M., Gignac, L., Hurd, J. L., Iggulden, R. C., Lin, C., Longo, R., Lu, G. Z., Ning, J., Nuetzel, J. F., Ploessl, R., Rodbell, K., Ronay, M., Schnabel, R. F., Tobben, D., Weber, S. J., Chen, L., Chiang, S., Guo, T., Mosley, R., Voss, S. and Yang, L. (1998). *Proc. of International Interconnect Technology Conference*, 137.

D'Heurle, F. M. and Gangulee, A. (1975). Electrotransport in copper alloy films and the defect mechanism in grain boundary diffusion. *Thin Solid Films* **25**: 531–544.

Ding, P. J., Lanford, W. A., Hymes, S. and Murarka, S. P. (1994). *Applied Physics Lett.* **64**: 2897.

Dion, M. J. (2001). Reservoir modeling for electromigration improvement of metal systems with refractory barriers. *IEEE International Reliability Physics Symposium*, 327–333.

Dixit, G., Gandikota, S., Yahalom, J., Parikh, S., Yoshida, N., Shankaranarayanan, K., Chen, J., Maity, N. and Yu, J. (2003). Enhancing the electromigration resistance of copper interconnects. *Proceedings of the IEEE 2003 International Interconnect Technology Conference*, 162–164.

Domenicucci, A. G., Filippi, R. G., Choi, K. W., Hu, C.-K. and Rodbell, K. P. (1996). Effect of copper on the microstructure and electromigration lifetime of Ti-AlCu-Ti fine lines in the presence of tungsten diffusion barriers. *Journal of Applied Physics* **80**: 4952.

Doyen, L., Federspiel, X., Arnaud, L., Terrier, F., Wouters, Y. and Girault, V. (2007). Electromigration multistress pattern technique for copper drift velocity and Black's parameters extraction. *IIRW Final Report*, 74–78.

Du, Y., Wang, G., Merrill, C. and Ho, P. S. (2002). Thermal stress and debonding in Cu/low-k damascene line structures. *52nd Electronic Components and Technology Conference*, 859–864.

Ernur, D., Iacopi, F., Carbonell, L., Struyf, H. and Maex, K. (2003). Influence of low-k dry etch chemistries on the properties of copper and a Ta-based diffusion barrier. *Microelectronic Engineering* **70**: 285–292.

Fang, H., Weidman, T., Shanmugasundram, A., Naik, M. and Kapoor, B. (2004). Electroless Co(W, P) capping application development. *Proc. of Advanced Metallization Conference*, Materials Research Society, 849–852.

Federspiel, X., Doyen, L. and Courtas, S. (2007). Use of resistance-evolution dynamics during electromigration to determine activation energy on single samples. *IEEE Trans. Device Mater. Reliab.* **7**(2): 236–241.

Field, D. P., Dornisch, D. and Tong, H. H. (2001). Investigating the microstructure-reliability relationship in Cu damascene lines. *Scripta Mater.* **45**: 1069–1075.

Filippi, R. G., Gribelyuk, M. A., Joseph, T., Kane, T., Suilivan, T. D., Clevenger, L. A., Costrini, G., Gambino, J., Iggulden, C. R., Kiewra, E. W., Ning, X. J., Ravikumar, R., Schnabel, R. F., Stojakovic, G., Weber, S. J., Gignac, L. M., Hu, C. K., Rath, D. L. and Rodbell, K. P. (2001). Electromigration in AlCu lines: Comparison of dual damascene and metal reactive ion etching. *Thin Solid Films* **388**: 303–314.

Frankovic, R. and Bernstein, G. H. (1996). *IEEE Trans. Electron Devices* **43**: 2233.

Gan, C. L., Thompson, C. V., Pey, K. L., Choi, W. L., Tay, H. L., Yu, B. and Radhakrishnan, M. K. (2001). Effect of current direction on the lifetime of different levels of Cu dual-damascene metallization. *Applied Physics Lett.* **79**(27): 4592–4594.

Gan, D., Ho, P. S., Huang, R., Leu, J., Maiz, J. and Scherban, T. (2005). Isothermal stress relaxation in electroplated Cu films. I. Mass transport measurement. *Journal of Applied Physics* **97**: 103531–103538.

Gan, D., Huang, R., Ho, P. S., Leu, J., Maiz, J. and Scherban, T. (2004). Effects of passivation layer on stress relaxation and mass transport in electroplated Cu films. *Proc. of International Workshop on Stress-induced Phenomena in Metallization*, 256–267.

Gan, Z. H., Shao, W., Mhaisalkar, S. G., Chen, Z., Li, H., Tu, K. N. and Gusak, A. M. (2006). Reservoir effect and the role of low current density regions on electromigration lifetimes in copper interconnects. *Journal of Mater. Res.* **21**(9): 2241–2245.

Gandikota, S., Padhi, D., Ramanathan, S., McGuirk, C., Naik, M., Parikh, S., K. Musaka, J. Yahalom and Dirish Dixit (2002). Cobalt alloy thin films for encapsulation of copper. *Proc. of Advanced Metallization Conference*, Materials Research Society, 329–335.

Gill, J., Sullivan, T., Yankee, S., Barth, H. and Glasow, A. (2002). Investigation of via-dominated multi-modal electromigration failure distributions in dual damascene Cu interconnects with a discussion of the statistical implications. *Proceeding of IEEE IRPS*, 298–304.

Gladkikh, A., Karpovski, M., Palevski, A. and Kaganovskii, Y. S. (1998). Effect of microstructure on electromigration kinetics in Cu lines. *J. Phys. D: Appl. Phys.* **31**: 1626–1629.

Glasow, A. V., Fischer, A. H., Bunel, D., Friese, G., Hausmann, A., Heitzsch, O., Hommel, M., Kriz, J., Penka, S., Raffin, P., Robin, C., Sperlich, H.-P., Ungar, F. and Zitzelsberger, A. E. (2003). The influence of the SiN cap process on the electromigration and stressvoiding performance of dual damascene Cu interconnects. *Proc. of IEEE IRPS*, 146–150.

Gleixner, R. J., Clemens, B. M. and Nix, W. D. (1997). Void nucleation in passivated interconnect lines: effects of site geometries, interfaces, and interface flaws. *J. Mater. Res.* **12**: 2081.

Glickman, E., Osipov, N., Ivanov, A. and Nathan, M. (1998). Diffusional Creep as a stress relaxation mechanism in electromigration. *Journal of Applied Physics* **83**(1): 100–107.

Glickman, E. E. and Klinger, L. M. (1995). *Proc. of MRS* **391**: 243.

Grone, A. R. (1961). Current-induced marker motion in copper. *J. Phys. Chem. Solids* **20**: 88–93.

Hau-Riege, C. S. (2004). An introduction to Cu electromigration. *Microelectronics Reliability* **44**: 195–205.

Hau-Riege, C. S., Hau-Riege, S. P. and Marathe, A. P. (2004). The effect of interlevel dielectric on the critical tensile stress to void nucleation for the reliability of Cu interconnects. *J. Appl. Phys.* **96**: 5792–5796.

Hau-Riege, C. S., Marathe, A. P. and Choi, Z. S. (2008). The effect of current direction on the electromigration in short-lines with reservoirs. *Proceeding of IEEE IRPS*, 381–384.

Hau-Riege, C. S. and Thompson, C. V. (2000a). *Journal of Applied Physics* **87**: 8467.

Hau-Riege, C. S. and Thompson, C. V. (2000b). The effects of microstructural transitions at width transitions on interconnect reliability. *Journal of Applied Physics* **87**(12): 8467–8472.

Hau-Riege, S. P. (2002). Probabilistic immortality of Cu damascene interconnects. *Journal of Applied Physics* **91**: 2014.

Hau-Riege, S. P. and Thompson, C. V. (2000c). The effects of the mechanical properties of the confinement material on electromigration in metallic interconnects. *J. Mater. Res.* **15**: 1797–1802.

Hau-Riege, S. P. and Thompson, C. V. (2000d). The effects of the mechanical properties of the confinement material on electromigration in metallic interconnects. *Journal of Mater. Res.* **15**(8): 1797–1802.

Haua-Riege, C. S. and Klein, R. (2008). The effect of a width transition on the electromigration reliability of Cu interconnects. *Proceeding of IEEE IRPS*, 377–380.

Hauschildt, M., Gall, M., Justison, P., Hernandez, R. and Ho, P. S. (2008). The influence of process parameters on electromigration lifetime statistics. *Journal of Applied Physics* **104**: 043503.

Hayashi, M., Nakano, S. and Wada, T. (2003). Dependence of copper interconnect electromigration phenomenon on barrier metal materials. *Microelectronics Reliability* **43**: 1545–1550.

Higashi, K., Yamaguchi, H., Omoto, S., Sakata, A., Katata, T., Matsunaga, N. and Shibata, H. (2004). Highly reliable PVDIALDPVD stacked barrier metal structure for 45 nm-node copper dual-damascene interconnects. *Proceedings of the IEEE 2004 International Interconnect Technology Conference*, 6–8.

Hinode, K., Hanaoka, Y., Tahed, K.-I. and Konda, S. (2001). *Japanese Journal of Applied Physics* **40**: L1097.

Hong, S. Q., Comrie, C. M., Russell, S. W. and Mayer, J. W. (1991). Phase formation in Cu-Si and Cu-Ge *Journal of Applied Physics* **70**(7): 3665–3660.

Hosaka, M., Kouno, T. and Hayakawa, Y. (1998). *Proc. of 36th IEEE IRPS*, 329.

Hou, Y. and Tan, C. M. (2007). Blech effect in Cu interconnects with oxide and low-k dielectrics. *Proceeding of IEEE IPFA*, 65–68.

Hou, Y. and Tan, C. M. (2009). Reservoir effect in Cu interconnects, *submitted to IEEE Trans. on Devices and Materials Reliability*.

Hsu, Y. L., Fang, Y. K., Chiang, Y. T., Chen, S. F., Lin, C. Y., Chou, T. H. and Chang, S. H. (2006). Failure mechanism of electromigration in via sidewall for copper dual damascene interconnection. *Journal of Electrochem. Soc.* **153**(8): G782–G786.

Hu, C. K. (1995). *Thin Solid Films* **260**: 124.

Hu, C. K. and Luther, B., *et al.* (1995). *Thin Solid Films* **262**: 84.

Hu, C. K. and Rosenberg, R., *et al.* (1999). Scaling effect on electromigration in on-chip Cu wiring. *Proceeding of IEEE International Interconnect Technology Conference*, 267–269.

Hu, C.-K., Gignac, L., Liniger, E., Herbst, B., Rath, D. L., Chen, S. T., Kaldor, S., Simon, A. and Tseng, W. T. (2003). Comparision of Cu electromigration lifetime in Cu interconnects coated with different caps. *Applied Physics Lett.* **83**: 869–871.

Hu, C.-K., Harper, J. M. E. (1998). Copper interconnections and reliability. *Mater. Chem. Phys.* **52**: 5–16.

Hu, C. K., Gignac, L. M., Neal, B. B.-O., Liniger, E., Yu, R., Flaitz, P. and Stamper, A. K. (2007). Electromigration reliability of advanced interconnects. *Proc. of 9th International Workshop on Stress-Induced Phenomena in Metallization.* S. Ogawa, P. S. Ho and E. Zschech, 27–41.

Hu, C. K., Gignac, L. M., Rosenberg, R., Herbst, B., Smith, S., Rubino, J., Canaperi, D., Chen, S. T., Seo, S. C. and Restanio, D. (2004). Atom motion of Cu and Co in Cu damascene lines with a CoWP cap. *Applied Physics Lett.* **84**: 4986–4988.

Hu, C. K. and Reynolds, S. (1997). CVD Cu interconnections and electromigration. *Proc. of Electrochem Soc. Symp.* **97–25**: 1514.

Hu, C. K., Rosenberg, R. and Lee, K. Y. (1999). Electromigration path in Cu thin-film lines. *Applied Physics Lett.* **74**(20): 2945–2947.

Igarashi, Y. and Ito, T. (1998). Electromigration properties of copper-zirconium alloy interconnects. *Journal of Vacuum Science and Technology B* **16**(5): 2745–2750.

Iggulden, R., Clevenger, L., Costrini, G., Dobuzinsky, D., Filippi, R., Gambino, J., Gignac, L., Lin, C., Longo, R., Lu, G., Ning, J., Nuetzel, J., Rodbell, K., Ronay, M., Schnabel, F., Stephens, J., Tobben, D. and Weber, S. (1998). *Proc. of the VLSI Multilevel Interconnection Conferernce*, 19.

Ishigami, T., Kurokawa, T., Kakuhara, Y., Withers, B., Jacobs, J., Kolics, A., Ivanov, I., Sekine, M. and Ueno, K. (2004). High reliability Cu interconnection utilizing a low contamination COW capping layer. *Proc. of IEEE*, 75–77.

Itabashi, T., Nakano, H. and Akahoshi, H. (2002). Electroless deposited CoWB for copper diffusion barrier metal. *Proceeding of 2002 IEEE International Interconnect Technology Conference*, 285–287.

Josell, D., Weihs, T. P. and Gao, H. (2002). *MRS Bulletin* **27**: 39.

Josell, D., Wheeler, D., Witt, C. and Moffat, T. P. (2003). *Electrochem. Solid-State Lett.* **6**: C143.

Jr., C. C., Harper, J. M. E., Holloway, K., Smith, D. A. and Schad, R. G. (1992). *Journal of vacuum Science and Technology A* **10**: 1706.

Kaanta, C. W., Bombardier, S. G., Cote, W. J., Hill, W. R., Kerszykowski, G., Landis, H. S., Poindexter, D. J., Pollard, C. W., Ross, G. H., Ryan, J. G., Wolff, S. and Cronin, J. E. (1991). *Proc. of IEEE VLSI Multilevel Interconnection Conference*, 144–152.

Kakuhara, Y., Kawahara, N., Ueno, K. and Oda, N. (2008). Electromigration lifetime enhancement of CoWP capped Cu interconnects by thermal treatment. *Japanese Journal of Applied Physics* **47**(6): 4475–4479.

Karimi, M., Tomkowski, T., Vidali, G. and Biham, O. (1995). Diffusion of Cu on Cu surfaces. *Phys. Rev. B: Condens. Matter* **52**: 5364.

Kim, C. and Morris, J. J. W. (1992). *Journal of Applied Physics* **72**: 1837.

Kim, H., Koseki, T., Ohba, T., Ohta, T., Kojima, Y., Sato, H. and Shimogaki, Y. (2005). Cu wettability and diffusion barrier property of Ru thin film for Cu metallization. *Journal of Electrochem. Soc.* **152**(8): G594–G600.

Kim, N.-H., Kim, S.-Y., Lee, S.-Y. and Chang, E.-G. (2007). Electromigration characteristics in dual-damascene copper interconnects by difference of via structures. *Microelectronic Engineering* **84**: 2663–2668.

Kirchheim, R. and Kaeber, U. (1991). *Journal of Applied Physics* **70**: 172.

Kisselgof, L. and Lloyd, J. R. (1997). Electromigration induced failure as a function of via interface. *Proceedings of Materials Research Society Symposium* **473**: 401–406.

Klinger, L. M., Glickman, E. E., Fradkov, V. E., Mullins, W. W. and Bauer, C. L. (1995a). *Proc. of MRS.* **391**: 295.

Klinger, L. M., Glickman, E. E., Fradkov, V. E., Mullins, W. W. and Bauer, C. L. (1995b). *Journal of Applied Physics* **78**: 3833.

Kohn, A., Eizenberg, M., Shacham-Diamond, Y. and Sverdlov, Y. (2001). *Mater. Sci. Eng. A* **302**: 18.

Koike, J., Haneda, M., Iijima, J. and Wada, M. (2006). Cu alloy metallization for self-forming barrier process. *Proc. of International Interconnect Technology Conference*, 161–163.

Koike, J., Wada, M., Usui, T., Nasu, H., Takahashi, S., Shimizu, N., Yoshimaru, M. and Shibata, H. (2006). Self-formed barrier with Cu-Mn alloy metallization and its effects on reliability. *AIP Conference Proceedings* **816**: 43–51.

Korhonen, M. A., Borgesen, P., Brown, D. D. and Li, C. Y. (1993). Microstructure based statistical model of electromigration damage in confined line metallizations in the presence of thermally induced stresses. *J. Appl. Phys.* **74**: 4995

Kouno, T., Hosaka, M., Niwa, H. and Yamada, M. (1998). *Journal of Applied Physics* **84**: 742.

Kwon, D., Park, H. and Lee, C. (2005). Electromigration resistance-related microstructural change with rapid thermal annealing of electroplated copper films. *Thin Solid Films* **475**: 58–62.

Kwon, O. K., Kwon, S. H., Park, H. S. and Kang, S. W. (2004). *Journal of Electrochem. Soc.* **151**: G109.

Lane, M. W., Linger, E. G. and Lloyd, J. R. (2003). Relationship between interfacial adhesion and electromigration in Cu metallization. *Journal of Applied Physics* **93**: 1417–1421.

Le, H. A., Ting, L., Tso, N. C. and Kim, C. U. (2002). Analysis of the reservoir length and its effect on electromigration lifetime. *Journal of Material Research* **17**: 167–171.

Lee, H. and Lopatin, S. D. (2005). The influence of barrier types on the microstructure and electromigration characteristics of electroplated copper. *Thin Solid Films* **492**: 279–284.

Lee, K. L., Hu, C. K. and Tu, K. N. (1995). In-situ scanning electron microscope comparison studies on electromigration of Cu and Cu(Sn) alloys for advanced chip interconnects. *Journal of Applied Physics* **78**(7): 4428–4437.

Lee, S. C. and Oates, A. S. (2006). Identification and analysis of dominant electromigration failure modes in copper/low k dual damascene interconnects. *Proceeding of IEEE International Reliability Physics Symposium*, 107–114.

Li, W., Tan, C. M. and Hou, Y. (2007). Dynamic simulation of electromigration in polycrystalline interconnect thin film using combined Monte Carlo algorithm and finite element modeling. *Journal of Applied Physics* **101**: 104314.

Li, Y., Bruynseraede, C., Groeseneken, G., Maex, K. and Tokei, Z. (2007). On the interaction between inter-metal dielectric reliability and electromigration stress. *Proceedings of IEEE International Reliability Physics Symposium*, 642–643.

Lin, M. H., Chang, K. P., Su, K. C. and Wang, T. (2007). Effects of width scaling and layout variation on dual damascene copper interconnect electromigration. *Microelectronics Reliability* **47**: 2100–2108.

Lin, M. H., Lin, Y. L., Chen, J. M., Tsai, C. C., Yeh, M. S., Liu, C. C., Hsu, S., Wang, C. H., Sheng, Y. C., Chang, K. P., Su, K. C. and Chang, Y. J. (2004). The improvement of copper interconnect electromigration resistance by cap/dielectric interface treatment and geometrical design. *Proc. of IEEE IRPS*, 229–233.

Lloyd, J. R. (1991). Electromigration failure. *Journal of Applied Physics* **69**(11): 7601–7604.

Lloyd, J. R., Clemens, J. J. and Snede, S. (1999). Copper metallization reliability. *Microelectronics Reliability* **39**: 1595–1602.

Lloyd, J. R. and Clement, J. J. (1995). Electromigration in copper conductors. *Thin Solid Films* **262**: 135–141.

Lu, L., Shen, Y., chen, X., Qian, L. and Lu, K. (2004). *Science* **304**: 422.

Marras, A., Impronta, M., De Munari, I., Valentini, M. G. and Scorzoni, A. (2007). Reliability assessment of multi-via Cu-damascene structures by wafer-level isothermal electromigration tests. *Microelectronics Reliability* **47**: 1492–1496.

Massalski, T. B. (1990). *Binary Alloy Phase Diagrams*. Pittsburgh, PA, Materials Research Society.

McBrayer, J. D., Swanson, R. M. and Sigmon, Y. W. (1986). Diffusion of metals in silicon dioxide. *Journal of the Electrochemical Society* **133**(6): 1242–1246.

McCusker, N. D., Gamble, H. S. and Armstrong, B. M. (2000). Surface electromigration in copper interconnects. *Microelectronics Reliability* **40**: 69–76.

Meyer, M. A. (2007). Effects of advanced process approaches on electromigration degradation of Cu on-chip interconnects. Cottbus, Germany, Brandenburg Univ. Technol.

Meyer, M. A., Herrmann, M., Langer, E. and Zschech, E. (2002). In-situ SEM observation of electromigration phenomena in fully embedded copper interconnect structures. *Microelectronic Engineering* **64**: 375.

Meyer, M. A. and Zschech, E. (2007). New microstructure-related EM degradation and failure mechanisms in Cu interconnects with CoWP coating. *Proc. 9th Int. Workshop Stress Induced Phenom. Metallization.* **945**: 107–114.

Michael, N. L. and Kim, C. U. (2001). Electromigration in Cu thin films with Sn and Al cross strips. *Journal of Applied Physics* **90**(9): 4370–4376.

Moreno-Armenta, M. G., Martinez-Ruiz, A. and Takeuchi, N. (2004). *Solid State Science* **6**: 9.

Muppidia, T., Fielda, D. P., J. E. S. Jr and Woo, C. (2005). Barrier layer, geometry and alloying effects on the microstructure and texture of electroplated copper thin films and damascene lines. *Thin Solid Films* **471**: 63–70.

Nakano, H., Itabashi, T. and Akahoshi, H. (2005a). *Journal of Electrochem. Soc.* **152**: C163.

Nakano, H., Itabashi, T. and Akahoshi, H. (2005b). Electroless deposited cobalt-tungsten-boron capping barrier metal on damascene copper interconnection. *Journal of the Electrochemical Society* **152**(3): C163–166.

Nguyen, H. V., Salm, C., Wenzel, R., Mouthaan, A. J. and Kuper, F. G. (2002). Simulation and experimental characterization of reservoir and via layout effects on the electromigration lifetime. *Microelectronics Reliability* **42**: 1421.

Nogami, T., Y.-C. Joo, Lopatin, S., Romero, J., Bernard, J., Blum, N., H.-J. Engelmann, Gray, J., Tracy, B., Chen, S., Lukanc, T., Brown, D., Besser, P., Morales, G. and Cheung, R. (1998). Graded Ta/TaN/Ta barrier for copper interconnects for high electromigration resistance. *Proc. of Advanced Metallization Conference*, 313–319.

Nogami, T., Romero, J., Dubin, V., Brown, D. and Adem, E. (1998). Characterization of the Cu/barrier metal interface for copper interconnects. *Proceedings of the IEEE 1998 International Interconnect Technology Conference*, 298–300.

O'Sullivan, E. J., Schrott, A. G., Paunovic, M., Sambucetti, C. J., Marino, J. R., Bailey, P. J., Kaja, S. and Semkow, K. W. (1998). *IBM J. Res. Dev.* **42**: 607.

Onishi, T., Fujii, H., Yoshikawa, T., Munemasa, J., Inoue, T. and Miyagaki, A. (2003). Effects of the high-pressure annealing process on the reflow phenomenon of copper interconnections for large scale integrated circuits. *Thin Solid Films* **425**: 265.

Orain, S., Barbé, J.-C., Federspiel, X., Legallo, P. and Jaouen, H. (2004). FEM-based method to determine mechanical stress evolution during process flow in microelectronics application to stress — voiding. *Proceedings of the 5th International Conference on Thermal and Mechanical Simulation and Experiments in Microelectronics and Microsystems, EuroSimE 2004*, 47–52.

Orio, R. L. D., Ceric, H., Carniello, S. and Selberherr, S. (2008). Analysis of electromigration in redundant vias. *Proceeding of International Conference on Simulation of Semiconductor Processes and Devices (SISPAD 2008)*, 237–240.

Padhi, D. and Dixit, G. (2003). Effect of electron flow direction on model parameters of electromigration-induced failure of copper interconnects. *Journal of Applied Physics* **94**(10): 6463–6467.

Padhi, D. and Dixit, G. (2003). Effect of electron flow direction on model parameters of electromigration-induced failure of copper interconnects. *Journal of Applied Physics* **94**: 6463–6467.

Parikh, S., Educato, J., Wang, A., Zheng, B., Wijekoon, K., Chen, J., Rana, V., Cheung, R. and Dixit, G. (2001a). Defect and electromigration characterization of a two-level copper interconnect. *Proceedings of the IEEE 2001 International Interconnect Technology Conference*, 183–185.

Parikh, S., Educato, J., Wang, A., Zheng, B., Wijekoon, K., Chen, J., Rana, V., Cheung, R. and Dixit, G. (2001b). *Proceedings of International Interconnect Technology Conference*, 183–185.

Park, C. W. and Vook, R. W. (1993). Electromigration-resistant Cu-Pd alloy films. *Thin Solid Films* **226**(2): 238–247.

Park, Y.-B. and Jeon, I. S. (2003). *Microelectronics Engineering* **69**: 26.

Park, Y. J., Lee, K. D. and Hunter, W. R. (2005). Observation and restoration of negative electromigration activation energy behavior due to thermo-mechanical effects. *Proc. of IEEE IRPS*, 18–23.

Paunovic, M., Schlesinger, M. and Weil, R. (2000). *Modern Electroplating*. New York, John Wiley & Sons Inc.

Petrov, N., Valverde, C., Chen, Q., Xu, C., Paneccasio, V., Stritch, D. and Witt, C. (2005). Process control and material properties of thin electroless Co-based capping layers for copper interconnects. *Proc. of SPIE* **6002**: O1–O11.

Proost, J., Maex, K. and Delaey, L. (1998). *Applied Physics Lett.* **73**: 2748.

Reimbold, G., Sicardy, O., Arnaud, L., Fillot, F. and Torres, J. (2002). Mechanical stress measurements in damascene copper interconnects and influence on electromigration parameters. *IEDM Tech. Digest* 745–748.

Rodbell, K. P., Ficalora, P. J. and Koch, R. (1987). *Applied Physics Lett.* **50**: 1415.

Rodbell, K. P. and Ficalora, P. J. (1985). *Applied Physics Lett.* **47**: 1010.

Rosenberg, R. and Berenbaum, L. (1968). Resistance monitoring and effects of non-adhesion during electromigration in aluminum films. *Applied Physics Lett.* **12**: 201–204.

Roy, A., Hou, Y. and Tan, C. M. (2009). Electromigration in width transition copper interconnect. *Microelectronics Reliability*.

Roy, A., Kumar, R., Tan, C. M., Wong, K. S. and Tung, C.-H. (2006). Electromigration in damascene copper interconnects of line width down to 100 nm. *Semicond. Sci. Technol.* **21**: 1369–1372.

Roy, A. and Tan, C. M. (2006). Experimental investigation on the impact of stress free temperature on the electromigration performance of copper dual damascene submicron interconnect. *MIcroelectronics Reliability* **46**(9–11): 1652–1656.

Roy, A. and Tan, C. M. (2007). Probing into the asymmetric nature of electromigration performance of submicron interconnect via structure. *Thin Solid Films* **515**: 3867–3874.

Roy, A. and Tan, C. M. (2008). Very high current density package level electromigration test for copper interconnects. *Journal of Applied Physics* **103**: 093707.

Roy, A., Tan, C. M., Kumar, R. and Chen, X. T. (2005). Effect of test condition and stress free temperature on the electromigration failure of Cu dual damascene submicron interconnect line-via test structure. *Microelectronics Reliability* **45**: 1443–1448.

Ryu, C., Kwon, K. W., Loke, A. L. S., Lee, H., Nogami, T., Dubin, V. M., Kavari, R. A., Ray, G. W. and Wong, S. S. (1999). Microstructures and reliability of copper interconects. *IEEE Trans. on Electron Devices* **46**: 1113–1120.

Sato, H. and Ogawa, S. (2001). Mechanism of dependency of EM properties on linewidth in dual damascene copper interconnects. *Proceedings of the IEEE 2001 International Interconnect Technology Conference*, 186–188.

Schindler, G. N., Penka, S., Steinlesberger, G., Traving, M., Steinhogl, W. and Engelhardt, M. (2005). Reliability studies of narrow Cu lines. *Microelectronic Engineering* **82**: 645–649.

Sekiguchi, A., Koike, J. and Maruyama, K. (2003). Microstructural influences on stress migration in electroplated Cu metallization. *Applied Physics Lett.* **83**(10): 1962–1964.

Sekiguchi, A., Koike, J., Kamiya, S., Saka, M. and Maruyama, K. (2001). Void formation by thermal stress concentration at twin interfaces in Cu thin films. *Applied Physics Lett.* **79**(9): 1264.

Shao, W., Vairagar, A. V., Tung, C. H., Xie, Z. L., Krishnamoorthy, A. and Mhaisalkar, S. G. (2005). Electromigration in copper damascene interconnects: Reservoir effects and failure analysis. *Surface and Coatings Technology* **198**(1–3): 257–261.

Shatzkes, M. and Lloyd, J. R. (1986). A model for conductor failure considering diffusion concurrently with electromigration resulting in a current exponent of 2. *Journal of Applied Physics* **59**: 3890.

Shen, Y.-L. and Ramamurty, U. (2003a). Temperature-dependent inelastic response of passivated copper films: Experiments, analyses, and implications. *Journal of Vacuum Science and Technology B* **21**(4): 1258–1264.

Shen, Y. L. and Ramamurty, U. (2003b). Temperature dependent inelastic response of passivated copper films: Experiments, analyses and implications. *Journal of Vacuum Science and Technology B* **21**(4): 1258–1264.

Stangl, M., Acker, J., Oswald, S., Uhlemann, M., Gemming, T., Baunack, S. and Wetzig, K. (2007). Incorporation of sulfur, chlorine, and carbon into electroplated Cu thin films. *Microelectronic Engineering* **84**(1): 54–59.

Stangl, M., Liptak, M., Acker, J., Hoffmann, V., Baunack, S. and Wetzig, K. (2009). Influence of incorporated non-metallic impurities on electromigration in copper damascene interconnect lines. *Thin Solid Films* **517**(8): 2687–2690.

Stangl, M., Lipták, M., Fletcher, A., Acker, J., Thomas, J., Wendrock, H., Oswald, S. and Wetzig, K. (2008). Influence of initial microstructure and impurities on Cu room-temperature recrystallization (self-annealing). *Microelectronic Engineering* **85**(3): 534–541.

Steinhogl, W., Schindler, G., Steinlesberger, G., Traving, M. and Engelhardt, M. (2005). *Journal of Applied Physics* **7**: 023706.

Stoebe, T. G., Gulliver, R. D., Ogurtani, T. O. and Huggins, R. A. (1965). *Acta Metall.* **13**: 701.

Surholt, T. and Herzig, C. (1997). Grain boundary self-diffusion in Cu polycrystals of different purity. *Acta Mater.* **45**(9): 3817–3823.

Tada, M., Abe, M., Furutake, N., Ito, F., Tonegawa, T., Sekine, M. and Hayashi, Y. (2007). Improving reliability of copper dual-damascene interconnects by impurity doping and interface engineering. *IEEE Trans. on Electron Devices* **54**(8): 1867–1877.

Tan, C. M., Hou, Y. and Li, W. (2007). Revisit to the finite element modeling of electromigration for narrow interconnects. *Journal of Applied Physics* **102**: 033705.

Tan, C. M. and Roy, A. (2006). Investigation of the effect of temperature and stress gradients on accelerated EM test for Cu narrow interconnects. *Thin Solid Films* **504**: 288.

Tan, C. M. and Roy, A. (2007). Electromigration in ULSI interconnects. *Materials Science and Engineering Review* **58**: 1–75.

Tan, C. M., Roy, A., Vairagar, A. V., Krishnamoorthy, A. and Mhaisalkar, S. G. (2005). Current crowding effect on copper dual damascene via bottom failure for ULSI applications. *IEEE Trans. Device Mater. Reliab.* **5**: 198–205.

Tan, C. M. and Zhang, G. (2004). Overcoming intrinsic weakness of ULSI metallization electromigration performances. *Thin Solid Films* **462–463**: 263–268.

Tan, C. M., Zhang, G. and Gan, Z. H. (2004). Dynamic study of the physical process in the intrinsic line electromigration of deep-submicron copper and Aluminum interconnects *IEEE Trans. Device Mater. Reliab.* **4**: 450.

Taubenblat, P. W., Marino, V. J. and Batra, R. (1979). *Journal of Wire* **12**: 114.

Usui, T., Miyajima, H., Masuda, H., Tabuchi, K., Watanbe, K., Hasegawa, T. and Shibata, H. (2006). Effect of plasma treatment and dielectric diffusion barrier on electromigration performance of copper damascene interconnects. *Japanese Journal of Applied Physics Part I* **45**: 1570.

Usui, T., Nasu, H., Koike, J., Wada, M., Takahashi, S., Shimizu, N., Nishikawa, T., Yoshimaru, A. and Shibata, H. (2005). Low resistive and highly reliable Cu dual-damascene interconnect technology using self-formed MnSixOy barrier layer. *Proceedings of the IEEE 2005 International Interconnect Technology Conference*, 88–90.

Usuni, T., Oki, T., Miyajima, H., Tabuchi, K., Watanabe, K., Hasegawa, T. and Shibata, H. (2004). Identification of electromigration dominant diffusion path for Cu damascene interconnects and effect of plasma treatment and barrier dielectrics on electromigration performance. *Proc. of IEEE IRPS*, 246–250.

Vaidya, S., Sheng, T. T. and Sinha, A. K. (1980). Linewidth dependence of electromigration in evaporated Al-0.5%Cu. *Applied Physics Lett.* **30**: 464–466.

Vairagar, A. V., Gan, Z., Shao, W., Mhaisalkar, S. G., Li, H., Tu, K. N., Chen, Z., Zschech, E., Engelmann, H. J. and Zhang, S. (2006). Improvement of electromigration lifetime of submicrometer dual-damascene Cu interconnects through surface engineering. *Journal of Electrochemical Society* **153**(9): G840–845.

Vairagar, A. V., Mhaisalkar, S. G. and Krishnamoorthy, A. (2004). Electromigration behavior of dual-damascence Cu interconnect structures — structure, width and length dependences. *Microelectronics Reliability* **44**(5): 747–754.

Vairagar, A. V., Mhaisalkar, S. G., Meyer, M. A., Zschech, E. and Krishnamoorthy, A. (2005). Reservoir effect on electromigration mechanisms in dual-damascene Cu interconnect structures. *Microelectronic Engineering* **82**: 675–679.

Wang, C. P., Lopatin, S., Marathe, A., Buynoski, M., Huang, R. and Erb, D. (2001). Binary Cu-alloy layers for Cu-interconnections reliability improvement. *Proceedings of the IEEE 2001 International Interconnect Technology Conference*, 86–88.

Wang, H., Bruynserae, C. and Maex, K. (2004). The influence of surface fluctuations on early failures in single-damascene Cu wires: A weakest link approximation analysis. *Proceedings of IRPS*, 625–626.

Wang, Q., Ekerdt, J. G., Gay, D., Sun, M. Y. and White, J. M. (2004). *Applied Physics Lett.* **84**: 1380.

Wang, R. C. J., Su, D. S., Yang, C. T., Chen, D. H., Doong, Y. Y., Shih, J. R., Lee, S. Y., Chiu, C. C., Peng, Y. K. and Yue, J. T. (2002). Investigation of electromigration properties and plasma charging damages for plasma treatment process in Cu interconnects. *Proc. of 7th International symposium on plasma and process-induced damage*, 166–168.

Wang, R. C. J., Su, D. S., Yang, C. T., Chen, D. H., Doong, Y. Y., Shih, J. R., Lee, S. Y., Chiu, C. C., Peng, Y. K. and Yue, J. T. (2004). Selective formation of CoWP capping barriers onto Cu interconnects by electroless plating. *Proc. of Advanced Metallization Conference*, Materials Research Society, 809–814.

Wei, L., Tan, C. M., *et al.* (2009). Dynamic simulation of void nucleation during electromigration in narrow integrated circuit interconnects. *Journal of Applied Physics* **105**: 14305.

Weide-Zaage, K., Zhao, J., Ciptokusumo, J. and Aubel, O. (2008). Determination of migration effects in Cu-via structures with respect to process-induced stress. *Microelectronics Reliability* **48**: 1393–1397.

Wendrock, H., Mirpuri, K., Menzel, S., Schindler, G. and Wetzig, K. (2005). Correlation of electromigration defects in small damascene Cu interconnects with their microstructure. *Microelectronic Engineering* **82**: 660–664.

Yang, C.-C., Spooner, T., Ponoth, S., Chanda, K., Simon, A., Lavoie, C., Lane, M., Hu, C.-K., Liniger, E., L.Gignac, Shaw, T., Cohen, S., McFeely, F. and Edelstein, D. (2006). Physical, electrical, and reliability characterization of Ru for Cu interconnects. *Proceedings of the IEEE 2006 International Interconnect Technology Conference*, 187–189.

Yang, Z. W., Zhang, D. H., Li, C. Y., Tan, C. M. and Prasad, K. (2004). Barrier layer effects on reliabilities of copper metallization. *Thin Solid Films* **462–463**: 288–291.

Yokogawa, S. and Tsuchiya, H. (2007). Impurity doping effects on electromigration performance of scaled-down Cu interconnects. *Proc. of the 9th International workshop on Stress-induced Phenomena in Metallization*. S. Ogawa, P. S. Ho and E. Zschech, 82–97.

Yokogawa, S., Tsuchiya, H., Kakuhara, Y. and Kikuta, K. (2008). Analysis of Al doping effects on resistivity and electromigration of copper interconnects. *IEEE Trans. Devices and Materials Reliability* **8**(1): 216–221.

Zaporozhets, T. V., Gusak, A. M., Tu, K. N. and Mhaisalkar, S. G. (2005). Three-dimensional simulation of void migration at the interface between thin metallic film and dielectric under electromigration. *Journal of Applied Physics* **98**: 103508.

Zhang, W., Brongersma, S. H., Li, Z., Li, D., Richard, O. and Maex, K. (2007). *Journal of Applied Physics* **101**: 063703.

Zschech, E., Ho, P. S., Schmeisser, D., Meyer, M. A., Vairagar, A. V., G. Schneider, Hauschildt, M., Kraatz, M. and Sukharev, V. (2009). Geometry and microstructure effect on EM-induced copper interconnect degradation. *IEEE Trans. Devices and Materials Reliability* **9**(1): 20–30.

Zschech, E., Meyer, M. A. and Langer, E. (2004). Effect of mass transport along interfaces and grain boundaries on copper interconnect degradation. *Materials, Technology and Reliability for Advanced Interconnects and Low-k dielectrics*. R. J. Carter, C. S. Hau-Riege, G. M. Kloster, T. M. Lu and S. E. Schulz. PA.

CHAPTER 5
Numerical Modeling of Electromigration

We can see from Chapters 3 and 4 that the kinetic of EM in metal thin film is complicated and it is a mass transportation process controlled by various driving forces such as the electron wind, temperature gradients, stress gradients and the surface tension (Huntington 1975; Shewmon 1989; Xia, Bower, Suo and Shih 1997). In addition, self-healing effect such as the mass backflow or Blech effect also plays an important role on the retardation of the EM process (Blech 1998).

Beside the kinetics of EM, EM is also a thermodynamics favorable process in which the entropy is increased and the increased system energy due to the EM driving forces is dispersed during the mass transport through various diffusion paths, such as the grain boundaries, void surface, Cu/barrier layer interface, Cu/cap-layer interface and lattice. During the EM failure in ULSI interconnections, the above-mentioned mass flow mechanisms continously interact with each other to influence the behaviors of void nucleation, void growth, void movement and void shape change.

Besides the physical mechanisms of the mass flow and the void dynamic as mentioned above, the EM dynamic is also influenced by the metal thin film materials, their microstructures, the surrounding materials in an interconnect system and the interconnect processing technology as line width decreases. As a result, the complexity of the physics of EM increases, rendering the inadequacy of the empirical EM equations such as the Black's equation although it has been widely used successfully in the industry to study the EM reliability of the interconnect system.

On the other hand, the demand for higher speed ULSI with more complex functionality is driving the continuous change in the interconnects' structures and materials as well as the interconnects processing technology, but without sacrificing their reliability requirements. Thus, implementation of reliability simulation capabilities at the design rules generation step becomes necessary to avoid either unnecessary conservative approach or approach that degrades its reliability without awareness. In fact, it is ironic that while it is impossible to design an integrated circuit without detail circuit modeling, and even detail process modeling is being done using T-Supreme for example, the absence of reliability modeling performed on the design is being practice today. This could be due to the fact that such a modeling tool is not available.

However, to be able to reach the target of reliability modeling, even only for interconnect reliability requires a detailed 3D modeling. This is especially true for copper metallization, than for aluminum metallization, because copper metallization involves complicated geometry of the DD interconnect segments and much lower critical stress for void nucleation. The latter fact clarifies an important role of non-uniform stress, current and temperature distributions in failure development (Sukharev 2004). Numerous experimental data and a number of theoretical results as described in detail in Chapters 3 and 4 have created a basis for understanding of the major mechanisms governing the reliability of the Cu interconnect. The problem that was not addressed yet is the availability of the practical and robust simulation model capable of capturing these major mechanisms of failure. Such model should be employed for optimization of the interconnect design in order to increase its lifetime (Sukharev 2004). Specially designed experimental investigations of IC reliability can be too costly and too slow to respond to today's technological advances. Physics-based modeling of IC reliability becomes necessary to complement experimental investigations.

In this chapter, I am not attempting to provide a detail on the numerical modeling on EM as this can be found elsewhere (Tan, Gan, Li and Hou 2010). Instead, I will give an overview of the progression of EM modeling.

5.1 1D Continuum Electromigration Modeling

Physics-based EM modeling begins from 1D analytical modeling by solving the following continuity equation (Clement 2001) derived from the time evolution of the vacancy concentration along an interconnect line:

$$\frac{\partial C_v}{\partial t} + \frac{\partial J_v}{\partial x} + r = 0. \tag{5.1}$$

Here C_v is the instantaneous vacancy concentration at position x and time t, J_v is the vacancy flux due to EM driving forces and r is the sink/source term which allows for the recombination or generation of vacancies at sites such as grain boundary, dislocations, or surface (Clement 2001).

In this category of simulation methodology, the focus is on the EM kinetics. The net vacancy flux along the length of an interconnect can be due to the electron wind force, vacancy diffusion as a result of concentration gradient (Fickian diffusion) and temperature gradient (Soret diffusion), hence the flux can be written as

$$J_v = -D\frac{\partial C_v}{\partial x} + \frac{DC_v Z^* eE}{k_B T} + \frac{Q^* DC_v}{k_B T^2} \cdot \frac{\partial T}{\partial x}, \tag{5.2}$$

where D is the vacancy diffusivity, $Z^* e$ is the effective charge of the diffusing species, E is the electric field and Q^* is the heat of transport.

The sink/source term r can be expressed as

$$r = \frac{C_v - C_{ve}}{\tau}, \tag{5.3}$$

where C_{ve} is the equilibrium concentration of vacancies within the grain, and τ is the average lifetime of a vacancy.

With the shrinking metal line dimension, it is found that, in the confined metal interconnects deposited on an oxidized silicon substrate and covered by a dielectric passivation layer, stress evolution occur in the metal interconnect during EM, and hence modeling of EM incorporating the effect of the transient stress buildup was developed by several workers such as Ross (Ross 1991), Kirchheim (Kirchheim 1992), Korhonen *et al.* (Korhonen, Borgesen, Tu and Li 1993) etc. The modeling of the EM driving forces was modified to account for the effect of the migration due to stress gradient as given below (Korhonen, Borgesen, Tu and Li 1993)

$$J_v = -D\frac{\partial C_v}{\partial x} + \frac{DC_v Z^* eE}{k_B T} + \frac{Q^* DC_v}{k_B T^2} \cdot \frac{\partial T}{\partial x} - \frac{DC_v}{kT} f\Omega \frac{\partial \sigma}{\partial x}. \quad (5.4)$$

With the development of the analytical modeling and interconnect technology, Korhonen's model received further review and continuous development in order to explain the physics of EM in the new advanced interconnect systems.

Based on the improved model, Park, Andleigh and Thompson reported another EM model to simulate the reliability of Al and Al-Cu interconnects (Park, Andleigh and Thompson 1999). In their work, the effect of the impurity Cu atoms in Al interconnects on stress evolution and lifetime was investigated in various structures as Cu impurity was found to improve the EM performance of Al interconnects. In addition, the significance of the effect of the mechanical stress on the diffusivity of both the Al and Cu was determined. More importantly, the application of the model can be further extended to the investigation of other interconnection materials such as Cu and its alloys by modifying the input material properties. The detail mathematical formulation used in the analysis can be found in their report (Park and Thompson 1997). Their analytic EM model was developed into an EM simulation package, MIT/EmSim (Andleigh, Fayad, Verminski and Thompson http://nirvana.mit.edu/emsim/) which was extensively employed in their many other EM studies (Hau-Riege and Thompson 2000; Hau-Riege 2002) and provided theoretical support to various experimental observation successfully.

5.2 2D EM Modeling

Besides the electron wind force, vacancy concentration gradient, stress and temperature gradients, the EM performance of an interconnect was found to be greatly affected by the microstructural inhomogeneities of the line, such as the grain size and texture distribution, the triple points of the grain boundaries, the barrier layer interface and the surrounding materials of the metal line as mentioned in the previous chapters. While the EM equation is generally easy to solve in one dimension, the simplifications fail to account for the details of the effect of microstructures. One step further into the reality is achieved by using two-dimensional models including the consideration of the microstructure of the metal conductor.

The addition of this second dimension complicates the governing EM equations significantly, and numerical approaches are generally required to find the solution. A 2D model was proposed by Kirchheim and Kaeber (Kirchheim and Kaeber 1991). As shown in Fig. 5.1, a 2D array of grains is produced by distributing the central points of the grains over an area randomly or according to a Poisson distribution. The grain boundaries are constructed by the Voronoi method (Frost and Thompson 1988). A line of width W is cut out by two parallel straight lines of distance W.

Such a numerical approach to solve the EM equation in a 2D computer generated EM model with the microstructure was also reported by Trattles *et al.* (Trattles, O'Neill and Mecrow 1994).

Besides studying the evolution of the vacancy concentration along a metal thin film, models were also developed primarily focusing on the kinetics of EM on the void surface, namely the size evolution and stability of a void due to the atomic migration along its surface under the various driving forces.

The simulation methodology for this class of model begins in the mid 1990's, and there are two different models developed. One is the *sharp interface model* that attempts to model the void surface as a sharp interface between the conductor material and empty space in the void.

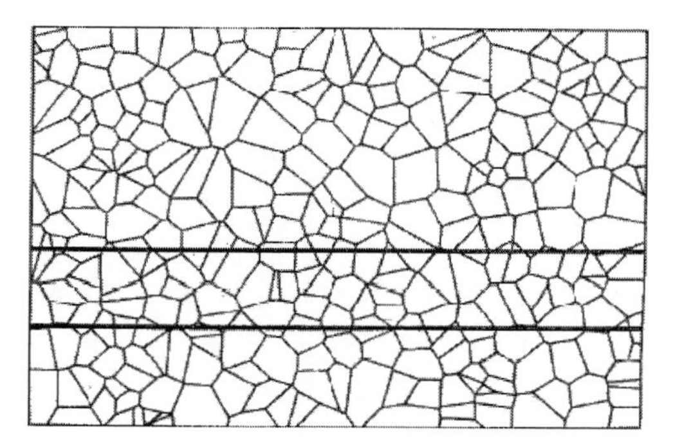

Fig. 5.1. Simulated distribution of grains from which an "Al line" has been selected by the two parallel straight lines (Kirchheim and Kaeber 1991). Reprinted with permission from Kircheim and Kaeber, 'Atomistic and computer modelling of metallization failure of integrated circuits by electromigration', Journal of Applied Physics, vol. 70, pp. 172–181. Copyright © 1991, American Institute of Physics.

The other method is the *phase field model*. This model defined the level of the presence of the conductor material based on a phase field scalar variable. The value of the variable defines the phase of the material at any point on a fixed grid that defining the simulated specimen. The transaction from metal material to an empty void is not sharp as modeled by the first approach, but gradual in terms of the presence of the material. It requires less book-keeping and seems to be more computational efficiency.

5.2.1 *Sharp interface model*

In 1994, Arzt and Kraft *et al.* (Arzt, Kraft, Nix and Sanchez 1994) reported their work on the investigation of the behavior of EM-induced voids in narrow, unpassivated aluminum interconnects. Experimentally, it was found that the fatal voids have a specific asymmetric shape with respect to the electron flow direction. They proposed, to our best knowledge, the first model which attempted to simulate the void shape changes on the basis of atomic diffusion along the void surface. In their model, they considered an isolated two-dimensional void which extends through the thickness of the line, and atomic diffusion on the void surface was assumed to be the primary transport mechanism. Depending on the balance of the arriving and departing atoms at each point on the surface, the movement and the shape change of a void relative to a fixed coordinated system in the line are tracked.

The driving forces for such a behavior are modeled using the surface mass flux through a surface element, and they considered two types of mass fluxes due to the electron wind force and the curvature of the surface respectively.

In their model, the current density distribution was calculated based on the assumption of a steady flow of an incompressible, non-viscous, circulation-free liquid, of which the expressions are mathematically identical to the equations of electrostatics. With the knowledge of current density and mass fluxes, the normal velocity at each point of the surface can be computed and the resulting equation of motion for the void surface can be obtained.

The simulation methodology was further elaborated in their subsequent report (Kraft and Arzt 1997). A finite element method was used in the calculation of the current density and the temperature distribution. A finite difference method was employed for the void motion and the shape changes

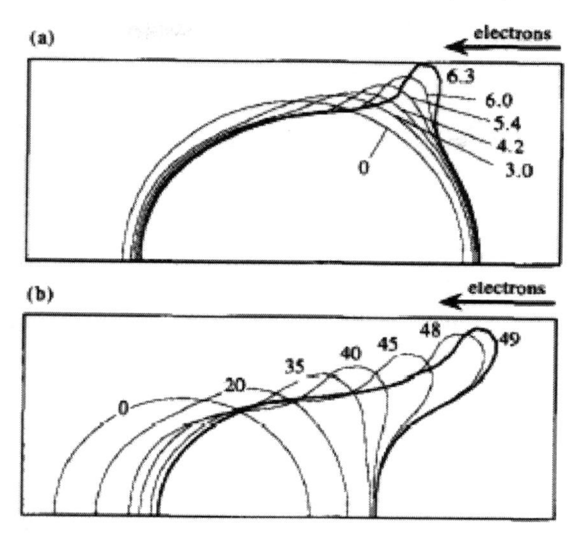

Fig. 5.2. Simulation of the development of initially semi-circular voids with radii of (a) 0.8 um (b) 0.6 um in a 1 um wide Al metal line. The applied current density was 2 MA/cm^2 and the time is given in units of 100 s (Kraft and Arzt 1997). Reprinted from Acta Mater, 45, Kraft and Arzt, 'Electromigration mechanisms in conductor lines: Void shape changes and slit-like failure', pp. 1599, Copyright © 1997, with permission from Elsevier.

computation. Figure 5.2 shows some of their simulation results of the void motion and the shape changes in an Al metal line.

The commercial finite element software ANSYS was used in their modeling. Also included in their work was the consideration of the crystallographic orientation leading to a surface diffusivity change for different angles of the surface with respect to the crystal. To model this effect, they considered the surface diffusivity to be no longer constant and the anisotropy were incorporated in the finite element analysis by describing the surface diffusivity as a function of the angle θ between the surface tangent and the conductor line length direction.

In their study, they found that slit voids that form within grains are likely to be caused by EM induced surface diffusion. Under sufficiently high current density, a rounded void is unstable and will spontaneously collapse into a slit.

Another similar model was proposed by Wang *et al.* (Wang, Suo and Hao 1996). In their early studies (Suo, Wang and Yang 1994; Yang, Wang and Suo 1994), they showed that atomic diffusion on the void surface, driven by the electrical current, can cause a circular void to translate into a slit void. During this translation process, two forces compete in determining the void shape, one is the surface tension force and the other is the electron wind force due to the electrical current. Surface tension force favors the formation of a rounded void, while electrical current favors a slit void; a rounded void will collapses and becomes a slit when the electron wind force dominates.

In their later work (Wang, Suo and Hao 1996), they reviewed the experimental and theoretical findings, and provided a numerical simulation of the void shape change. Similar to the Kraft's model, they adopted the sharp interface approach and assumed that the void shape change was due to the surface diffusion only, considering all other transport processes as negligibly slow during the void shape change. This approach utilizes the same equations as in Kraft's model (Arzt, Kraft, Nix and Sanchez 1994) to model the mass transport. By approximating the void perimeter by many short straight segments, they formulated a finite element procedure for the shape evolution, as had been done by Kraft (Arzt, Kraft, Nix and Sanchez 1994).

The difference between the above-mentioned two models lies in the implementation of the physical system analysis. Wang *et al.* uses a conformal mapping technique to determine the electric field around the circular void. Compared with the finite element method by Kraft, Wang's approach, while mathematically elegant, is not versatile and is hard to yield more complex geometry evolutions (Atkinson 2003). Other similar works are reported by Gungor *et al.* (Gungor and Maroudas 1998; Gungor and Maroudas 1999) and Schimschak *et al.* (Schimschak and Krug 2000).

In the sharp interface models developed, the common assumption is that the atomic surface diffusion is the dominant diffusion mechanism, and the surface diffusion is driven by the electron wind force and the surface tension force due to the curvature of the void surface only. The anisotropy of the void surface diffusivity is emphasized, and slit voids will only form in grains with certain crystallographic orientations. They studied moving boundary problems entail explicit tracking of the boundary and the interface is described by specifying a large number of points on it. This requires a tracking and accurate book-keeping of the surface elements and their geometry, and renders the sharp interface model very complicated and also tends to have rather poor numerical stability (Karma and Rappel 1996; Mahadevan and Bradley 1999), even though it is able to explain some experimental observations quite well.

5.2.2 *Phase field model*

If the entire domain is described by a continuously varying scalar order parameter ϕ which has a value of $+1$ for region well within the metal "phase" and -1 for region well within the void "phase", and ϕ has a value between $+1$ to -1 for the metal-void interface as shown in Fig. 5.3, we have the phase field model (Mahadevan and Bradley 1999).

This model was introduced independently by Fix (Fix 1983) and Collins & Levine (Collins and Levine 1985), and it received considerable attention in the context of phenomena associated with evolving interface. The early attempt to use the phase field model in line interconnect failure simulation was reported by Mahadevan and Bradley (Mahadevan and Bradley 1999). In their study, they simulated the time evolution of a perturbation to the edge of a current carrying, single-crystal, unpassivated metal line. Surface electron wind force migration, surface self-diffusion due to the current crowding, and the curvature of the void surface are all taken into account. They adopted the same formula that model the diffusion mechanisms along the void surface used by Kraft (Arzt, Kraft, Nix and Sanchez 1994), but the idea of a sharp interface between metal and void is abandoned.

The phase field model was used for an isolated void in an infinite thin film, and this method can be easily extended so that it applies to the time

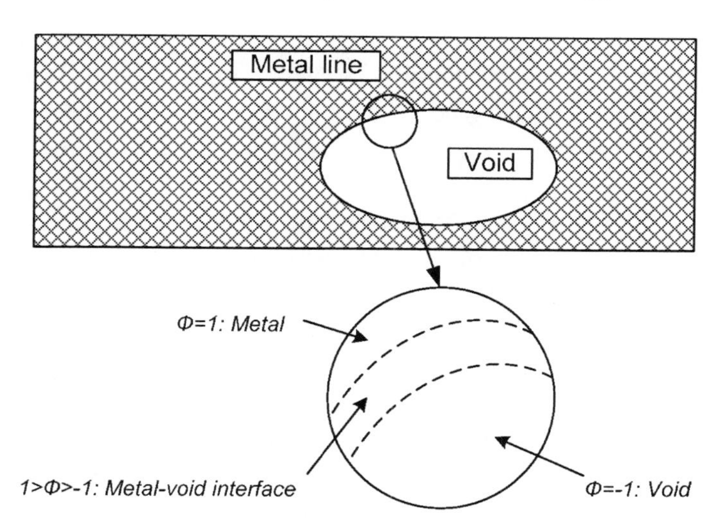

Fig. 5.3. The void simulation in a phase field model.

evolution of an edge perturbation in a metal line of finite extent. By solving the phase field equations and electric field equations numerically, the model provides the time evolution of a small notch at the edge of a current-carrying single crystal metal line. The model is able to predict a threshold value of the applied current so that the edge perturbation will grow into a slit-shaped void that spans the wire. They also explained the physical origin of this instability and pointed out the importance of the crystalline anisotropy and mass transport along the edge of the line.

At nearly the same time as Mahadevan and Bradley, Bhate *et al.* (Bhate, Kumar and Bower 2000) reported their own version of the phase field model for simulating the process of electron wind force migration, curvature driven surface migration, and the stress driven migration. In their work, Bhate *et al.* (Bhate, Kumar and Bower 2000) briefly discussed the theory of the sharp interface model and its limitations or disadvantages, and hence they proposed their own phase field model. Their approach is based on the introduction of an order parameter field to characterize the damaged state of an interconnect. The order parameter takes on distinct uniform values within the material and the void, varying rapidly from one to the other over narrow interfacial layers associated with the void surface. They derived the field equations for the order parameters based on the micro-force balance principle of Gurtin (Gurtin 1996).

The atomic diffusion paths through the bulk and grain boundary are assumed to be negligible, and the only mode of transport for atoms is the diffusion along the void surfaces in their model. This is the same assumption as in most sharp interface models except the surface diffusivity in their model is treated as isotropic. However, this assumption is not justifiable since it has been proved that the anisotropic diffusivity of atoms along the void surface has a strong effect on the evolution of void during EM (Arzt, Kraft, Nix and Sanchez 1994).

The assumption of the sole void surface path in 2D model is unsuitable in the new Cu interconnect technology. The fast diffusion path along the interface of the capping layer and Cu can dramatically change the behavior of the void during the EM. Experimental observations found that the void can migrate, coalescent and interact with grain boundaries under the influence of the various driving forces (Lee, Hu and Tu 1995; Meyer, Herrmann, Langer and Zschech 2002; Zaporozhets, Gusak, Tu and Mhaisalka 2005; Sukharev,

Kteyan, Zschech and Nix 2009). In fact, the consideration of both the EM kinetics (multiple driving forces) and its thermodynamic (multiple diffusion paths) are imperative for the EM simulation in the interconnect systems.

Recently Sukharev *et al.* (Sukharev, Kteyan, Zschech and Nix 2009) developed a comprehensive 2D finite element model which accounts for the different driving forces that we have discussed as well as the concentration gradient and the microstructure of the line. Also, the high elastic anisotropy of copper, orientation dependence of the elastic constants are included in their modeling. Furthermore, they also included vacancy generation/annihilation kinetics which was found to affect the stress evolution in the line significantly. The stress dependence of the vacancies concentration is considered as well. With this model, they can simulate the void nucleation and migration and coalescence as observed in the in-situ EM studies. However, they neglected Joule heating as well as the initial thermal stress due to SFT which was found to be significant for EM. Also, the application of the model to 3D case is yet to be demonstrated.

To distinguish the difference between Al and Cu EM in term of the presence of interface diffusion in Cu EM, Sukharev (Sukharev 2004) considered not only the interface diffusion in his modeling, the atom exchange between the material bulk and interface as mentioned by Rosenberg and Ohring (Rosenberg and Ohring 1971) was also included. He found that if the liner's conductivity is very low, then void nucleates first at the via bottom. On the other hand, if the liner's conductivity is high, void will nucleates at the top left corner of the line above the via. His result was found to agree with his experimental result. Sukhareve *et al.* (Sukharev, Kteyan, Zschech and Nix 2009) also include the stress re-distribution due to vacancy exchange between interface and Cu grains in their model.

Other simulation techniques based on atomic flux divergence was introduced, making the study of the EM physics more realistically in a 3D system.

5.3 Electromigration Simulation Using Atomic Flux Divergence and Finite Element Analysis

The lifetime of a metal line is closely related with the void and hillock formation in the line. Divergence of the atomic flux due to EM gives rise to the formation of the void and hillock as we have discussed. As such, AFD method was developed to model the EM process.

5.3.1 *Computation methods for Atomic Flux Divergence (AFD)*

5.3.1.1 *Formulation of AFD*

Sasagawa *et al.* reported their detailed study in AFD in 1998 (Sasagawa, Nakamura, Saka and Abe 1998; Sasagawa, Naito, Saka and Abe 1999). They proposed a calculation method of AFD due to EM by considering the various factors on void formation for both cases of unpassivated polycrystalline line and bamboo line. Later on, the calculation method of AFD was also applied to passivated polycrystalline line including the effect of the back flow stress in their subsequent work (Sasagawa, Nakamura, Saka and Abe 2002). In their study, the AFD is identified as a parameter governing the void formation, and the distributions of the current density and temperature as well as the material properties of the metal thin film was considered in the computation of the AFD. Lifetime and failure site in a passivated/unpassivated metal line were predicted by means of the numerical simulation of the processes of the void initiation and void growth based on the AFD, and the change in the distributions of the current density and temperature with void growth was taken into account.

In 2001, Dalleau and Weide-Zaage reported their study on 3D voids simulation in ULSI metallization structures (Dalleau and Weide-Zaage 2001). In their study, the different mass flow such as the electron wind force migration, the thermo-migration due to temperature gradients and the stress-migration due to thermal induced stress gradients were modeled respectively as follows.

$$\vec{J}_{EWM} = \frac{N}{k_B T} eZ^* \vec{j} \rho D_0 \exp\left(-\frac{E_A}{k_B T}\right), \tag{5.5}$$

$$\vec{J}_{th} = -\frac{NQ^* D_0}{K_B T} \exp\left(-\frac{E_A}{K_B T}\right) \nabla T, \tag{5.6}$$

$$\vec{J}_{str} = \frac{N\Omega D_0}{K_B T} \exp\left(-\frac{E_A}{K_B T}\right) \nabla \sigma_H, \tag{5.7}$$

where E_{a} is the measured activation energy, Q^* is the heat of transport, σ_H is the local hydrostatic stress value ($\sigma_H = (\sigma_{xx} + \sigma_{yy} + \sigma_{zz})/3$). Based on the above atomic flux equations, they derived an approximated Atomic Flux divergence (AFD) for the respective driving forces as follows:

$$div(\vec{J}_{EWM}) = \left(\frac{E_A}{k_B T} - \frac{1}{T} + \alpha\frac{\rho_0}{\rho}\right) \cdot \vec{J}_{EWM} \cdot \nabla T, \tag{5.8}$$

$$\begin{aligned} div(\vec{J}_{th}) = & \left(\frac{E_A}{k_B T} - \frac{3}{T} + \alpha\frac{\rho_0}{\rho}\right) \cdot \vec{J}_{th} \cdot \nabla T \\ & + \frac{NQ^* D_0}{3k_B^3 T^3} j^2 \rho^2 e^2 \exp\left(-\frac{E_A}{k_B T}\right), \end{aligned} \tag{5.9}$$

$$\begin{aligned} div(\vec{J}_{str}) = & \left(\frac{E_A}{k_B T} - \frac{1}{T}\right) \vec{J}_{str} \nabla T \\ & + \frac{2EN\Omega D_0 \alpha_l}{3(1-\upsilon)k_B T} \exp\left(-\frac{E_A}{k_B T}\right)\left(\frac{1}{T} - \alpha\frac{\rho_0}{\rho}\right) \nabla T^2 \\ & + \frac{2EN\Omega D_0 \alpha_l}{3(1-\upsilon)k_B T} \exp\left(-\frac{E_A}{k_B T}\right) \frac{j^2 \rho^2 e^2}{3k_B^2 T}. \end{aligned} \tag{5.10}$$

The coupled field analyses, namely electrical-thermal analysis and thermal-mechanical analysis, is performed using a commercial finite element analysis (FEA) package ANSYS. Based on their analysis results, the distributions of the current density, the temperature, the hydrostatic stress and their respective gradients are obtained and the value of AFD due to these three driving forces are calculated according to the formula above. Both the failure site of the test structure and the lifetime of the test structure are predicted by their method.

In their subsequent work, the formula were also applied to different test structures, such as meander structure, pad structure (Dalleau and Weide-Zaage 2001), and dual-damascene structure (Weide-Zaage, Dalleau, Danto and Fremont 2007) for both the Al and Cu technologies (Dalleau, Weide-Zaage and Danto 2003). Adopting the similar AFD formula, Tan *et al.* studied the intrinsic EM performance of deep submicron Cu and Al interconnects, including the effect of surface migration and void nucleation (Tan, Zhang and Gan 2004). In their subsequent work, the method was also successfully used to explain many experimental observations (Tan, Li, Tan and Low 2006; Tan and Roy 2006; Roy and Tan 2007).

In the derivation of the Eqs. (5.9) and (5.10), the following energy balance equation in a thin film conductor at steady state is used (Shirley 1985):

$$j^2\rho + K\nabla^2 T - \frac{k(T - T_s)}{th} = 0, \tag{5.11}$$

where K and k are the thermal conductivity of the metallic film and the dielectrics, T and T_S are the temperature of the metallic film and the substrate, t is the metallic film thickness, and h is the dielectric film thickness. The three terms in Eq. (5.11) represent Joule heating due to electrical current, lateral heat conduction along the plane of a film, and vertical heat conduction through the dielectrics into the substrate, respectively. The contributions by the three terms were calculated for Cu thin film with different thicknesses, and Tan *et al.* (Tan, Hou and Li 2007) found that as the line thickness decreases, heat conduction through the dielectrics into the substrate increases dramatically. Therefore, it is inaccurate to ignore the last term as it was done in the conventional derivations. Also, in the derivations of Eq. (5.10), the coupling effect between the hydrostatic stress and the temperature is considered and formulated as follows

$$\sigma_H = \frac{1}{3}(\sigma_{11} + \sigma_{22} + \sigma_{33}) = -\frac{2E\Delta\alpha_1}{3(1-\upsilon)}(T - T_{SFT}), \tag{5.12}$$

where σ_H is the hydrostatic stress in the interconnect, σ_{11}, σ_{22}, σ_{33} are stresses along the principal axes, T_{SFT} is the stress free temperature of the interconnect, and $\Delta\alpha$ is the difference of the CTE between the interconnect metallization and the surrounding materials. Equation (5.12) is derived based on Eshelby's inclusion model (Niwa, Yagi, Tsuchikawa and Masaharu 1990) and the cross section of the metal film was assumed to be an elongated ellipsoid in their analysis (Tan, Hou and Li 2007).

In reality, the cross section of a typical Cu interconnect is rectangular with cap layer on the top and diffusion barrier layers on the sidewalls and bottom, so Eq. (5.12) is insufficient in describing the coupling effect between the stress and temperature. In fact, the stress distribution is highly nonuniform due to the material and the structure inhomogeneities in interconnects. Through simulations, the thermomechanical stress distribution is strongly dependent on the structure geometry such as the line width, the aspect ratio, the presence of via, etc. The discrepancy will be even higher for narrow interconnects due to the increased thermomechanical hydrostatic stress as shown by (Tan, Hou and Li 2007).

In order to remove the above-mentioned assumptions, Tan *et al.* (Tan, Hou and Li 2007) derived the atomic flux divergence based on the Green's theorem. With this formulation, they were able to explain the reservoir effect observed experimental which cannot be done using Eqs. (5.8)–(5.10).

5.3.1.2 *Voiding mechanism simulation*

With the knowledge of AFD, the mass transport through a given controlled volume at any location in the test structure can be obtained, and thus the process of voiding can be simulated. In the work of Sasagawa (Sasagawa, Naito, Saka and Abe 1999; Sasagawa, Nakamura, Saka and Abe 2002), the metal line is divided into elements with a smaller size of the element to give more realistic results. The thickness of the element is changed by the procedure shown in Fig. 5.4. Through this procedure, the growth of voids in the metal line is simulated.

Based on the finite element analysis and the AFD formula, the distribution of AFD in each element can be calculated. Each calculation step is assigned to realistic time. The volume decreasing in each element per a calculation step in the simulation is calculated based on the volume of the element, the realistic time corresponding to one calculation step, the atomic volume and the value of AFD. The finite element analysis of the current

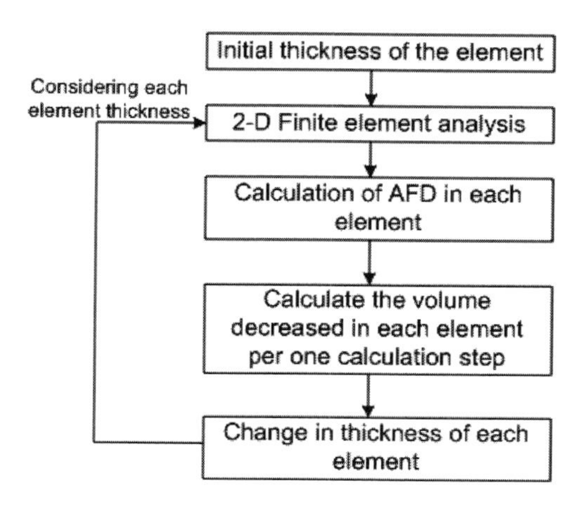

Fig. 5.4. Simulation flow chart in Sasagawa's study.

density and temperature in the metal line is implemented again according to the new thickness of each element and the decrement of thickness of the element is regarded as the void that is being formed. Following this procedure, a 2D simulator is able to simulate the growth of the voids and predict the failure time when the void becomes fatal to the interconnects.

However, in the work by Dalleau and Weide-Zaag (Dalleau and Weide-Zaage 2001; Dalleau, Weide-Zaage and Danto 2003; Weide-Zaage, Dalleau, Danto and Fremont 2007) as well as Tan (Tan, Zhang and Gan 2004), the process of void growth is simulated in a different procedure as shown in Fig. 5.5.

Different from Sasagawa's procedure, the elements in the meshed structure is physically deleted to simulate the void growth. In Dalleau's study, the deletion is realized by changing the corresponding material properties to negligible electrical and thermal conductivities, as well as a negligible Young modulus value for stress analysis. In Tan's study, those selected elements which have the highest AFD value is erased from the finite element model. In both simulations, the structure is automatically modified, and the finite element analysis is repeated until the failure condition is reached.

5.3.1.3 *Lifetime prediction*

The prediction of the lifetime of the EM test structure depends on the method to determine the realistic time in the simulation and the specified condition/criterion beyond which the elements do not exist anymore. In the work by Sasagawa *et al.* (Sasagawa, Naito, Saka and Abe 1999; Sasagawa, Nakamura, Saka and Abe 2002), the calculation process of the numerical

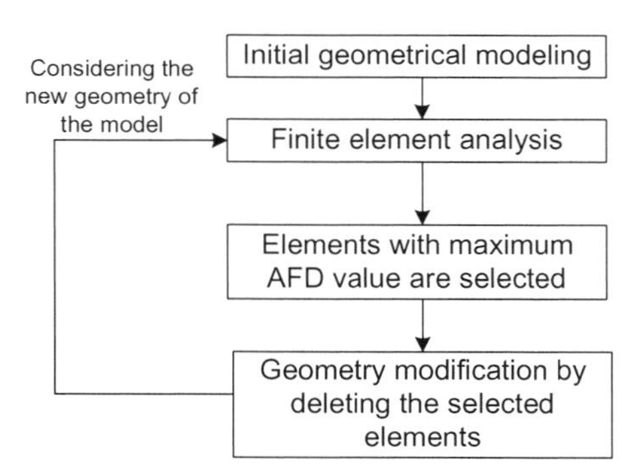

Fig. 5.5. The simulation flowchart adopted by Dalleau *et al.* and Tan *et al.*

simulation for the lifetime prediction is carried out repeatedly until the metal line fails where the entire line width is either occupied by the elements whose temperature exceeds the melting point of the metal due to Joule heating and/or the thickness of the elements become smaller than a pre-defined infinitesimal threshold value. Based on their experimental measurement and the hypothesis of the effective width of the slit void, they determined the threshold value should be 2×10^{-3} times the initial film thickness. By assigning one calculation step to a realistic time, the lifetime of the test structure can thus be predicted. The method of prediction was found to have a good agreement with their experimental observations.

Different prediction approaches is adopted in Dalleau et al's study (Dalleau and Weide-Zaage 2001; Dalleau, Weide-Zaage and Danto 2003; Weide-Zaage, Dalleau, Danto and Fremont 2007). The time dependent evolution of the local atomic concentration is expressed as

$$div(J_{EWM} + J_{th} + J_{str}) + \frac{\partial N}{\partial t} = 0. \tag{5.13}$$

According to Eqs. (5.8)–(5.10). Equation (5.13) is equivalent to the following equation

$$Nf + \frac{\partial N}{\partial t} = 0, \tag{5.14}$$

where f is a function including different physical parameters. The theoretical evolution of the atomic density is then given by Eq. (5.15)

$$N = N_0 e^{-ft}, \tag{5.15}$$

and N_0 is the initial value of the atomic concentration. In their simulation, the element is considered to be empty of material when the atomic concentration reduced to 10% of the initial concentration ($N/N_0 = 10\%$). By solving Eq. (5.15), the time required to delete one element can be calculated. As the void growth is simulated by the deletion of groups of element, time to failure of the structure can be extracted by accumulating the time needed for the deletion of all the elements. However, large difference is found between lifetime measurements and calculations.

One drawback of AFD based modeling is its incapability of simulating the activity of vacancies/atoms during EM from a thermodynamics aspect. The redistribution of vacancies/atoms through various diffusion paths during the void nucleation at the early stage of EM failure is essential for the narrow interconnect system. This is because the process of EM failure after the void is nucleated can be very abrupt with the small dimension and the high current density in the narrow interconnect system (Tan, Raghavan and Roy 2007). Under such circumstance, the method to retard the void nucleation at the early stage of EM failure will be a primary concern as this period of time could be almost the entire EM lifetime for narrow interconnections.

However, the redistribution of vacancies/atoms will not disturb much the physical environment of the entire EM test structure, and the process is nearly impossible to be simulated using EM AFD simulator since the calculation of AFD is based on the finite element analysis on the physical environment of the entire EM model for the interconnect. Including the simulation of EM void nucleation from the thermodynamic aspect in the EM simulation is important, and Monte Carlo method was employed for the purpose.

5.4 Monte Carlo Simulation of Electromigration

Monte Carlo method is a class of non-deterministic computational algorithms for simulating the behavior of various physical and mathematical systems. There are only few publications on the simulation of EM using Monte Carlo method, and they are different from other simulation methods by being stochastic as opposed to deterministic algorithms as introduced in the above sections.

5.4.1 *Monte Carlo simulation of the movement of atoms during EM*

Smy *et al.* reported their work on a Monte Carlo computer simulation of EM in 1993 (Smy, Winterton and Brett 1993). In the model, the motion of atoms are simulated according to two processes during EM, a drift of activated atoms in response to the electron wind force and the annealing due to the atoms movement to minimize the curvature of the grain boundaries. The grain structure is modeled by disks, each comprising some 2000 atoms. This is an attempt to describe events at the atomic level limited by the restrictions on computer memory. However, the physical laws used are not those obeyed by an individual atom, but rather are those appropriate to describe the average behavior of the atoms.

Annealing is a motion of disks to minimize surface curvature, and is carried out at a fraction of the rate at which disks are migrated. The disks are selected at random along the line, and are moved from sites of positive curvature to sites of negative curvature for reducing the surface energy.

This disk manipulation routines are collectively referred to as SIMBAD code reported in their previous reports, which have been applied to the studies in areas as diverse as the microstructure of deposited metals (Dew, Smy and Brett 1992a; Dew, Smy and Brett 1992b), dielectrics (Tait, Dew, Smy and Brett 1990) and the formation of hailstones (Lozowski, Brett, Tait and Smy 1991).

Another atomic Monte Carlo simulation of EM in polycrystalline thin films was reported by Bruschi (Bruschi, Nannini and Piotto 2000). In their study, they proposed an atomic Monte Carlo EM simulations based on a model of atom diffusion that includes the effect of electron wind force on the activation barrier. The samples are represented by a 3D array of cubic cell. Each cell is a data structure which includes the following objects: (i) a list of sites, (ii) a list of atoms and (iii) an electron wind force. The electron wind force is calculated by simulating the injection of a constant current and solving an equivalent resistor network extracted from the sample. The atomic Monte Carlo simulator calculates the interatomic interactions and seeks the possible destination for atom hopping which can be either a substrate site or an induced site.

Substrate sites simulate the vacancies in a grain, and induce sites simulate the vacancies along the grain boundary. Incorporation of atoms into the induced sites results in grain growth. The sample then evolves through a series of transitions, each one consists of an atom jump from the initial position to an empty site (substrate site or induced site) providing that the site lies within a maximum jump length from the initial position, and the model can simulate the void growth, the hillock formation and the grain growth on the sample.

Basically, the atomistic Monte Carlo simulation is a promising approach that can be applied to various problems related to diffusion and nucleation of atoms on a surface (Amar, Family and Amar 1996; Bruschi, Cagoni and Nannini 1997).

5.4.2 *Monte Carlo simulation of void movement during EM*

As the interconnect technology shifting to Cu, there are new experimental observations of EM reported in the literature, such as void migration at the interface between thin metallic film and dielectric (Vairagar, Krishnamoorthy, Tu, Mhaisalkar, Gusak and Meyer 2004). Zaporozhets *et al.* (Zaporozhets, Gusak, Tu and Mhaisalka 2005) proposed an atomistic Monte Carlo methodology to simulate the void migration in a Cu interconnect as observed in the experiments of their research group.

The simulator models the shape evolution, atomic displacement in the frame of a stochastic Monte Carlo method, with probability of displacement depending exponentially on the change of the energy as a result of the displacement. The electron wind force energy and the pair interaction energy are taken into consideration. The metallic film material is modeled at an atomic level, and the electric field is solved by using finite difference method with a time-dependent boundary condition. The interaction of bulk, grain boundary, dielectric and vacancy with the moving atoms is described by a matrix of interaction coefficients for the different pair interaction energy.

Based on their model, they simulated the behavior of the void under several conditions, namely (i) Surface void migration along the interface metal/dielectric in the absence of grain boundary; (ii) void migration along grain boundaries of different orientations with respect to current; and (iii) behavior of the voids at triple junctions of grain boundaries.

The use of Monte Carlo method to model physical problems allows us to examine more complex systems than we otherwise can. In this method, the EM theory formulation is based on a single atom or a single vacancy. With the understanding of the basic theory of EM physics, Monte Carlo method randomly generates numbers of scenarios for each atom/vacancy to simulate its behavior. The EM process is modeled by tracking the behavior of all atoms, vacancies or their clusters. In fact, the accuracy of a Monte Carlo simulation is proportional to the square root of the number of scenarios used. Unfortunately, this also means that it is computation intensive and should be avoided if simpler solutions are possible. While solving equations which describe the interactions between two atoms is fairly simple, solving the same equations for hundreds and thousands of atoms is nearly impossible for a standard computer system. For this reason, Monte Carlo method does not become the main stream of EM simulation.

Recently, as demonstrated by Li (Li, Tan and Hou 2007), the Monte Carlo methodology can be modified to simulate the behavior of vacancy clusters without compromising the accuracy of the simulation. The Monte Carlo simulation can actually be incorporated with finite element analysis to simulate the void nucleation from the thermodynamics perspective during EM process. The attempt of combining the simulation of EM kinetics and its thermodynamics aspect will produce a more realistic EM model and help us understand the physical phenomenon of EM better.

5.4.3 *Holistic EM simulation*

Li *et al.* (Li, Tan and Hou 2007) developed a model that combines Monte Carlo method and Finite element analysis. The model includes the inhomogeneities of diffusivities of metal atoms along various diffusion paths such as grain boundary, void surface and interface under the various driving forces of atom migration. A Monte Carlo method is used to simulate the void nucleation process followed by the dynamic simulation of void evolution due to the driving forces of EM based on Finite element analysis. In the Monte Carlo simulation, the void nucleation is modeled through a series of atom displacement with vacancies, and the displacement is dependent exponentially on the energy change as the result of the displacement. The energy calculation includes electron wind force energy, strain energy (plastic and elastic), thermal energy and pair interaction energy. It simulates the thermodynamics process of EM.

In their simulation, the vacancies are shown to move through the grain boundary from the anode to cathode end, and they are immobilized at the intersection of barrier layer-grain boundary ultimately. Following the void nucleation, the void growth is simulated by calculating the atomic flux divergence (AFD) due to various driving forces of EM. The void shape is then modified according to the maximum value of AFD in each element in the finite element model, as illustrated in Fig. 5.7. After that, the coupled field finite element analysis calculates the redistribution of the physical parameters, such as current density, temperature and stress, and the resulting new AFD value will be used to modify the void shape for the next iteration.

The simulation procedure for void growth is similar to the AFD based simulation method presented in the previous section. Thus, Li *et al.* (Li, Tan and Hou 2007) demonstrate a methodology to incorporate the Monte

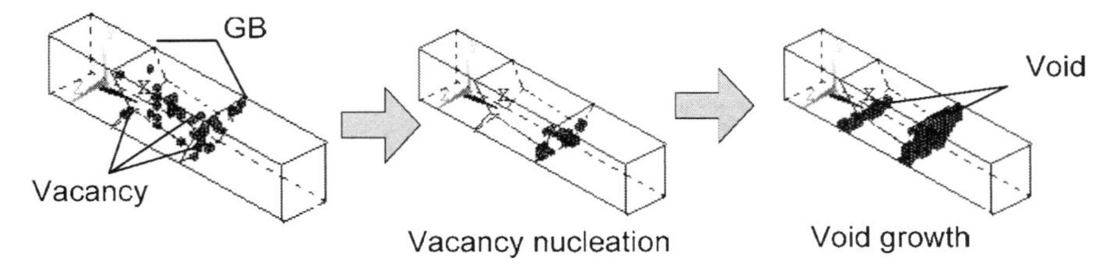

Fig. 5.6. The simulation of the vacancy nucleation and the void growth during EM.

Carlo method into Finite element analysis to simulate the EM process from void nucleation to void growth, considering the thermodynamics process of EM (various diffusion paths) and the kinetic of EM (various driving forces). With this model, it helps us to further understand the physics of EM and identify key material properties which can influence the reliability of interconnect for ULSI application. As an example, they used the method to derive an expression for the Median time to failure of EM failure governed by the void nucleation as in narrow interconnects (Li, Tan and Raghavan 2009), and their results agree well with the experimental data, even to the extrapolated data. This expression reveals the detail nature of the lump parameters in the Black's equation such as the material dependent constant, the activation energy and the current exponent. It even pointed out the inaccuracy in using the Black's equation for extrapolation. However, the expression is too complicate to use, and further simplification is needed.

5.5 Resistance Change Modeling

Resistance change with time is a direct measurement of EM performance of an interconnect. While we have discussed the void dynamics to determine the EM lifetime, Castro *et al.* (Castro, Hoofman, Michelon, Gravesteijn and Bruynseraede 2007) performed void growth modeling to understand the resistance change profile. The model starts with a small void located at the top interface between the copper and the cap layer or the barrier. The model explains the jump in the resistance profile and the progressive resistance change after the jump. They also studied the effect of barrier cross-sectional area as compared to the line cross-sectional area, line width/thickness, line length and the multiple voids on the resistance profiles. Although their results agree with experimental data qualitatively only, it helps to understand the resistance change profile better, shading insight to the possible EM dynamics.

References

Amar, J. G., Family, F. and Amar, G. (1996). *Thin Solid Films* **272**: 208.
Andleigh, Y. K., Fayad, W., Verminski, M. and Thompson, C. V. (http://nirvana.mit.edu/emsim/).
Arzt, E., Kraft, O., Nix, W. D. and Sanchez, J. J. E. (1994). Electromigration failure by shape change of voids in bamboo lines. *Journal of Applied Physics* **76**: 1563.
Atkinson, R. R. (2003). New Brunswick Rutgers, The State University of New Jersey. PhD.
Bhate, D. N., Kumar, A. and Bower, A. F. (2000). *Journal of Applied Physics* **87**: 1712.
Blech, I. A. (1998). Diffusion back flows during electromigration. *Acta Mater.* **46**(11): 3717.
Bruschi, P., Cagoni, P. and Nannini, A. (1997). *Physical Review B* **55**: 7955.
Bruschi, P., Nannini, A. and Piotto, M. (2000). *Computational Material science* **17**: 299.
Castro, D. T., Hoofman, R. J. O. M., Michelon, J., Gravesteijn, D. J. and Bruynseraede, C. (2007). Void growth modeling upon electromigration stressing in narrow copper lines. *Journal of Applied Physics* **102**: 123515.
Clement, J. J. (2001). Electromigration modeling for integrated circuit interconnectreliability analysis. *IEEE Trans. Device Mater. Reliab.* **1**(1): 33–42.
Collins, J. B. and Levine, H. (1985). Diffuse interface model of diffusion-limited crystal growth. *Physical Review B* **31**: 6119.
Dalleau, D. and Weide-Zaage, K. (2001). Three-dimensional voids simulation in chip metallization structures: a contribution to reliability evaluation. *Microelectronics Reliability* **41**(9–10): 1625–1630.
Dalleau, D., Weide-Zaage, K. and Danto, Y. (2003). *Microelectronics Reliability* **43**: 1821.
Dew, S. K., Smy, T. and Brett, M. J. (1992a). *J. Vac. Sci. Technol.* **B10**: 618.
Dew, S. K., Smy, T. and Brett, M. J. (1992b). *IEEE Trans. Electron. Dev* **39**: 1559.
Fix, G. (1983). Free boundary problems. *Research Notes in Mathematics*. A. Fasano and M. Primicero. New York, Pitman. **2**.
Frost, H. J. and Thompson, C. V. (1988). *Journal of Electronic Material* **17**: 447.
Gungor, M. R. and Maroudas, D. (1998). *Applied Physics Letter* **72**: 3452.
Gungor, M. R. and Maroudas, D. (1999). *Journal of Applied Physics* **85**: 2233.
Gurtin, M. (1996). *Physica D* **92**: 178.
Hau-Riege, C. S. and Thompson, C. V. (2000). *Journal of Applied Physics* **87**: 8467.
Hau-Riege, S. P. (2002). Probabilistic immortality of Cu damascene interconnects. *Journal of Applied Physics* **91**: 2014.
Hau-Riege, S. P. and Thompson, C. V. (2000). Electromigration saturation in a simple interconnect tree. *Journal of Applied Physics* **88**: 2382.
Huntington, H. B. (1975). Effect of driving forces on atom motion. *Thin Solid Films* **25**(2): 265–280.
Karma, A. and Rappel, W. (1996). *Physical Review E* **53**: R3017.
Kirchheim, R. (1992). Stress and electromigration in Al-lines of the integrated circuits. *Acta. Metall. Mater.* **40**: 309.
Kirchheim, R. and Kaeber, U. (1991). *Journal of Applied Physics* **70**: 172.

Korhonen, M. A., Borgesen, P., Tu, K. N. and Li, C.-Y. (1993). Stress evolution due to electromigration in confined metal lines. *Journal of Applied Physics* **73**: 3790–3799.

Kraft, O. and Arzt, E. (1997). *Acta. Mater.* **45**: 1599.

Lee, K. L., Hu, C. K. and Tu, K. N. (1995). *In-situ* scanning electron microscope comparison studies on electromigration of Cu and Cu(Sn) alloys for advanced chip interconnects. *Journal of Applied Physics* **78**(7): 4428–4437.

Li, W., Tan, C. M. and Hou, Y. (2007). Dynamic simulation of electromigration in polycrystalline interconnect thin film using combined Monte Carlo algorithm and finite element modeling. *Journal of Applied Physics* **101**: 104314.

Li, W., Tan, C. M. and Raghavan, N. (2009). Dynamic simulation of void nucleation during electromigration in narrow integrated circuit interconnects. *Journal of Applied Physics* **105**: 14305.

Lozowski, E. P., Brett, M. J., Tait, R. N. and Smy, T. (1991). *Quart. J. Royal Meteorological Soc.* **117**: 427.

Mahadevan, M. and Bradley, R. M. (1999). *Physical Review B* **59**: 11037.

Meyer, M. A., Herrmann, M., Langer, E. and Zschech, E. (2002). *In-situ* SEM observation of electromigration phenomena in fully embedded copper interconnect structures. *Microelectronic Engineering* **64**: 375.

Niwa, H., Yagi, H., Tsuchikawa, H. and Masaharu, K. (1990). *Journal of Applied Physics* **68**: 328.

Park, Y. J., Andleigh, V. K. and Thompson, C. V. (1999). *Journal of Applied Physics* **85**: 3546.

Park, Y. J. and Thompson, C. V. (1997). *Journal of Applied Physics* **82**: 4277.

Rosenberg, R. and Ohring, M. (1971). *Journal of Applied Physics* **42**: 5671.

Ross, C. A. (1991). Stress and electromigration in thin film metallization. *Mat. Res. Soc. Proc.* **225**: 35–46.

Roy, A. and Tan, C. M. (2007). Probing into the asymmetric nature of electromigration performance of submicron interconnect via structure. *Thin Solid Films* **515**: 3867–3874.

Sasagawa, K., Naito, K., Saka, M. and Abe, H. (1999). *Journal of Applied Physics* **86**: 6043.

Sasagawa, K., Nakamura, N., Saka, M. and Abe, H. (1998). *Trans. ASME, J. Elect. Pack* **120**: 360.

Sasagawa, K., Nakamura, N., Saka, M. and Abe, H. (2002). *Journal of Applied Physics* **91**: 1882.

Schimschak, M. and Krug, J., (2000). *Journal of Applied Physics* **87**: 695.

Shewmon, P. G. (1989). *Diffusion in solids*, Minerals, Metals and Materials Society.

Shirley, C. J. (1985). *Journal of Applied Physics* **57**: 777.

Smy, T. J., Winterton, S. S. and Brett, M. J. (1993). A Monte Carlo computer simulation of electromigration. *Journal of Applied Physics* **73**: 2821.

Sukharev, V. (2004). Physically-based simulation of electromigration induced failures in copper dual-damascene interconnect. *Proceedings. 5th International Symposium on Quality Electronic Design*, 225–230.

Sukharev, V., Kteyan, Zschech, E. and Nix, W. D. (2009). Microstructure effect on EM-induced degradations in dual inlaid copper interconnects. *IEEE Trans. Device Mater. Reliab.* **9**(1): 87–96.

Suo, Z., Wang, W. and Yang, M. (1994). *Applied Physics Letter* **64**: 1944.

Tait, R. N., Dew, S. K., Smy, T. and Brett, M. J. (1990). *SPIE International Symposium on Optical and Optoelectronic Applied Science and Engineering*. Bellingham, WA. **1324**.

Tan, C. M., Gan, Z., Li, W. and Hou, Y. (2010). *Applications of Finite element Methods for Reliability Study of ULSI Interconnections*, Springer Verlag.

Tan, C. M., Hou, Y. and Li, W. (2007). Revisit to the finite element modeling of electromigration for narrow interconnects. *Journal of Applied Physics* **102**: 033705.

Tan, C. M., Li, W., Tan, K. T. and Low, F. (2006). Development of highly accelerated electromigration test. *Microelectronics Reliability* **46**: 1638–1642.

Tan, C. M., Raghavan, N. and Roy, A. (2007). Application of gamma distribution in electromigration for submicron interconnects. *Journal of Applied Physics* **102**: 103703.

Tan, C. M. and Roy, A. (2006). Investigation of the effect of temperature and stress gradients on accelerated EM test for Cu narrow interconnects. *Thin Solid Films* **504**: 288.

Tan, C. M., Zhang, G. and Z. H. Gan (2004). Dynamic study of the physical process in the intrinsic line electromigration of deep-submicron copper and aluminum interconnects. *IEEE Trans. Device Mater. Reliab.* **4**: 450.

Trattles, J. T., O'Neill, A. G. and Mecrow, B. C. (1994). *Journal of Applied Physics* **75**: 7799.

Vairagar, A. V., Krishnamoorthy, A., Tu, K. N., Mhaisalkar, S. G., Gusak, A. M. and Meyer, M. A. (2004). *Applied Physics Letter* **85**: 2502.

Wang, W., Suo, Z. and Hao, T. H. (1996). *Journal of Applied Physics* **79**: 2394.

Weide-Zaage, K., Dalleau, D., Danto, Y. and Fremont, H. (2007). *Microelectronics Reliability* **47**: 319.

Xia, L., Bower, A. F., Suo, Z. and Shih, C. F. (1997). A finite element analysis of the motion and evolution of voids due to strain and Electromigration induced surface diffusion. *J. Mech. Phys. Solids* **45**(9): 1473–1493.

Yang, W., Wang, W. and Suo, Z. (1994). *J. Mech. Phys. Solids* **42**: 897.

Zaporozhets, T. V., Gusak, A. M., Tu, K. N. and Mhaisalka, S. G. (2005). Three-dimensional simulation of void migration at the interface between thin metallic film and dielectric under electromigration. *Applied Physics Letter* **98**: 103508.

CHAPTER 6
Future Challenges

Although extensive researches have been undertaken on interconnect electromigration for the past few decades as we can see from the previous chapters, there are several challenges remain that deserve further research effort, and they are described as follows.

6.1 Electromigration Modeling

As we have seen, the underlying physical processes during electromigration are complex and dynamics, and there is a need for the physics based modeling in order to examine the effect of various design and process parameters on the electromigration performance of interconnections. Such a model must be comprehensive to include the following, and it must be 3D as mentioned in Chapter 5:

(a) Micro-structure and its evolution during electromigration, e.g. grain rotation, grain growth etc. Sukharev *et al.* (Sukharev, Kteyan, Zschech and Nix 2009) have developed model that accounts for the micro-structure effect on EM. However, grain rotation, grain orientation, grain texture etc. have not been included, and its 3D demonstration is yet to be made. Also, initial stress in the film has not yet included. Lastly, how to create a realistic micro-structure in the modeling is a major subject that needs to be investigated.

(b) The comprehensive model must include all the driving forces for electromigration. The consideration of the various driving forces in EM modeling has been made by Sukharev (Sukharev 2004) as well as Tan *et al.* (Tan, Hou and Li 2007) and Li *et al.* (Li, Tan and Hou 2007; Li, Tan and Raghavan 2008). To include all the driving forces and micro-structure effect as mentioned above will be a challenge to the model developers.

(c) When all the driving forces and micro-structure are included, the diffusion of vacancies and atoms through the different diffusion paths available must be simulated. Li *et al.* (Li, Tan and Hou 2007) have combined the driving forces and diffusion paths modeling using the combination of Monte Carlo and finite element methods. While Monte Carlo method is good for void nucleation, it cannot simulate void migration, coalescence and growth because it is too computation-time consuming. Li *et al.* used finite element method to model the void growth, but the void migration and coalescence will require the method developed by Sukharev *et al.* (Sukharev, Kteyan, Zschech and Nix 2009) to compute.

(d) Stress evolution during EM is important in determining the EM performance of a line. Many works have been on the stress modeling during EM such as Sukharev (Sukharev 2004), Lloyd *et al.* (Lloyd 1991; Clement and Thompson 1995), Filippi *et al.* (Filippi, Wang, Wachnik, Chidambarrao, Korhonen, Shaw, Rosenberg and Sullivan 2002), Flinn (Flinn 1995), Gan *et al.* (Gan, Huang, Ho, Leu, Maiz and Scherban 2004), Korhonen *et al.* (Korhonen, Borgesen, Tu nd Li 1993), Orain *et al.* (Orain, Barbé, Federspiel, Legallo and Jaouen 2004), and Shen *et al.* (Shen, Guo and Minor 2000) etc. The incorporation of this stress evolution into the model with microstructure, all the driving forces and all the diffusion paths will be needed.

(e) As line width becomes narrower, stress relaxation will be dominated by plastic deformation. The diffusional creep and plastic deformation for stress relaxation will also need to be modeled in.

(f) The residual film stress will have a tremendous effect on the EM performance of the line, and the modeling of this residual stress during the back end process is needed. The effect of stress free temperature on EM performance and also the void location have been shown by Roy *et al.* (Roy, Tan, Kumar and Chen 2005; Roy and Tan 2006).

Weide-Zaage *et al.* (Weide-Zaage, Zhao, Ciptokusumo and Aubel 2008) have developed a model to compute the total accumulated stress after the entire back end process. However, grain growth stress is not considered in their work. This residual stress will need to be included in the comprehensive model.

(g) The resulting micro-structure of the line due to interconnect geometry, process parameters, barrier material type etc. will need to be modeled so that realistic micro-structure can be incorporated into the comprehensive model.

One can see that the challenge to develop such a comprehensive model is huge, and it really requires extensive collaborative efforts among researchers working in this area. On the other hand, with such a comprehensive model, one can predict the EM performance of the interconnect fabricated, examine the size effect on EM performance, validate the extrapolation of the EM test results to normal operating condition, assist in the understanding of the failure physics of EM failed samples and thus provide reliability improvement.

Fortunately, the required knowledge for the comprehensive model is mostly available, and the power of modeling tools as well as computation power are also advancing to take up this challenging task. Like the maturity of TCAD tool such as T-supreme, Medici, Cadence etc, it is hoped that the tool for interconnect reliability will also become available to aid in the design-in reliability for IC.

6.2 EDA Tool Development

While it is good to have a comprehensive model, but it remains at the evaluation of the EM of interconnect lines. The incorporation of the model into an EDA tool will be needed in order to assist IC designers for the proper design and layout of the interconnections used in their designs.

Various simulators and modeling tools have been developed to predict circuit reliability due to electromigration. For examples, BERT (Hu 1992), iTEM (Teng, Cheng, Rosenbaum and Kang 1997), and ERNI (Cherry, Hau-Riege, Alam, Thompson and Troxel 1999) were developed for reliability analysis in Al-based interconnect.

For Cu based interconnect, SysRel, a layout reliability prediction tool based on the critical (jL) product, was built by Alam *et al.* (Alam, Lip, Thompson and Troxel 2004); RELIANT was also developed for predicting failure rate of circuits due to electromigration by Frost *et al.* (Frost and Poole 1989).

However, all the above mentioned models are 2D in nature. It is found that 2D simulation will underestimate the peak current density when the track and via width are different, which may result in an electromigration susceptible process design as shown by Trattle *et al.* (Trattles, Neill and Mecrow, 1991). 2D circuit modeling is also unable to represent the entire circuit geometry due to layers overlapping. Complex geometries include junctions and segments of varying shape and width cannot be built in 2D model. Some important factors on electromigration, such as high stress due to sharp corner, current crowding due to the change of shape of current path, and change in the void initiation location due to the change in the aspect ratio of the cross section of the metal line (Li, Tan and Hou 2007) are difficult to simulate in 2D modeling.

As metallization level and via number increases due to scaling, it becomes increasingly difficult to simply use 2D layout for an efficient and effective EM awareness design.

In fact, in actual IC implementation on a wafer, it is 3D in nature, and the temperature and thermo-mechanical stress in the various parts of the IC are highly dependent on the surrounding materials and their materials properties, including their thermal conductivities, thermal expansivities, Young's modulus, poisson ratio etc. Also, the architectural of the 3D layout will also affect the current density, temperature and thermal-mechanical stress distributions in the IC (Li, Tan and Hou 2007).

Owing to the above considerations, many electromigration studies have been undertaken using 3D modeling. Lin *et al.* (Lin, Chang, Su and Wang 2007) studied the effect of line width on electromigration in dual damascene Cu interconnects; Atakov *et al.* (Atakov, Sriram, Dunnell and Pizzanello 1998) characterized the reliability of multiple-via contacts; Rzepka *et al.* (Rzepka, Banerjee, Meusel and Hu 1998) showed the effect of self-heating based on a 4-level interconnect test structure and presented a thermal simulator for the design process; the impact of different test structure design on electromigration lifetime had also been investigated by Guo *et al.* (Guo, Foo, Xu, Pei and Ping 1999); Jerke *et al.* (Jerke and Lienig 2004) built a quasi 3D model of the interconnect via region to verify the irregularities for current densities determination in arbitrarily shaped custom-circuit layouts.

However all these reported electromigration studies are typically conducted on either a simple straight line or two level line-via structure which is just part of the entire circuit. As electron wind force due to the current density is no longer the dominate driving force for electromigration when interconnect line width drops below 200 nm, and driving force due to thermo-mechanical stress gradient is found to be at least two order of magnitude higher than that from the electron wind force (Li, Tan and Hou 2007), the surrounding materials and the interconnect structures are critical to the electromigration performance of interconnections, and thus a complete 3D interconnect structure with the consideration of the surrounding materials properties has become necessary in order to accurately assess the interconnect reliability in ULSI.

Furthermore, temperature is a key factor in determining the degradation rate of the various materials, including ULSI interconnections, and the heat dissipation of an ULSI depends on the actual 3D layout of an integrated circuit. Thus a complete 3D circuit model for an ULSI will be needed.

Presently, EDA tool inclusion for electromigration awareness IC design is still limited to 2D. An automatic wire widening method had been developed by Zhan *et al.* (Zhan and Heng 1998) for EM reliability enhancement in 2D layout; Lienig *et al.* (Lienig and Jerke 2005) described a 2D physical design methodology for current density verification and current-driven layout de-compaction to address electromigration; Alam *et al.* (Alam, Gan, Thompson and Troxel 2007) also showed a method for full-chip electromigration analysis using SysRel. However, as current density is no longer the sole driving force for electromigration in interconnects when the line width decreases below 200 nm (Tan, Hou and Li 2007), and their electromigration reliability depends critically on their surrounding materials (Li, Tan and Hou 2007), such a 2D modeling with the current density as the main consideration is no longer adequate.

In order to have a full chip 3D interconnect electromigration modeling, the various level of mask layout must be able to convert into a 3D structure of the IC that resembles its actual implementation. Recently, Tan *et al.* (Tan and He 2009) presented a method to do such a conversion from Cadence mask layout to 3D finite element model. With this 3D finite element model, they computed the atomic flux divergence of the entire interconnect structure for all the driving forces for EM under different loading condition of the circuit

(He and Tan 2009), and the effect of layout on the void locations can be clearly seen.

However, while the full chip 3D modeling is promising, the demonstrated work (He and Tan 2009) is only limited to simple inverter circuit. To apply the method to the complex chip will require huge memory size and time for the computation, and thus its incorporation into EDA tool will not be possible. Further work is needed to simplify the modeling so that an accurate EDA tool for electromigration-awareness design can be possible.

6.3 Physics of Electromigration

Besides the future work needed for the electromigration modeling, and despite the many reported works on the physics of electromigration, the physical nature of the very basic electron wind force remain unclear. The presence of the electron wind force is in doubt from a physical point of view as mentioned in Chapter 2, and electromigration phenomena can also be explained without the use of the electron wind force (Sah and Jie 2008). This fundamental question, though it may not hinder the development of practical EM formulation presently, may create problem when line width is down to a few nm range where classical electrical conductivity theory is no longer applicable.

Another question need to be clarified is the nature of the critical stress for void nucleation. Although the yield strength of Cu and the bonding strength of Cu are much higher than that of Al, the critical stress for EM voiding is found to be below 50 MPa in Cu interconnects whilst that for Al is about 200 MPa (Hau-Riege 2004). The factors that determine the critical stress for voiding are yet to be known. As thermo-mechanical stress is increasingly important as the line width is shrinking, the identification of the factors for critical stress is necessary to improve the reliability of Cu interconnections.

Sukhareve *et al.* (Sukharev, Kteyan, Zschech and Nix 2009) also observed that void nucleation depends on the grain size, but the dependence is presently unclear.

6.4 Electromigration Testing

As stress free temperature (SFT) is crucial for EM performance of interconnects, one must be able to accurately determine the SFT experimentally without resolving to the use of synchrotron X-ray which is not readily available.

Also, the use of the Black's equation to extrapolate the EM test results to normal condition is getting increasingly more inaccurate due to the increasing importance of the thermo-mechanical stress and decreasing importance of the electron wind force for EM. Li *et al.* (Li, Tan *et al.* 2009) has developed an analytical expression for the median time to failure based on the physics of the EM kinetics and thermodynamics, and the accuracy of the expression in the median time to failure extrapolation to normal operating condition is better than that from the Black's equation. However, for IC design, further simplification of the expression is needed in order to be useful for semiconductor industry, or at least an algorithm must be provided to determine the parameters values from some experimental measurements.

With the demand of higher interconnect reliability, shorter available reliability test time, and accuracy in extrapolation, we need to properly evaluate highly accelerated EM test methods, in order to ensure invariant of the failure mechanisms when extrapolated to other stress conditions. Tan *et al.* (Tan, Li, Tan and Low 2006) have developed a method that combine modeling and testing for such evaluation. As EM becomes more microstructural dependent, a comprehensive model discussed earlier will be required instead, and the non-linear dependences of the materials degradation with applied stress levels must be taken into account. This call for a detail understanding of these dependences, and such a model will be very complex and yet to be developed.

6.5 New Failure Mechanism for Interconnects

As the line spacing dimension between adjacent interconnects is approaching 100 nm and below, the intrinsic reliability in BEOL low-k dielectrics show some similarity to the gate oxide reliability concerns that existed during the early years of integrated circuit development. This leads to the formulation of a number of basic and technologically important questions. First, which time-dependent model of reliability is necessary for accurate assessment of BEOL intrinsic reliability? Second, which defect mechanisms cause serious erosion of intrinsic reliability and what will be their impact on the intrinsic time-dependent dielectric breakdown (TDDB) mechanism? (Hartfield, Ogawa, Park, Chiu and Guo 2004).

A separate but equally important line of reasoning that motivates such inquiry is the fact that these new low-k dielectrics generally possess inferior thermo-mechanical and electrical materials properties in comparison with those of a silicon dioxide dielectric. Consequently, their use provides a challenge to ensuring the robustness of the now-standard damascene process in modern interconnect fabrication (Hartfield, Ogawa, Park, Chiu and Guo 2004).

Because there are numerous and generally nontrivial integration steps used in the fabrication of Cu-based interconnects, a number of potentially new reliability and defectivity issues may need to be understood and characterized. So, additional important questions arise. What are the important process and integration steps that will have great impact on intrinsic reliability performance? Where are the most susceptible locations that will act as weak-links for dielectric reliability? (Hartfield, Ogawa, Park, Chiu and Guo 2004).

Cu ion migration in semiconductors and insulators under an electric field is a well-known phenomenon that can limit Cu/low-k interconnect reliability. Even as fabrication of dual-damascene interconnects has matured, confinement of Cu was not always a simple matter because the damascene process steps could leave Cu residue material outside the trench before barrier confinement and capping (Kim, Cho and Ho 1999; Bersuker, Blaschke, Choi and Wick 2000; Noguchi, Ohashi, Yasuda, Jimbo, Yamaguchi, Owada, Takeda and Hinode 2000; Tsu, McPherson and McKee 2000).

Furthermore, defect control with respect to the post-CMP trench interface was vital (Balakumar, Tsang and Matsuki 2003; Chou, Wu, Huang, Lin, Chang, Jang and Liang 2003; Kato, Sekiyama, Takeshiro, Fukuda and Yanazawa 2003; Konishi, Yamada, Noguchi and Tanaka 2003; Noshioka, Ariga, Inoue, Tokushige and Tsujimura 2003; Seo, Kawano, Satake, Hattori, Harada, Itoh and Ueda 2003). The impact of ambient exposure time due to Cu migration and corrosion is shown to affect dielectric breakdown (Noguchi, Miura, Kubo, Tamaru, Yamaguchi, Hamada, Makabe, Tsuneda and Takeda 2003). Another process factor is the presence of line-edge roughness (LER) effects, which can seriously degrade the TDDB distribution (Noguchi, Oshima, Tanaka, Sasajima, Aoki, Sato, Ishikawa, Saito, Konishi, Hotta, Uno and Kikushima 2003). For 22 nm and beyond, line-edge roughness (LER) and its effects are expected to increase (Peters Jul 14, 2009). Thus, intermetal dielectric (IMD) configurations — rather than interlevel dielectrics (ILDs) — apparently exhibit potential weak-link characteristics from an integration and reliability standpoint.

Another approach observes that intrinsic failures of the IMD dielectric stack are associated with crack formation and argues that the electric field can be strong enough to trigger interface delamination. Consequently, intrinsic adhesion strength rather than Cu ion migration will be the intrinsic limiter of dielectric reliability (Alers, Sanganeria, Shaviv, Kooi, Jow and Ray 2003). Another work also notes that narrower leads show decreased breakdown strength and argues that field strength enhancements at the top of the Cu trench will also limit scaling (Noguchi, Miura, Kubo, Tamaru, Yamaguchi, Hamada, Makabe, Tsuneda and Takeda 2003).

In short, the issue of dielectric reliability in BEOL Cu/low-k systems continues to evolve with scaling. The detail investigation of this failure mechanism and associated solutions are needed in order to enhance interconnect reliability, and this can be challenging.

6.6 Alternative Interconnect Structure

To further reduce the RC delay, partial air-gap interconnect is proposed. IMEC claimed that beyond 20 nm half pitch, air gaps become a promising solution (Peters Jul 14, 2009). How will this new interconnect structure affect EM remains to be investigated. To remove the TDDB impact as mentioned earlier, complete air-gap interconnect structure may be one alternative. But to have the Cu interconnect stand-alone will be challenging, and nano-twinned Cu might be a promising choice for the complete air-gap interconnect because of its higher mechanical strength. How the EM performance of such an interconnect will be interesting to know. However, as a number of material properties of the nano-twinned Cu are yet to be known, EM modeling of such interconnect remain to be highly speculative, if not on hold.

As low-k dielectric is a poor thermal conductor, the heat dissipation will be retarded for Cu/low-K system. To improve the thermal conductivity of the low-k dielectric, attempt is made to incorporate carbon nanotube in the dielectric as carbon nanotube has excellent thermal and mechanical properties. However, successful implementation requires a method to disperse the nanotube uniformly in the dielectric and also to ensure covalent bonding between the nanotube and the dielectric material so that phonon can be transport through the nanotube for better heat conduction, and also mechanical enhancement can be achieved. Recently, a method of functionalization of carbon nanotube to achieved the above-mentioned method is demonstrated for epoxy (Baudot and Tan 2009). The application of such a method to low-k dielectric is yet to be done.

6.7 Alternative Interconnect System

For better and more reliable interconnects, carbon nanotube (CNT), with excellent electrical, thermal and mechanical properties are considered. CNTs are considered as an alterative interconnect material due to their high aspect ratio, tremendous current carrying capacity and large electromigration tolerance (Kreupl, Graham, Duesberg, Steinhoegl, Liebau, Unger and Hoenlein 2002; Kreupl, Graham, Liebau, Duesberg, Seidel and Unger 2004; Zhu, Sun, Xu, Zhang, Hess and Wong 2005; Zhu, Hess and Wong 2006). The electrical conductance of an ideal ballistic nanotube is $4e^2/h \approx 155\,\mu S$ or about $6.5\,\mathrm{k}\Omega$. (Dattta 1997) A bundle consisting of several parallel nanotubes exhibit lower resistance compared with equivalent area copper wire at nanometer scale (Kreupl, Graham, Liebau, Duesberg, Seidel and Unger 2004). The carbon-carbon bonding between the neighbouring atoms in the CNT is one of the strongest bonding in the world (Wong, Sheehan and Lieber 1999). Migration of carbon atoms will not happen even under very high current density. Therefore, the electromigraion tolerance of CNTs should be much better than that of other traditional interconnect material, such as, copper and aluminum. Current carrying capacity and reliability of individual CNT has been studied under both electrical and thermal stress, which shows no observable failure in the single nanotube structure under high current densities (10^9 A/cm^2) and 250°C (Wei, Vajtai and Ajayan 2001).

However, the condition in reported study is rather different from the silicon CMOS environment. Chai *et al.* (Chai, Zhang, Gong and Chan 2007) performed reliability evaluation of CNT interconnect in a silicon CMOS environment, and they found that the conductivity of CNT interconnect decreases with time rapidly due to the oxidation of the outer shell as a result of the Joule heating of the tube at high power dissipation. Although the conductivity of bundle of CNT in a via is stable due to the heat dissipation through the surrounding SiO_2 dielectric, a sharp decrease in the conductance was observed when the temperature increases beyond 340K. Therefore,

CNT as an interconnect may not be a good choice. On the other hand, embedding CNT into Cu shows promising results as was discussed in Chapter 5.

Other type of interconnect system include optical interconnects (Singer Nov 1, 2006), and atomic waves communication through spintronic (Hellemans 2006). The feasibility of using these methods remain to be demonstrated.

In fact, as the line width decreases to nm range, the size effect on the Cu line resistivity causes the resistivity of Cu to be higher than that of Al. In such situation, will the future interconnect going back to Al? Or will another metal be chosen for the interconnect? An answer yet to be seen. In fact, the consideration on using different metal for interconnect will not rely only on its EM reliability. Complexity in manufacturing processing as well as cost of the material will be more important issues to address, and these are the reasons that Au, though having lower electrical conductivity than Al, was not considered as VLSI interconnects.

From the above discussion, we can see that there are indeed several huge challenges in the interconnect reliability study, in particular its electromigration performance. It is my hope that readers of this book will have a better understanding of the EM physics, and are motivated to continue their research in this field.

References

Alam, S. M., Gan, C. L., Thompson, C. V. and Troxel, D. E. (2007). Reliability computer-aided design tool for full-chip electromigration analysis and comparison with different interconnect metallizations. *Microelectronics Journal* **83**(4–5): 463–473.

Alam, S. M., Lip, G. C., Thompson, C. V. and Troxel, D. E. (2004). Circuit level reliability analysis of Cu interconnects. *Proceedings. 5th International Symposium on Quality Electronic Design*, 238–243.

Alers, G. B., Sanganeria, M., Shaviv, R., Kooi, G., Jow, K. and Ray, G. W. (2003). Failure mechanisms in dielectric barriers. *Proceeding of Advanced Metallization Conf. (AMC).*

Atakov, E. M., Sriram, T. S., Dunnell, D. and Pizzanello, S. (1998). Effect of VLSI interconnect layout on electromigration performance. *Proceeding of IEEE IRPS*, 348–355.

Balakumar, S., Tsang, C. F. and Matsuki, N. (2003). CMP process development and post-CMP defects studies on Cu/ultra low k materials with single damascene scheme. *Proceeding of Advanced Metallization Conf. (AMC)*, 613–620.

Baudot, C. and Tan, C. M. (2009). Solubility, dispersion and bonding of functionalized carbon nanotubes in epoxy resins. *International Journal of Nano-Technologies* **6**: 618–627.

Bersuker, G., Blaschke, V., Choi, S. and Wick, S. (2000). Conduction processes in Cu/low-k interconnection. *Proceeding of IEEE IRPS*, 344–347.

Chai, Y., Zhang, M., Gong, J. and Chan, P. C. H. (2007). Reliability evaluation of carbon nanotube interconnect in a silicon CMOS environment. *Proceeding of International Conference on Electronic Materials and Packaging*, 589–593.

Cherry, Y., Hau-Riege, S., Alam, S. M., Thompson, C. V. and Troxel, D. E. (1999). ERNI: A tool for technology-generic circuit-level reliability projections. *Interconnect Focus Center Annual Review*.

Chou, T. J., Wu, Z. C., Huang, Y. L., Lin, S. H., Chang, W., Jang, S. M. and Liang, M. S. (2003). Post Cu CMP surface modification for reliability improvement in 300 mm CLN90 BEOL Cu/low-k interconnect. *Proceeding of Advanced Metallization Conf. (AMC)*, 433–437.

Clement, J. J. and Thompson, C. V. (1995). Modeling electromigration-induced stress evolution in confined metal lines. *Journal of Applied Physics* **78**: 900.

Dattta, S. (1997). *Electronic Transport in Mesoscopic Systems*, Cambridge University Press.

Filippi, R. G., Wang, P.-C., Wachnik, R. A., Chidambarrao, D., Korhonen, M. A., Shaw, T. M., Rosenberg, R. and Sullivan, T. D. (2002). The electromigration short-length effect in AlCu and Cu interconnects. *Stress-Induced Phenomena in Metallization: 6th International Workshop on Stress-Induced Phenomena in Metallization* **612**: 33–48.

Flinn, P. A. (1995). Mechanical stress in VLSI interconnections: Origins, effects, measurement, and modeling. *Mater. Res. Bull.* **20**: 70–73.

Frost, D. F. and Poole, K. F., (1989). Reliant: A reliability analysis tool for VLSI interconnects. *IEEE Journal on Solid-State Circuits* **24**(2): 458–462.

Gan, D., Huang, R., Ho, P. S., Leu, J., Maiz, J. and Scherban, T. (2004). Effects of passivation layer on stress relaxation and mass transport in electroplated Cu films. *Proc. of International Workshop on Stress-induced Phenomena in Metallization*, 256–267.

Guo, Q., Foo, L. K., Xu, Z., Pei, Y. and Ping, N. S. (1999). Impact of test structure design on electromigration of metal interconnect. *Proceeding of IEEE IPFA*, 39–43.

Hartfield, C. D., Ogawa, E. T., Park, Y.-J., Chiu, T.-C. and Guo, H. (2004). Interface reliability assessments for copper/low-k products. *IEEE Trans. Device Mater. Reliab.* **4**(2): 129–141.

Hau-Riege, C. S. (2004). An introduction to Cu electromigration. *Microelectronics Reliability* **44**: 195–205.

He, F. F. and Tan, C. M. (2009). *Circuit level interconnect reliability study using 3D circuit model*, to be submitted.

Hellemans, A. (2006). Spin doctoring. *IEEE Spectrum*, 13.

Hu, C. M. (1992). BERT: an IC reliability simulator. *Microelectronics Journal* **23**: 97–102.

Jerke, G. and Lienig, J. (2004). Hierarchical current-density verification in arbitrarily shaped metallization patterns of analog circuits. *IEEE Trans. on Computer-Aided Design of Integrated Circuits and Systems* **23**(1): 80–90.

Kato, I., Sekiyama, S., Takeshiro, S., Fukuda, T. and Yanazawa, H. (2003). New method to improve electrical characteristics of low-k dielectrics in Cu-damascene interconnections. *Proceeding of Advanced Metallization Conf. (AMC)*, 699–704.

Kim, S. U., Cho, T. and Ho, P. S. (1999). Leakage current degradation and carrier conduction mechanisms for Cu/BCB damascene process under bias-temperature stress. *Proceeding of IEEE IRPS*, 277–282.

Konishi, N., Yamada, Y., Noguchi, J. and Tanaka, U. (2003). Improvement in Cu-CMP technology for 90-nm nodes. *Proceeding of Advanced Metallization Conf. (AMC)*, 127–132.

Korhonen, M. A., Borgesen, P., Tu, K. N. and Li, C.-Y. (1993). Stress evolution due to electromigration in confined metal lines. *Journal of Applied Physics* **73**: 3790–3799.

Kreupl, F., Graham, A. P., Duesberg, G. S., Steinhoegl, W., Liebau, M., Unger, E. and Hoenlein, W. (2002). Carbon nanotube in interconnect application. *Microelectronic Engineering* **64**: 399–408.

Kreupl, F., Graham, A. P., Liebau, M., Duesberg, G. S., Seidel, R. and Unger, E. (2004). Carbon nanotubes for interconnect applications. *Proceeding of IEEE IEDM*, 683–686.

Li, W., Tan, C. M. and Hou, Y. (2007). Dynamic simulation of electromigration in polycrystalline interconnect thin film using combined Monte Carlo algorithm and finite element modeling. *Journal of Applied Physics* **101**: 104314.

Li, W., Tan, C. M. and Raghavan, N. (2008). Predictive dynamic simulation for void nucleation during Electromigration in ULSI interconnects. *Journal of Applied Physics*, under review.

Li, W., Tan, C. M., *et al.* (2009). Dynamic simulation of void nucleation during electromigration in narrow integrated circuit interconnects. *Journal of Applied Physics* **105**: 14305.

Lienig, J. and Jerke, G. (2005). Electromigration-aware physical design of integrated circuits. *Proceeding of International Conference on VLSI Design*, 78–82.

Lin, M. H., Chang, K. P., Su, K. C. and Wang, T. H. (2007). Effects of width scaling and layout variation on dual damascene copper interconnect electromigration. *Microelectronics Reliability* **47**(12): 2100–2108.

Lloyd, J. R. (1991). Mechanical stress and electromigration failure. *Material research society symposium* **225**: 47.

Noguchi, J., Miura, N., Kubo, M., Tamaru, T., Yamaguchi, H., Hamada, N., Makabe, K., Tsuneda, R. and Takeda, K.-I. (2003). Cu-ion-migration phenomena and its influence on TDDB lifetime in Cu metallization. *Proceeding of IEEE IRPS*, 287–292.

Noguchi, J., Ohashi, N., Yasuda, J.-I., Jimbo, T., Yamaguchi, H., Owada, N., Takeda, K.-I. and Hinode, K. (2000). TDDB improvement in Cu metallization under bias stress. *Proceeding of IEEE IRPS*, 339–343.

Noguchi, J., Oshima, T., Tanaka, U., Sasajima, K., Aoki, H., Sato, K., Ishikawa, K., Saito, T., Konishi, N., Hotta, S., Uno, S. and Kikushima, K. (2003). Integration and reliability issues of Cu/SiOC interconnect for ArF/90 nm node SoC manufacturing. *Proceeding of IEEE Int. Electron Devices Meeting (IEDM)*, 527–530.

Noshioka, Y., Ariga, Y., Inoue, T., Tokushige, K. and Tsujimura, M. (2003). The development of defect free post-CMP cleaning in Cu/low-k damascene wiring. *Proceeding of Advanced Metallization Conf. (AMC)*, 645–650.

Orain, S., Barbé, J.-C., Federspiel, X., Legallo, P. and Jaouen, H. (2004). FEM-based method to determine mechanical stress evolution during process flow in microelectronics. application to stress avoiding. *Proceedings of the 5th International Conference on Thermal and Mechanical Simulation and Experiments in Microelectronics and Microsystems, EuroSimE 2004*, 47–52.

Peters, L. (Jul 14, 2009). IMEC Reveals Interconnect Roadmap to 10 nm. *Semiconductor International.*

Roy, A. and Tan, C. M., (2006). Experimental investigation on the impact of stress free temperature on the electromigration performance of copper dual damascene submicron interconnect. *MIcroelectronics Reliability* **46**(9–11): 1652–1656.

Roy, A., Tan, C. M., Kumar, R. and Chen, X. T. (2005). Effect of test condition and stress free temperature on the electromigration failure of Cu dual damascene submicron interconnect line-via test structure. *Microelectronics Reliability* **45**: 1443–1448.

Rzepka, S., Banerjee, K., Meusel, E. and Hu, C. M. (1998). Characterization of self-heating in advanced VLSI interconnect lines based on thermal finite element simulation. *IEEE Trans. on Components, Packaging, and Manufacturing Technology, Part A* **21**(3): 406–411.

Sah, C. T. and Jie, B. (2008). The Driftless electromigration theory (diffusion-generation-recombination-trapping theory). *Journal of Semiconductors* **29**(5): 815–821.

Seo, S., Kawano, H., Satake, M., Hattori, T., Harada, T., Itoh, Y. and Ueda, T. (2003). Improvement of short defect density by controlling surface height of Cu line. *Proceeding of Advanced Metallization Conf. (AMC)*, 121–125.

Shen, Y.-L., Guo, Y. L. and Minor, C. A. (2000). Voiding induced stress redistribution and its reliability implications in metal interconnects. *Acta Materialia* **48**(8): 1667–1678.

Singer, P. (Nov 1, 2006). Copper/low k challenges for 45 and 32 nm. *Semiconductor International.*

Sukharev, V. (2004). Physically-based simulation of electromigration induced failures in copper dual-damascene interconnect. *Proceedings. 5th International Symposium on Quality Electronic Design*, 225–230.

Sukharev, V., Kteyan, A., Zschech, E. and Nix, W. D. (2009). Microstructure effect on EM-induced degradations in dual inlaid copper interconnects. *IEEE Trans. Device Mater. Reliab.* **9**(1): 87–96.

Tan, C. M. and He, F. F. (2009). 3D circuit model for 3D IC reliability study. *Proceeding of IEEE EuroSimE.*

Tan, C. M., Hou, Y. and Li, W. (2007). Revisit to the finite element modeling of electromigration for narrow interconnects. *Journal of Applied Physics* **102**: 033705.

Tan, C. M., Li, W., Tan, K. T. and Low, F. (2006). Development of highly accelerated electromigration test. *Microelectronics Reliability* **46**: 1638–1642.

Teng, C., Cheng, Y., Rosenbaum, E. and Kang, S. (1997). iTEM: A temperature-dependent electromigration reliability diagnosis tool. *IEEE Trans. on Computer-Aided Design of Integrated Circuits and Systems* **16**: 882–893.

Trattles, J. T., Neill, A. G. O. and Mecrow, B. G. (1991). Three dimensional finite element determination of current density and temperature distributions in pillar like vias. *Proceeding of IEEE VMIC*, 343–345.

Tsu, R., McPherson, J. W. and McKee, W. R. (2000). Leakage and breakdown reliability issues associated with low-k dielectrics in a dual-damascene Cu process. *Proceeding of IEEE IRPS*, 348–353.

Wei, B. Q., Vajtai, R. and Ajayan, P. M. (2001). Reliability and current carrying capacity of carbon nanotubes. *Applied Physics Lett.* **79**(8): 1172–1174.

Weide-Zaage, K., Zhao, J., Ciptokusumo, J. and Aubel, O. (2008). Determination of migration effects in Cu-via structures with respect to process-induced stress. *Microelectronics Reliability* **48**: 1393–1397.

Wong, E. W., Sheehan, P. E. and Lieber, C. M. (1999). Nanobeam mechanics: Elasticity, strength, and toughness of nanorods and nanotubes. *Science* **277**: 1971–1975.

Zhan, C. and Heng, F. L. (1998). A fast minimum layout perturbation algorithm for electromigration reliability enhancement. *Proceeding of IEEE International Symposium on Defect and Fault Tolerance in VLSI Systems*, 56–63.

Zhu, L. B., Hess, W. and Wong, C. P. (2006). *In-situ* opening aligned carbon nanotube film/arrays for multichannel ballistic transport in electrical interconnect. *Proceeding of 56th Electronic Components and Technology Conf.*, 171–176.

Zhu, L. B., Sun, Y. Y., Xu, J. W., Zhang, Z. Q., Hess, W. and Wong, C. P. (2005). Aligned carbon nanotubes for electrical interconnect and thermal management. *Proceeding of 55th Electronic Components and Technology Conf.*, 44–50.

Index

$1/f$ noise measurement, 132
1D heat flow, 69
3D model, 31

accelerated life test, 67
activation energy, 17, 19, 43, 79
AFD, 21, 215, 216, 260
AFD distribution, 186
Al based metallization, 143
Al plugged via, 52
ALD, 196
alloying, 29
analytical modeling, 246
anisotropy, 14
annealing, 165
antibonding, 16
applied stress, 43
atom displacement, 263
atom hopping, 16, 261
atom-vacancy complexes, 13
atomic diffusion coefficient, 17
atomic diffusion paths, 252
atomic displacement, 262
Atomic Flux divergence, 20, 28, 254
atomic flux divergence distribution, 212
atomic migration, 247
atomic surface diffusion, 250
atomistic Monte Carlo simulation, 261

Babal tower, 99
Back flow, 29
back stress, 221
ballistic model, 12, 13, 15
bamboo structure, 40, 77, 169
bamboo-like Cu grain structure, 162
barrier material, 166, 195
barrier metal, 24, 55, 191
BEM, 80, 87, 100
bimodal distribution, 116
Black equation, 15, 23–25, 27, 29, 92, 112, 131, 264
Blech effect, 29, 45
Blech length effect, 23, 58
bond breaking, 15
Born approximation, 14
Born-Oppenheimer adiabatic approximation, 14

cap layer materials, 156
carrier density modulation, 15
cathode thinning, 208
censored points, 116
chamber vacuum break, 56
chemical–mechanical-polishing, 144
CMP, 199
CNT, 179, 180
compliant dielectric polymer, 44
concentration gradient, 253
conformal mapping technique, 250
contact pitting, 30
contact resistance, 57
continuity equation, 245
copper precipitate, 42
copper silicide formation, 154
corrosion, 199
coupled field analyses, 255
CoWB, 161, 162
CoWP, 158–160

cracking, 185
crystalline anisotropy, 252
crystallographic orientation, 57, 249, 250
Cu agglomerates, 194
Cu based interconnects, 143
Cu drift velocity, 179
Cu in Al, 42
Cu precipitation, 109
Cu silicidation, 157
Cu surface diffusivity, 169
Cu(Al), 180
Cu(Al) alloy film, 178
Cu(Sn), 180
Cu(Sn) alloy, 177
Cu-Mg alloy film, 176
Cu-Zr film, 176
Cu/cap interface engineering, 145
Cu/cap layer, 145, 217
Cu/cap layer interface, 218
Cu/CNT composite via, 180
Cu/etch stop layer, 218
current crowding, 47, 61, 75, 208
current density exponent, 23
current density gradient, 47
current exponent, 79
curvature driven surface migration, 252

damage map, 99
Damascene, 152, 153
Damascene method, 144
Damascene processing method, 144
Damascene structure, 165
defect, 45
defect density, 13
design induced failure mechanisms, 202
design rules generation, 244
diffusing path, 145
diffusion, 261
diffusion flux, 39
diffusion path, 21, 39, 41, 217, 259, 263
drift velocity, 17, 18, 45, 170, 204
drift velocity measurement, 143
driving force, 13–15, 31, 64, 146, 245, 247, 253, 263, 270
Dual Damascene, 145, 146, 149–151

EBICH, 126
effective blocking boundary, 221
effective boundary width, 40
effective bulk modulus, 45
effective charge number, 17
electrical-thermal analysis, 255
electric field, 13, 14
electron density, 12, 14
electron scattering, 12
electron wind, 65
electron wind force, 11, 13, 15, 16, 248, 250, 252
electron wind force energy, 263
electron wind force migration, 254
EM design rule, 31
EM driving force, 39
EM kinetics, 263
EM test monitoring, 105
end-contact segment, 71
enhancement effect, 213
enhancement factor, 216
enhancement of EM lifetime, 157
equilibrium vacancy concentration, 62
etch stop layer, 200
extrapolate, 22
extrapolation, 24, 25, 125
extrusion, 37

failure criteria, 103, 104
fast wafer level electromigration test, 80
Fermi-Dirac distribution, 13
Feynman-Hellman theorem, 14
FIB, 126
Fick's diffusion laws, 18
Fickian diffusion, 111, 245
finite difference method, 248, 262
finite element analysis, 28, 47, 51, 62, 212, 257, 263
finite element method, 248
finite element model, 253
Finite Element Modeling, 101
flux divergence, 40, 42, 45, 53, 108, 170, 182
free-electron approximation, 13
FWLR, 99

grain boundary, 247, 263
grain boundary diffusion, 208, 218
grain growth, 261
grain orientation, 40
grain size, 30, 38, 40, 57, 246
grain size distribution, 38
grain structure, 21, 30

hard surrounding materials, 20
heat conductivity, 54
hillock, 29, 37
hillock formation, 83, 115, 261
hillock growth, 60
hydrogen, 29, 60
hydrogen plasma treatment, 154
hydrostatic stress, 62, 256

IC layout, 221
IC reliability, 244
impurity, 14, 16, 42, 173
incubation, 150
inhomogeneity, 21
interatomic interactions, 261
interface, 263
interface diffusion, 146, 208, 253

jL product, 150
Joule heating, 27, 71, 73, 108, 148, 149

Kelvin-type measurements, 75
kinetic, 28
kink site, 172
Kubo's linear response formalism, 14

lattice defect, 15
lattice diffusion, 53, 143
length dependence, 46
lifetime, 246
lifetime distribution, 38
lifetime enhancement, 221
lifetime enhancement effect, 62
lifetime prediction, 258
line width, 61
line width dependence, 203
line-related EM, 124
line-via structure, 102, 208
line/stud contact, 42
liner's conductivity, 253
'log-normal' distribution, 23
low-k interlevel dielectric, 144

magnetic force microscopy, 128
material difference, 51
mechanical strength, 60
mechanical stress, 57, 62, 104, 181, 190, 195
mechano-diffusion fluxes, 63
median energy to fail, 88
median time to failure, 22, 23, 264
metal atom accumulation, 37
metallic cap-layer, 158
micro-crack, 60
micro-force balance principle, 252
microstructural changes, 104
micro-structural discontinuities, 41
microstructure, 42, 246, 253, 269
microstructure analysis, 198
microstructure transition, 212
migration energies, 15
mixture distribution, 124, 125
mixture model, 117
momentum transfer, 12, 13, 66
Monte Carlo method, 28, 260, 262, 263
moving boundary problems, 250
muffin-tin formalism, 14
multi-level metallization, 73
multi-model failure time distribution, 220
multiple censored data analysis, 118

narrow-to-wide transitions, 211
Nernst-Einstein equation, 17
NH_3 plasma treatment, 155, 156
NIST structure, 68, 98
non-fatal voids, 106
nucleation, 261
numerical modeling, 27

OBIRCH, 126
open failure, 113
optical microscope, 126
optical microscopy imaging, 130
outlier points, 116

outlier points analysis, 125
oxide thickness, 46, 68

Package-Level Electromigration Test, 77
pair interaction energy, 262, 263
passivation layer, 29, 60, 69
passivation materials, 44
passivation thickness, 60
phase field model, 248, 251
physical failure analysis, 125
physics-based EM modeling, 245
physics-based modeling, 244
pitfalls, 98
Pitfalls of SWEAT, 90
plasma-induced-damage, 156
plasma treatment, 153
polycrystalline, 21
polycrystalline structures, 146
polyimide, 71
precipitate distribution, 42
probability plot, 25
protrusions, 115
PVD, 196

quasibamboo microstructure, 146

rate of resistance change, 77
RC delay, 144
recovery of resistance, 104
refractory materials, 30
reliability modeling, 244
repeatability, 67
reservoir, 61, 212, 213, 221
reservoir effect, 257
reservoir length, 62
residual resistivity dipole, 15
resistance change, 220, 226, 228
resistance change profile, 265
resistance profile, 227
resistance saturation, 58
resistance saturation effects, 149
resistivity, 12, 13
RIE, 146, 148–152
RTA, 171

S-vacancy pairs, 174
saturation reservoir length, 215
Scanning electron microscope, 126
scattering, 15
self-diffusivity, 112
self-forming Mn barrier, 197
SFT, 187, 188
shape evolution, 262
sharp interface model, 247, 250
short-length effect, 29, 110, 111, 221
SiCN cap-layer, 157
signature of the resistance change, 115
Silicon oil baths, 78
SiN cap-layer, 157
slit voids, 249
slit-like crack, 64
slit-like void formation, 109
slit-shaped void, 252
Soret diffusion, 245
spatial variation, 38
standard NIST structure, 108
standard test structures, 67
statistical method, 25
step height, 41
strain energy, 263
stress driven migration, 252
stress evolution, 212, 245, 246
stress free temperature, 186, 256
stress gradient, 63, 146
stress-induced Al backflow, 52
stress-migration, 254
subtractive etching, 144
supersaturated, 104
surface contamination, 40
surface diffusion, 65
surface diffusivity, 249, 252
surface electromigration, 66
surface engineering, 153, 199
surface free energy, 194
surface mass flux, 248
surface migration, 255
surface tension force, 250
surface void migration, 262
SWEAT, 80, 83, 86, 87, 89, 94, 96–99

TCR, 94, 108
TDDB, 227
TEM, 131
Temperature Coefficient of Resistivity, 84
temperature gradient, 23, 47, 59, 71, 200
temperature profile, 46, 69
temperature ramping, 81
tensile stress, 43, 147
test data analysis, 115
test structure, 27
test structure design, 73
texture, 151
textured film, 41
thermal conductivity, 68
thermal energy, 263
thermal expansion coefficient, 184
thermal expansivities, 44
thermal expansivity coefficient, 208
thermal-mechanical analysis, 255
thermal mismatch, 44
thermodynamics process, 28
thermomechanical stress distribution, 256
thermo-migration, 254
thick passivation, 20
thinning, 66
time evolution of voids, 127
TRACE, 80, 83
transition metals, 14
triaxial stresses, 191
triple point, 30, 65, 172, 246
true linear acceleration, 25
twin boundary, 172

vacancies, 14, 15, 261
vacancies accumulation, 64
vacancy, 104
vacancy accumulation, 37, 111
vacancy flux, 18, 37, 38
vacancy-impurity complexes, 104
vacancy supersaturations, 108
vacuum break, 30
"via-fed" structure, 73, 77
Via-line structure, 24
via-related EM, 124
void, 15, 29
void evolution, 128
void growth, 44, 109, 261, 263
void identification, 128
void migration, 66, 262
void nucleation, 29, 40, 44, 109, 111, 217, 255, 263
void shape, 44, 220, 263
void shape change, 64, 248, 250
void surface, 263
void surface diffusivity, 250
voltage tapping, 75
Voronoi method, 247

weak spots, 31
WIJET, 80, 85–87, 100
wind force, 14
W-plug via, 52, 73
W studs, 110

X-ray, 131

Young's modulus, 191

Biography

Cher Ming Tan was born in Singapore in 1959, and received his B.Eng. in Electrical Engineering from National University of Singapore in 1984. He received his M.Eng. and Ph.D. degrees in Electrical Engineering from the University of Toronto, Canada in 1988 and 1992 respectively. He has worked in Fairchild Semiconductor, Hewlett Packard, LiteOn Semiconductor Corporation (Taiwan), and Chartered Semiconductor Manufacturing Ltd before joining Nanyang Technological University (NTU), Singapore as faculty member in 1996. He is the past chair of IEEE Singapore Section, Distinguish Lecturer of IEEE Electronic Device Society, Founding Chair of IEEE Nanotechnology Chapter and Fellow of Singapore Quality Institute. He is a senior member of IEEE and American Society of Quality. He is currently a Senior Scientist in Singapore Institute of Manufacturing Technology, A*Star and Faculty Associate of Institute of Microelectronics, A*Star, Singapore.

His research interests include reliability and failure physics modeling of electronic components and systems, finite element modeling of materials degradation, statistical modeling of engineering systems, and nanomaterials and devices reliability.